Archimedes

Volume 5

Archimedes

NEW STUDIES IN THE HISTORY AND PHILOSOPHY OF SCIENCE AND TECHNOLOGY

VOLUME 5

Archimedes has three fundamental goals; to further the integration of the histories of science and technology with one another: to investigate the technical, social and practical histories of specific developments in science and technology; and finally, where possible and desirable, to bring the histories of science and technology into closer contact with the philosophy of science. To these ends, each volume will have its own theme and title and will be planned by one or more members of the Advisory Board in consultation with the editor. Although the volumes have specific themes, the series itself will not be limited to one or even to a few particular areas. Its subjects include any of the sciences, ranging from biology through physics, all aspects of technology, broadly construed, as well as historically-engaged philosophy of science or technology. Taken as a whole, *Archimedes* will be of interest to historians, philosophers, and scientists, as well as to those in business and industry who seek to understand how science and industry have come to be so strongly linked.

Archimedes

Volume 5
New Studies in the History and Philosophy of Science and Technology

---

Leadership and Creativity

A History of the Cavendish Laboratory, 1871–1919

by

DONG-WON KIM

Korea Advanced Institute of Science and Technology,
Taejon, Korea

KLUWER ACADEMIC PUBLISHERS
DORDRECHT / BOSTON / LONDON

A C.I.P. Catalogue record for this book is available from the Library of Congress.

ISBN 1-4020-0475-3

Published by Kluwer Academic Publishers,
P.O. Box 17, 3300 AA Dordrecht, The Netherlands.

Sold and distributed in North, Central and South America
by Kluwer Academic Publishers,
101 Philip Drive, Norwell, MA 02061, U.S.A.

In all other countries, sold and distributed
by Kluwer Academic Publishers,
P.O. Box 322, 3300 AH Dordrecht, The Netherlands.

Printed on acid-free paper

Printed in the Netherlands.

Dedicated to my mentors,
Erwin N. Hiebert and Silvan S. Schweber

TABLE OF CONTENTS

CHAPTER 4. THE EMERGENCE OF THE CAVENDISH SCHOOL, 1895-1900

CHAPTER 5. J.J. THOMSON'S LEADERSHIP AND THE DEVELOPMENT OF THE CAVENDISH SCHOOL, 1901-1914

CHAPTER 6. THE END OF AN ERA, 1914-1919

ACKNOWLEDGMENTS

The ambition to write a decent history of the Cavendish Laboratory at the University of Cambridge first struck me in the fall of 1985, my first semester at Harvard graduate school. As a foreign student from a non-Western developing country, South Korea, I was frightened and somewhat doubtful that I could survive in Harvard's competitive environment. Although the history of physics had long been my favorite subject for study, my experience and knowledge were naturally quite limited. For a course taught by Erwin N. Hiebert on the history of physical sciences during the twentieth century, I read an article by George P. Thomson about his father's discovery of the electron. This article, "J.J. Thomson and the Discovery of the Electron (*Physics Today 9* (1956): 19-23)," focused on Joseph John Thomson's greatness as a physicist and his charisma as a teacher. Fascinated by this account of J.J. Thomson's charming character, I devoted my term paper to an examination of his role as director of the Cavendish Laboratory. When Professor Hiebert encouraged me to delve further into the history of J.J. Thomson's achievements, I quickly discovered that the available histories of the Cavendish Laboratory depended heavily on the reminiscences and memoirs of former directors and researchers of the Cavendish and that these histories lacked systematic analysis. I was especially bothered by the apparent consensus that an 1895 regulation change at Cambridge permitting non-Cambridge graduates to enter the University for post-graduate research was the chief cause of the Cavendish's sudden success at the turn of the twentieth century. I simply could not accept this idea. Thus began a long research project. The subject for a term paper developed into a doctoral dissertation (in 1991) and finally matured into this book, which represents a thorough condensation and revision of my dissertation along with the addition of two new chapters.

My deepest gratitude is directed to Erwin N. Hiebert and to Silvan S. Schweber. Erwin led me to this wonderful subject and has given my research efforts considerable attention ever since. I am very proud of the fact that he accepted me as his last doctoral candidate. He and Mrs. Elfrieda Hiebert offered myself and my family unfailing kindness, which was my secret source of strength as I worked to overcome many difficulties I encountered as a graduate student in the United States as well as a scholar and teacher in Korea. Sam Schweber, who generously took over the role of my dissertation advisor when Erwin retired in 1989, has looked after me ever since we met in a departmental colloquium in 1987. He read almost every word I wrote about the Cavendish and offered me incisive critiques. His encouragement was invaluable to me. It was he who urged me to publish my first paper about the Cavendish (which appeared in the *British Journal for the History of Science* in 1995) and who pushed me to extend my dissertation into a book. Sam also offered me wise counsel when I was confronted with personal difficulties after returning to

Korea in 1991. It was truly my great good fortune to have encountered—at the same point in time—two such exceptional mentors as Erwin Hiebert and Sam Schweber.

A number of other scholars contributed valuable criticism and advice to the writing of this book. Among them, I would like to especially thank Simon Schaffer, Andrew Warwick, Isobel Falconer, Peter Harman, and Jeff Hughes. I am also deeply grateful to Jed Z. Buchwald, who carefully read the entire draft of this book and recommended that I submit the manuscript to the Kluwer Academic Publishers for publication. I am very happy to see that this manuscript survived his sharp scrutiny. I also would like to express special thanks to professors and graduate students at Johns Hopkins University, where I spent my sabbatical year of 1998-1999 writing Chapters 5 and 6. Their criticisms during departmental colloquia were most helpful to me in revising and improving these two chapters. Among my friends and colleagues at Johns Hopkins, my particular thanks go to Robert Kargon, Stuart W. Leslie, and Buhm Soon Park. In addition, my heartfelt appreciation goes to several Korean colleagues who offered me encouragement to continue this project. To Judy Hardesty, who read and revised the entire draft, I offer my deep gratitude.

My sincere thanks naturally go to the institutes and libraries that made this book possible. The Korea Advanced Institute of Science and Technology partly financed my research trips to England and the United States. The American Institute of Physics provided me with a research grant to access its Niels Bohr Library. The Cavendish Laboratory, Cambridge University Library, the Imperial College Library, and the Niels Bohr Library generously permitted me to use their collections during the preparation of this book and to reproduce items from their collections. I owe a special debt of gratitude to Spencer Weart (American Institute of Physics) and Keith Papworth (Cavendish Laboratory). Cambridge University Press granted me permission to use and quote from my 1995 paper on the topic of the Cavendish Laboratory ("J.J. Thomson and the emergence of the Cavendish School, 1885-1900," *British Journal for the History of Science 28* (1995): 191-226). My sincere thanks also go to Jolanda Voogd and Helen van der Stelt at the Kluwer Academic Publishers.

Last, but not least, I offer my loving thanks to my wife, Soo-Yeon, to my daughters, Daye and Jinsol, and to my parents, Mr. Bo-Jung and Mrs. Soon-Young Kim, who always encouraged me to continue my research and writing and gave me heart to continue when I became exhausted. Their reaffirmation to me of the importance of family is one of the most important fruits that I harvested during the process of writing this book.

Dong-Won Kim
Korea Advanced Institute of Science and Technology
Korea

LIST OF ABBREVIATIONS

AIP	American Institute of Physics (Niels Bohr Library)
B. A. Report	*British Association Report*
BJSH	*British Journal for the History of Science*
CUL MSS	Cambridge University Library Manuscripts Collection
	ADD 7653 (E. Rutherford)
	ADD 7654 (J.J. Thomson)
	ADD 7655 (J. C. Maxwell)
CUR	*Cambridge University Reporter*
DSB	*Dictionary of Scientific Biography*
DNB	*Dictionary of National Biography*
HSPS/HSPBS	*Historical Studies in the Physical (and Biological) Sciences*
MT	Mathematical Tripos
NST	Natural Sciences Tripos
Not. Proc. Roy. Inst.	*Notices of the Proceedings at the Meetings of the Members of the Royal Institution*
Phil. Mag.	*Philosophical Magazine*
Phil. Trans.	*Philosophical Transactions (series A)*
Proc. Roy. Soc.	*Proceedings of the Royal Society of London*
Proc. Camb. Phil. Soc.	*Proceedings of the Cambridge Philosophical Society.*

INTRODUCTION

Historical accounts of successful laboratories often consist primarily of reminiscences by their directors and the eminent people who studied or worked in these laboratories. Such recollections customarily are delivered at the celebration of a milestone in the history of the laboratory, such as the institution's fiftieth or one-hundredth anniversary. Three such accounts of the Cavendish Laboratory at the University of Cambridge have been recorded. The first of these, *A History of the Cavendish Laboratory, 1871-1910*, was published in 1910 in honor of the twenty-fifth anniversary of Joseph John Thomson's professorship there. The second, *The Cavendish Laboratory, 1874-1974,* was published in 1974 to commemorate the one-hundredth anniversary of the Cavendish. The third, *A Hundred Years and More of Cambridge Physics,* is a short pamphlet, also published at the centennial of the Cavendish.[1] These accounts are filled with the names of great physicists (such as James Clerk Maxwell, Lord Rayleigh, J.J. Thomson, Ernest Rutherford, and William Lawrence Bragg), their glorious achievements (for example, the discoveries of the electron, the neutron, and DNA) and interesting anecdotes about how these achievements were reached. But surely a narrative that does justice to the history of a laboratory must recount more than past events. Such a narrative should describe a living entity and provide not only details of the laboratory's personnel, organization, tools, and tool kits, but should also explain how these components interacted within their wider historical, cultural, and social contexts.[2]

---

[1] J.J. Thomson et al., *A History of the Cavendish Laboratory, 1871-1910* (London: Longmans, Green, 1910); J. G. Crowther, *The Cavendish Laboratory, 1874-1974* (New York: Science History Publications, 1974); Cambridge University Physics Society, *A Hundred Years and More of Cambridge Physics* (Cambridge, 1974, 1980, 1995). Other histories are: A. Wood, *The Cavendish Laboratory* (Cambridge: Cambridge University Press, 1946); E. Larson, *The Cavendish Laboratory: Nursery of Genius* (London: Ward, 1962); G. P. Thomson, *J.J. Thomson and the Cavendish Laboratory* (London: Thomas Nelson & Sons, 1964).

[2] Some notable examples of recent studies of the histories of various laboratories (and research schools) are: J. B. Morrell, "The Chemist Breeders: The Research Schools of Liebieg and Thomson," *Ambix 19* (1972): 1-46; Gerald Geison, *Michael Foster and the Cambridge School of Physiology: The Scientific Enterprise in Late Victorian Society* (Princeton: Princeton University Press, 1978); Bruno Latour and Steve Woolgar, *Laboratory Life: The Social Construction of Scientific Facts* (Beverly Hills, Calif.: Sage, 1979); Frederic L. Holmes, *Lavoisier and the Chemistry of Life: An Exploration of Scientific Creativity* (Madison: University of Wisconsin Press, 1984); John L. Heilbron and Robert W. Seidel, *Lawrence and His Laboratory: A History of the Lawrence Berkeley Laboratory*, vol. 1 (Berkeley: University of California Press, 1989); Joseph S. Fruton, *Contrasts in Scientific Style: Research Groups in the Chemical and Biochemical Sciences* (Philadelphia: American Philosophical Society, 1990); Kathryn M. Olesko, *Physics as a Calling: Discipline and Practice in the Königsberg Seminar for Physics* (Ithaca, New York: Cornell University Press, 1991); and Robert E. Kohler, *Lords of the Fly: Drosophila Genetics and the Experimental Life* (Chicago: University of Chicago Press, 1994). Valuable discussions about research schools are found in Gerald L. Geison and Frederic L. Holmes (ed.), *Research Schools: Historical Reappraisals, Osiris 8* (1993).

My goal in writing this book was to deliver a critical history of the Cavendish Laboratory in its early years. At the core of this book is my belief that the evolution of the Cavendish Laboratory was not as smooth as the reader of previous historical accounts might expect. Contrary to the suggestions of earlier accounts, the prestige of the Cavendish's first directors, Maxwell, Lord Rayleigh, J.J. Thomson, and even Rutherford, did not automatically guarantee the Laboratory smooth sailing and a rosy future. Similarly, the 1895 change in University policy that permitted non-Cambridge graduates to study at the Cavendish did not promise the Laboratory's sudden elevation to the status of important research center. Instead, the Cavendish evolved from a relatively small university teaching laboratory to the world center of experimental physics through a *slow but steady* accumulation of knowledge and human resources. To present the history of the Cavendish Laboratory from this perspective, I have described and analyzed the participants in the Laboratory's development, as much as possible, in terms of their contributions to that development.

This book answers the following questions. What made it possible to create the Cavendish Laboratory in the 1870s? What was the Laboratory's principal role within Cambridge University and how did this role change over time? Who performed research at the Cavendish, when did they work there, and what topics did they investigate? In what ways and to what extent did the Laboratory's directors influence the work of Cavendish researchers? How did the Cavendish become the mecca of experimental physics during the first third of the twentieth century? In short, why was the Cavendish Laboratory so successful?

Throughout this book, special attention has been given to Cambridge University's influence on the Cavendish Laboratory. Not only did the Cavendish develop within this venerable University, but it also advanced in tandem with the University's educational system. Despite this close relationship, a struggle between the University's old traditions and the Laboratory's new values could not be avoided because the Cavendish propagated new modes of doing science. The early history of the Cavendish, therefore, illustrates the manner in which conflicts between traditional and new values were negotiated in late Victorian Cambridge. In concentrating on the Cavendish's context within Cambridge University, however, I have somewhat minimized the Laboratory's relationship with the outside world.

From 1860 to 1930, higher education in Europe was undergoing transformation from systems that were "small, homogeneous, elite and pre-professional" to systems that were "large, diversified, middle-class and professional."[3] The engine of this change was the second industrial revolution, during which science and technology became revered as the keys to society's goals. As Germany and the United States transformed their educational systems to meet the challenge of industrialization,

---

[3] Konrad H. Jarausch, "Higher Education and Social Change: Some Comparative Perspectives," in Konrad H. Jarausch (ed.), *The Transformation of Higher Learning 1860-1930: Expansion, Diversification, Social Opening, and Professionalization in England, Germany, Russia, and the United States* (Chicago: University of Chicago Press, 1983), 10.

efforts in Britain to adapt education to the new technological challenges were largely unsuccessful until the Paris International Exhibition of 1867, which made it obvious that industrial development was flourishing on the Continent and in the United States much more than on the British Isles. This realization intensified the efforts of British educational reformers to advance their pro-scientific program, even in the tradition-steeped halls of Oxford and Cambridge. However, the most important educational changes in England did not occur in existing universities (as they did in Germany and the United States), but in newly established regional colleges and universities.[4] Oxford and Cambridge remained as aloof as possible to public pressure for change, and whatever small concessions the two universities made to educational reformers were made at a foot-dragging pace.

Cambridge University, in a "least possible" response to pressure for greater educational emphasis on science and technology, in the 1870s established a professorship of experimental physics and a professorship of engineering, at the same time instituting the Cavendish Laboratory and Michael Foster's physiological laboratory. The Cavendish was quickly absorbed into the University's existing system, and the Laboratory's development was hampered by that system's painfully slow evolution. As late as 1873, the Royal Commission on Scientific Instruction and the Advancement of Science (better known as the Devonshire Commission) considered it necessary to remind Oxford and Cambridge about the importance to the universities of research in the sciences. To persuade Oxford and Cambridge to establish more teaching positions in the sciences, the Commission included in its report comparisons of the University of Berlin and the two British universities that showed the Oxbridge system to be wanting.[5] Nevertheless, Oxford and Cambridge delayed for a full decade before reluctantly establishing a few new lectureships in the sciences in 1883.

Thus, to understand the Cavendish Laboratory, one first must understand Cambridge University and the radical changes it experienced during the nineteenth century. In 1800, the University was a bastion of traditions (some of which could be traced back to the Middle Ages) and a closed, male-only community which allowed only its own graduates to pursue advanced study within its halls or to serve as fellows or officers. Its primary goal was to produce educated gentlemen, sound in mind and body, according to the tenets of the Church of England, to which the University had been tightly bound by the Religious Test Act since the seventeenth century. Along with mathematics, the classics, and philosophy, a Cambridge education emphasized rowing, fencing, swimming, and social gathering.[6] Professors

---

[4] Graeme Gooday, "Precision Measurement and the Genesis of Physics Teaching Laboratories in Victorian Britain," *BJHS 23* (1990): 25-51 on 25-31.

[5] Royal Commission on Scientific Instruction and the Advancement of Science, "The Report of the Science Commission of the Old Universities," *Nature 8* (1873): 317-319, 337-341. See also Roy M. MacLeod, "Resources in Science in Victorian England: The Endowment of Science Movement, 1868-1900," in Peter Mathias (ed.), *Science and Society 1600-1900* (Cambridge: Cambridge University Press, 1972), 111-166.

[6] For more about student life at Cambridge at the turn of the nineteenth century, see Sheldon Rothblatt,

were few, and most teaching was done by tutors affiliated with the various colleges in Cambridge. Research was regarded as a purely private activity. By 1900, however, students who had graduated from colleges and universities outside Cambridge were permitted to enter the University as "advanced students." Women had entered the University, and its connections with the Church of England had been relaxed. The number of Cambridge professors, university lecturers, and university demonstrators had greatly increased, and they had become chiefly responsible for educating students. Research had taken firm root in Cambridge's educational system, and a few of its research centers, including the Cavendish Laboratory, had become world famous. By 1900, Cambridge University was a leading institution for the study of the humanities, social sciences, medicine, and modern science.

Such rapid change had not been altogether welcome at Cambridge. Resistance, clashes, delays, and compromise were inevitable, and the result was a unique mixture of old and new that distinguished Cambridge from other universities. In 1908, Karl Breul, Cambridge's Schröder Professor of German, accurately characterized the University's intellectual atmosphere as:

> a happy blending of tradition and freedom, of ancient customs and new methods; a careful adapting [of] old institutions to modern needs, of training the intellect and moulding the character . . . The old humanistic tradition of classical studies in Cambridge is, by the best of her sons, successfully applied to the more modern studies.[7]

The typical nineteenth-century Cambridge undergraduate pursued his bachelor's degree for three academic years, each of which consisted of three terms: the Michaelmas term (October to December), the Lent term (January to March), and the Easter term (May to June). To "keep his term," the undergraduate was required to reside in Cambridge for a specified number of days. Most students stayed an additional term to take the January Senate House examination (also known as the tripos examination) necessary to earn a Cambridge "bachelor's" degree. Some stayed even longer to prepare for college fellowships. During the "Long Vacation" of summer, only the most diligent students remained in Cambridge to prepare for examinations, perform research, or study in a quiet atmosphere. Most undergraduates came to Cambridge from elite "public" schools like Eton or Harrow, but some scholarship students were graduates of small country grammar schools.

The University was a federation of colleges in which each member college enjoyed independent administration and traditions and, according to its size and wealth, contributed financially to the University. The position of Vice-Chancellor of the University was filled, in rotation, by a head (Master) of one of the colleges. Each college selected its own students, and what mattered most to the daily life

---

"The Student Sub-Culture and the Examination System in Early 19th Century Oxbridge," in Lawrence Stone (ed.), *The University in Society, Volume 1: Oxford and Cambridge from the 14th to the Early 19th Century* (Princeton: Princeton University Press, 1974), 247-303.

[7] Karl Breul, *Students' Life and Work in the University of Cambridge: Two Lectures*, revised edition (Cambridge: Bowes and Bowes, 1910), 5.

of the typical Cambridge student was not affiliation with the University, but affiliation with a college: the place where one studied, dined, played, and slept, ultimately to become a "Trinity man," a "King's man," or a "Newnham lady (or woman)." A small minority of Cambridge University students had no specific college affiliation; because they were "non-collegiate," their social status was lower than that of other students. For most Cambridge students, the University became a palpable force in daily life only when they matriculated into the University, when they took University examinations, and when they received their Cambridge University degrees.

Cambridge colleges managed and educated their students through a unique system in which every undergraduate was required to have a tutor:

> a graduate of experience and great influence in the college, who is appointed to stand him 'in loco parentis' [in place of a parent], and who not only looks after his interests in everything concerning the University and the lectures, but is always prepared to advise him in any serious case of difficulty and doubt that may trouble him.[8]

Thus, Cambridge tutors, whose first loyalty was directed to their respective colleges, were responsible for the education of each student from entrance to a college to graduation from the University. Tutors assigned readings, helped students prepare for some examinations, and advised ambitious students who wanted to excel at examinations in the choice of private coach. Courses taught by University professors, lecturers, readers, and demonstrators were not compulsory and, in fact, competed with the education provided by college lecturers. Tutors who were also college lecturers, as was common, usually advised their students to attend their own lectures rather than courses taught by University professors and other educators, for the quite understandable reason that their incomes were based on the number of students attending their lectures. Clearly, the increase in the size of the University teaching staff in the last quarter of the nineteenth century presented college lecturers with a serious challenge. J.J. Thomson, a Cambridge student who would become the third director of the Cavendish Laboratory, considered himself lucky to have as his tutor a classicist because "he let me choose the mathematical lectures I attended, whereas if he had been a mathematician he would have made me go to his own lectures."[9]

Cambridge University's control over the quality of its undergraduates was exerted through the University examination which, starting in the early eighteenth century, was systematized and held each January in the Senate House. This examination was called the "tripos," after the three-legged stool on which early Cambridge students sat during disputations to prove their competence.[10] During the

---

[8] Ibid., 21. Brackets added.

[9] J.J. Thomson, *Recollections and Reflections* (New York: MacMillan, 1937), 34.

[10] For the history of the Mathematical Tripos (and Senate House Examination), see W. W. Rouse Ball, *History of the Mathematical Tripos* (Cambridge, 1880) and *History of the Study of Mathematics at Cambridge* (Cambridge, 1889). A short version, "The History of the Mathematical Tripos," can be found in W. W. Rouse Ball, *Cambridge Papers* (London: Macmillan and Co., 1918), 252-316. See also J. W. L. Glaisher, "The Mathematical Tripos," *Proceedings of the London Mathematical Society 18*

eighteenth century, the tripos was systematized into an examination orally dictated by examiners and, at the turn of the nineteenth century, it evolved into a mixture of oral and paper tests, finally becoming a printed examination in 1827: "From a thing of wood to a man, from a man to a speech, from a speech to sets of verses, from verses to a sheet of coarse foolscap paper, from a paper to a list of names, and from a list of names to a system of examination."[11]

Tripos questions were primarily mathematical but also covered some philosophy and theology. As the years went by, the length of the examination increased. In 1800, it lasted eighteen hours and took three days; from 1828 on, it lasted twenty-three hours and took four days; from 1833 on, it took more than twenty-seven hours and lasted five days; from 1839 on, it took thirty-three hours and lasted six days; and from 1848 on, it took over forty-four hours and lasted eight days. The University Calendar of 1802 delivered a vivid description of the Senate House examination:

> On the Monday morning, a little before eight o'clock, the students, generally about a hundred, enter the Senate-House, preceded by a master of arts, who on this occasion is styled the father of the College to which he belongs. On two pillars at the entrance of the Senate-House are hung the classes and a paper denoting the hours of examination of those who are thought most competent to contend for honours. Immediately after the University clock has struck eight, the names are called over, and the absentees, being marked, are subject to certain fines. The classes to be examined are called out, and proceeded to their appointed tables, where they find pens, ink and paper provided in great abundance. In this manner, with the utmost order and regularity, two-thirds of the young men are set to work within less than five minutes after the clock has struck eight. There are three chief tables, at which six examiners preside. At the first, the senior moderator of the present year and the junior moderator of the preceding year. At the second, the junior moderator of the present, and the senior moderator of the preceding year. At the third, two examiners appointed by the Senate. The two first tables are chiefly allotted to the six first classes; the third, or the largest, to the [poll men].
>
> The young men hear the propositions or questions delivered by the examiners; they instantly apply themselves; demonstrate, prove, work out and write down, fairly and legibly (otherwise their labour is of little avail) the answers required. All is silence; nothing heard save the voice of the examiners; or the gentle request of some one, who may wish a repetition of the enunciation. It requires every person to use the utmost dispatch; for as soon as ever the examiners perceive anyone to have finished his paper and subscribed his name to it another question is immediately given . . .
>
> The examiners are not seated, but keep moving round the tables, both to judge how matters proceed and to deliver their questions at proper intervals. The examination, which embraces arithmetic, algebra, fluxions, the doctrine of infinitesimals and increments, geometry, trigonometry, mechanics, hydrostatics, optics and astronomy, in all their various gradations, is varied according to circumstances: no one can anticipate a question, for in the course of five minutes he may be dragged from Euclid to Newton, from the humble arithmetic of Bonnycastle to the abstract analytics of Waring. While this examination is proceeding at the three tables between the hours of eight and nine, printed problems are delivered to each person of the first and second classes; these he

---

(1886): 4-38.

[11] Ball, *Cambridge Papers*, 314.

takes with him to any window he pleases, where there are pens, ink, and paper prepared for his operations. It is needless to add that every person now uses his utmost exertion, and solves as many problems as his abilities and time will allow.[12]

From the mid-eighteenth century on, tripos results were published according to class rank. Graduates with the best scores received honour degrees and were divided into three classes; wranglers (first class), senior optimes (second class), and junior optimes (third class). The remaining candidates, called "poll" men (the many), received bachelor degrees without distinction. In 1824, the introduction of a new examination, the Classical Tripos, caused the old Senate House examination to be distinguished as the "Mathematical Tripos."[13] In 1851, two more examinations were established: the Moral Sciences Tripos and the Natural Sciences Tripos. In 1858, a General Examination for poll men finally was introduced, and the various triposes were reserved for honour degree candidates. At the beginning of the twentieth century, about half of Cambridge's graduating class took at least one tripos examination. The Cambridge graduates whose names appear in this book nearly all distinguished themselves as honour graduates.

The intentions and lives of students who aspired to earn honours at Cambridge University differed considerably from those who did not. Poll men went to Cambridge to pass "a few enjoyable years between their school time and the beginning of their life's work; to make themselves proficient in sports and all manly exercises; to acquire a certain polish of manners and ease in social intercourse, but not to make themselves proficient in any special branch of learning."[14] Poll men focused on passing two examinations, the "Little Go" (Previous Examination) and the General Examination (Special Examination). Honour candidates, also known as "reading men," following a very different path, studied for seven to ten hours each day to get high marks in the Senate House examination (or, starting in the mid-nineteenth century, in one or two tripos examinations).

Of the four tripos examinations, the most important was the Mathematical Tripos, which produced many outstanding nineteenth-century mathematicians and natural philosophers, among them John William Herschel, William Whewell, William Thomson (Lord Kelvin), George Gabriel Stokes, James Clerk Maxwell, John William Strutt (Lord Rayleigh), and J.J. Thomson, as well as a large number of eminent economists, civil servants, and Anglican bishops. Those who earned the title of wrangler were highly respected both within and outside the University. The first wrangler was often known as the "senior wrangler," and other wranglers were also identified by their examination rank, such as "second wrangler" or "fifth wrangler." Although the title itself did not guarantee reward, high wranglers often

---

[12] J. R. Tanner (ed.), *The Historical Register of the University of Cambridge: being a supplement to the Calendar with a record of University offices, honours and distinction to the year 1910* (Cambridge: Cambridge University Press, 1917), 352-353. Brackets added.

[13] The Mathematical Tripos, however, remained the only examination for the B.A. degree until 1850. The Classical Tripos, first held in 1824, could be taken only by those who had been at least junior optimes in the preceding Mathematical Tripos. This restriction was abolished in 1850.

[14] Breul, *Students' Life and Works in the University of Cambridge*, 8.

Figure 1. Two kinds of Cambridge undergraduates: the left represents the honour candidate while the right two the poll man. ("There was a sharp Scholar who read an hour every day. Till they said—'Your head man won't stand it.' He said, 'To be candid, if I read on much more I'd be dead.') [From Unknown author ("Naughty Boy"), Nonsense Scribbles at Cambridge (Cambridge, na.)].

Figure 2. Presentation of the senior wrangler to the Vice-Chancellor (1842) [From S.C. Roberts, Introduction to Cambridge (Cambridge: Cambridge University Press, 1934), 18].

were elected college fellows after receiving their bachelor degrees.[15]

As Cambridge University changed during the nineteenth century, so did the focus of the Mathematical Tripos. In 1800, the Mathematical Tripos consisted of "arithmetic, algebra, fluxions, the doctrine of infinitesimals and increments, geometry, trigonometry, mechanics, hydrostatics, optics, and astronomy, in all their various gradations."[16] Honour candidates focused their study on Euclid's *Elements*, Newton's *Principia*, a few books on algebra, and manuscripts prepared by previous wranglers.[17] During the 1810s and 1820s, three rebellious Cambridge men, John Herschel, George Peacock, and Charles Babbage, launched a radical reform of mathematical education by introducing into Cambridge the Continental style of algebra. As their analytical methods gradually gained stature in Cambridge, the Mathematical Tripos became more and more a written examination because "mastering analytical mathematics required years of tough progressive study" with paper and pencil.[18] In 1848, the Mathematical Tripos was divided into two parts, a change with far-reaching effects. For the first three days of the examination, candidates tackled elementary subjects, including the first three sections of the *Principia*, elementary parts of statics, dynamics, hydrostatics, optics and astronomy. At the end of the three days, after a short interval, the examiners issued a list of the honour candidates, namely those who would be permitted to attend the next five days of the examination, which was devoted to the higher parts of mathematics and physics such as "Statics, Dynamics of particles and of rigid bodies, hydrostatics and hydrodynamics, optics, astronomy, the lunar theory tested both analytically and by Newton's method, and planetary theory, precession and nutation, and the undulatory theory of light."[19] Shortly after completing the second stage of the Mathematical Tripos, a few very ambitious wranglers also took the examination for the Smith Prize. During the 1870s, additional changes were made to the contents of the Mathematical Tripos.

The Mathematical Tripos of the nineteenth century was a very difficult examination indeed. David B. Wilson summarized its rigors as follows:

> Because of the great importance of a student's place in the results and because of the resultant stiff competition, examination papers were always long enough so that even

---

[15] John Gascoigne, "Mathematics and Meritocracy: The Emergence of the Cambridge Mathematical Tripos," *Social Studies of Science 14* (1984): 547-584, especially 560-565.

[16] Ball, *Cambridge Papers*, 281, and 284-285.

[17] Ibid.

[18] Andrew Warwick, "Exercising the Student Body: Mathematics and Athleticism in Victorian Cambridge," in Christopher Lawrence and Steven Shapin (ed.) *Science Incarnate: Historical Embodiments of Natural Knowledge* (Chicago: University of Chicago Press, 1998), 288-326 on 290. See also Warwick, "A Mathematical World on Paper: Written Examinations in Early 19th Century Cambridge," *Studies in History and Philosophy of Modern Physics, 29* (1998): 295-319; "The World of Cambridge Physics," in R. Staley (ed.), *The Physics of Empire: Public Lectures* (Cambridge: Whipple Museum of the History of Science, 1994), 57-86.

[19] David B. Wilson, "Experimentalists among the Mathematicians: Physics in the Cambridge Natural Sciences Tripos, 1851-1900," *HSPS 12:2* (1982): 325-371 on 337. Sections 1 and 2 of the paper (pp. 327-340) contain a fine explanation of the Mathematical Tripos during the nineteenth century.

> the best students would not finish early. Also, to allow more specialization at the advanced level, stage II papers contained numerous questions in each area. A senior wrangler would typically obtain no more than fifty to sixty per cent of the total possible score for stages I and II, while a low wrangler would earn ten to twenty per cent. Two men taking the tripos the same year could answer quite different examinations.[20]

Thus, "for the lower wranglers, the Tripos was an enigma. For those seeking to become junior optimes, the minimum honours and the prerequisite for competing in the Classical Tripos, all was chaos."[21] While preparing for the examination, some candidates lost their physical health; some suffered nervous breakdowns. Even the most successful wranglers, like William Thomson or Maxwell, experienced serious stress as a result of the Mathematical Tripos. For example, when Maxwell entered the Senate House to take the examination, he "felt his mind almost blank." When he left the Senate House following the examination, he was "dizzy and staggering, and was some time in coming to himself."[22]

To alleviate the inevitable stress of daily studies, most Cambridge students took some regular physical exercise in the afternoon by rowing, swimming, walking, or, in the late nineteenth century, playing golf. "This exercise," Andrew Warwick pointed out, "became the recognized complement of hard study, and students experimented with different regimes of working, exercising, and sleeping until they found what they believed to be the most productive combination."[23]

As mentioned previously, ambitious Cambridge undergraduates prepared for the Mathematical Tripos by hiring private coaches who, more often than not, had been wranglers. Private coaches were not faculty members of the university; instead, they specialized in teaching techniques for answering the greatest number of examination questions in the shortest possible time. John William Strutt (third Lord Rayleigh), senior wrangler of 1865, used to relate "how he had answered one question during the time that the answers were being collected from other candidates. It was an advantage to be low down in alphabetical order!"[24] Private coaching was relatively expensive: in the 1870s the going rate was about £35 per year, making "mathematics a more expensive subject than classics or history, where private tuition was not nearly so general, and students were content with the lectures given by the Professor and College lecturers."[25] Two legendary coaches of the nineteenth century were William Hopkins and Edward John Routh, each of whom trained hundreds of wranglers. Between 1828 and 1849, Hopkins produced 175 wranglers, including seventeen senior wranglers. Among his pupils were William Thomson

---

[20] Wilson, "Experimentalists among the Mathematicians," 337.
[21] H. W. Becher, "Voluntary Science in Nineteenth Century Cambridge University to the 1850's," *BJHS 19* (1986): 57-87 on 83.
[22] Lewis Campbell and William Garnett, *The Life of James Clerk Maxwell* (London: Macmillan & Co., 1882), 176.
[23] Warwick, "Exercising the Student Body," 294-295.
[24] Robert John Strutt (fourth Baron Rayleigh), *Life of John William Strutt, Third Lord Rayleigh, O.M., F.R.S.*, an augmented edition (Madison: University of Wisconsin Press, 1968), 34. Chapter 2 includes a vivid description of Routh and his class during the 1860s.
[25] J.J. Thomson, *Recollections*, 42.

and Maxwell. More than 600 pupils of Routh—himself the senior wrangler of 1854—became wranglers, including John William Strutt and J.J. Thomson. Between 1858 and 1888, Routh coached 41 winners of the Smith Prize. The following description of Routh's class illustrates his recipe for success:

> He [Routh] gave catechetical lectures three times a week to classes of eight or ten men of approximately equal knowledge and ability. The work to be done between two lectures was heavy, and included the solution of some eight or nine fairly hard examples on the subject of the lectures. Examination papers were also constantly set on tripos lines (bookwork and riders), while there was a weekly paper of problems set to all pupils alike. All papers sent up were marked in public, the comments on them in class were generally brief, and, to save time, solutions of the questions were circulated in manuscript. Teaching also was supplemented by manuscripts on the subjects. Finally to the more able students he was accustomed, shortly before their tripos, to give memoirs or books for analyses and commentaries. The course for the first three years and the two earlier long vacations covered all the subjects of the examination—the last long vacation and the first term of the fourth year were devoted to a thorough revision.[26]

Although the Mathematical Tripos produced many celebrated nineteenth-century mathematical physicists, such as William Thomson, Maxwell, Lord Rayleigh, and J.J. Thomson, preparing for it was an imperfect way to gain training in physics because it did not test for experimental knowledge or skills.[27] As J.J. Thomson pointed out:

> The [tripos] question asked is often to find a relation between a number of mathematical symbols representing various physical quantities; nothing is asked as to what are the physical consequences resulting from this relation. This, however, is just the thing which is of interest to the physicist; it is as if he received a message in cipher and made no attempt to decode it.[28]

Beginning in the mid-nineteenth century, questions about emerging experimental areas, such as heat, electricity, and magnetism, occasionally were added to the Mathematical Tripos, but these were included only in the advanced part of the examination. Few candidates attempted to answer these less familiar questions; the new subjects were not taught in the lecture rooms of either the colleges or the University, which lacked both experienced teachers in these subjects and necessary apparatus for teaching them.

These drawbacks of the Mathematical Tripos were not corrected in the Natural Sciences Tripos established in 1851. Physics was classified with chemistry, and although knowledge of heat, electricity, magnetism, or sound sometimes was necessary to answer a question of mineralogy, geology, or physiology, the knowledge required was largely qualitative, not quantitative. The Natural Sciences Tripos explicitly excluded the mathematical subjects of mechanics and optics, as

---

[26] Ball, *Cambridge Papers*, 309-310.

[27] For the strength and weakness of the Mathematical Tripos in the nineteenth century, see P. M. Harman (ed.), *Wranglers and Physicists: Studies on Cambridge Mathematical Physics in the Nineteenth Century* (Manchester: Manchester University Press, 1985).

[28] J.J. Thomson, *Recollections*, 60. Brackets added.

well as two laws of thermodynamics. Indeed, the Natural Sciences Tripos barely survived its first two decades of existence. The idea of combining the Mathematical Tripos and the Natural Sciences Tripos to produce a balanced physics examination emerged slowly during the last quarter of the nineteenth century, beginning with the creation of the new professorship of experimental physics and the establishment of the Cavendish Laboratory in the beginning of the 1870s.[29] This development is discussed in greater detail in the first few chapters of this book.

---

[29] Wilson, "Experimentalists among the Mathematicians," 340-366.

CHAPTER 1

THE BEGINNING OF THE CAVENDISH TRADITIONS, 1871-1879

> Our principal work, however, in the Laboratory must be to acquaint ourselves with all kinds of scientific methods, to compare them, and to estimate their values
>
> Maxwell in his Introductory Lecture[1]

1.1. Preparing the Way

In 1868 the University of Cambridge, after much debate, finally added the subjects of heat, electricity, and magnetism to the new scheme of testing for Honours in the Mathematical Tripos. To facilitate teaching of the newly required subjects, on November 25, 1868, a Physical Science Syndicate was appointed "to consider the best means of giving instruction to students in physics, especially in Heat, Electricity and Magnetism, and the methods of providing apparatus for this purpose." After three months of thorough study, the Syndicate issued a detailed report on February 27, 1869 confirming the necessity for incorporating these subjects into the curriculum.[2] "No reason can be assigned," the Syndicate concluded, why the University should not become "a great school of physical and experimental as it is already of mathematical and classical instruction." To this end, the Syndicate recommended that the University found a new professorship and establish a "well appointed Laboratory" "to render the Professor's teaching practical." The chief duties of the new professor were "to teach and illustrate the laws of Heat, Electricity and Magnetism, to apply himself to the advancement of the knowledge of such subjects and to promote their study in the University."

Certain aspects of the Syndicate's report are noteworthy. First, in recommending the establishment of a new professorship, the Syndicate strongly emphasized the new professor's role as lecturer, not as researcher. The new professor was to provide "the large amount of additional teaching" needed to train the candidates for several examinations: the Mathematical Tripos (MT), the Natural Sciences Tripos (NST), and the ordinary degree examinations in chemistry and in mechanism and applied

---

[1] J. C. Maxwell, "Introductory Lecture on Experimental Physics," in W. D. Niven (ed.), *Scientific Papers of James Clerk Maxwell*, 2 vols. (Cambridge: Cambridge University Press, 1890), vol. 1, 241-255 on 250.

[2] *CUR* (16 November 1870): 93-96.

Figure 1.1. James Clerk Maxwell, the first Cavendish Professor of Experimental Physics (1871-1879) [Courtesy of the Cavendish Laboratory].

science. The new professor also would be responsible for providing additional training for candidates for the first examination for the medical degree (M. B.).

Second, the Syndicate stressed that the foundation of the new professorship "would be incomplete" without the establishment of a properly equipped laboratory. Building such a facility, the Syndicate estimated, would require the sum of £6,300: £5,000 for a new building and £1,300 for apparatus, cases, and furniture. The Syndicate also recommended that the annual salaries for the men who filled the new laboratory posts should be £500 for the professor, £100 for the professor's demonstrator, and £60 for a lecture-room attendant.

Third, the Syndicate recommended simultaneous establishment of the new professorship and laboratory. This emphasis on simultaneity indicated that the future laboratory was intended to be used as a property of the University for practical examinations, teaching, and occasional research rather than functioning as the private possession of its lead professor. As a University property, the future laboratory would have the benefit of automatic official recognition. By comparison, the laboratory of William Thomson at the University of Glasgow had not received official university recognition for its first sixteen years.[3]

The realization of this visionary scheme, however, depended on the University's ability to finance it. On May 13, 1869, another syndicate was appointed "to consider the means of raising the necessary funds for establishing a Professor and Demonstrator of Experimental Physics, and for providing buildings and apparatus required for that department of Science." The new Syndicate first asked several wealthy colleges "whether they would be willing to make contributions from their corporate funds," but the colleges approached were reluctant to share their resources for a University purpose. Next, the Syndicate investigated the possibility of funding the laboratory by raising the University's capitation tax and by using the financial resources of Cambridge's two building funds.[4] However, the financial problems raised by these possibilities were so complicated and proved so divisive that the new professorship would probably have to be delayed for a few years.[5] The future of the whole plan became uncertain.

Then, suddenly, after the Long Vacation, the Vice-Chancellor made the startling announcement of a "munificent offer of his Grace the Duke of Devonshire, the Chancellor of the University."

---

[3] Romualdas Sviedrys, "The Rise of Physical Science at Victorian Cambridge," *HSPS 2* (1970): 127-151 on 138. For more details about Thomson's laboratory, see David Murray, *Memories of the Old College of Glasgow: Some Chapters in its History of the University* (Glasgow: Jackson, Wylie and Co., 1927), 131-140; S. P. Thompson, *The Life of William Thomson,* 2 vols. (London: MacMillan & Co., 1910), vol. 1, chapter VII; Crosbie Smith & M. Norton Wise, *Energy & Empire: A Biographical Study of Lord Kelvin* (Cambridge: Cambridge University Press, 1989), 128-134.

[4] *CUR* (19 October 1870): 18-20. The original report was issued on May 31, 1870.

[5] *CUR* (26 October 1870): 49-51.

Holker Hall, Grange, Lancashire.

MY DEAR VICE-CHANCELLOR.

I have the honour to address you for the purpose of making an offer to the University, which, if you see no objection, I shall be much obliged to you to submit in such manner as you may think fit for the consideration of the Council and the University.

I find in the Report date Feb. 29 [27], 1869, of the Physical Science Syndicate, recommending the establishment of a Professor and Demonstrator of Experimental Physics, that the buildings and apparatus required for this department of Science are estimated to cost £6,300.

I am desirous to assist the University in carrying this recommendation into effect, and shall accordingly be prepared to provide the funds required for the building and apparatus, as soon as the University shall have in other respects completed its arrangements for teaching Experimental Physics, and shall have approved the plan of the building.

I remain,
My Dear Mr Vice-Chancellor,
Yours very faithfully
DEVONSHIRE.[6]

The Chancellor's generous offer inspired the colleges to move from stubborn reluctance to a new spirit of cooperation and they indicated readiness to provide remuneration for both the professor and his demonstrator. On November 28, 1870, two years after Cambridge had first considered the best means of giving physics instruction to its students, the Professorship of Experimental Physics was formally proposed in the Cambridge Senate. The proposal was approved on February 9, 1871.

The question of who would fill the new position now arose. Because the new chair would be very prestigious in Britain, a prominent figure was needed. After the new professorship was officially proposed in November of 1870, the Master of Peterhouse, H. W. Cookson, wrote to the most eminent British physicist at that time, William Thomson, with an offer of the proposed chair. Only one year previously, Thomson had turned down an offer of a Cambridge praelectorship in Science made to him by the Master of Trinity. Lady Thomson had been ill.[7] Now, Thomson rejected Cambridge's latest offer: he was too much involved in the economic, social and scientific life of Glasgow to consider a move to Cambridge.[8] Next, Hermann Helmholtz in Berlin was approached but, having just been appointed to the prestigious chair of physics once filled by Gustav Magnus, he was unable to leave Berlin.[9]

When it became clear that neither Thomson nor Helmholtz would accept the professorship, James Clerk Maxwell was pressed to be a candidate. Since 1865,

---

[6] *CUR* (19 October 1870): 13. Brackets added.

[7] Thompson, *Life of W. Thomson*, vol. 1, 558-562.

[8] See Smith & Wise, *Energy and Empire*.

[9] Thomson wrote Helmholtz to explain the offer and urged him to consider the invitation seriously because his acceptance would be "a great gratification and advantage to English scientific men." He himself, Thomson added, "would consider the difference of distance from Glasgow to Cambridge and Berlin a great gain." See Thompson, *Life of W. Thomson*, vol. 1, 564-566.

when he had resigned the chair of Natural Philosophy at King's College, London, Maxwell had lived at his estate, Glenlair, devoting himself to writing up the results of his investigations in heat, electricity and magnetism. His *Theory of Heat* would appear in 1871, and his famous *A Treatise on Electricity and Magnetism* would be published in 1873. Those who supported Maxwell's candidacy were uncertain whether he would abandon his comfortable life at Glenlair to come to Cambridge; nevertheless Stokes, John William Strutt, Rev. E. W. Blore (Vice-Master of Trinity), and others wrote to persuade him.[10]

Despite the hopes and the urgings of these eminent persons, Maxwell hesitated to stand for the professorship, giving as his reason his lack of experience in guiding students in experimental work.

> Glenlair, Dalbeattie,
> 15th February 1871
>
> MY DEAR BLORE
>
> Though I feel much interest in the proposed Chair of Experimental Physics, I had no intention of applying for it when I got your letter, and I have none now, unless I come to see that I can do some good by it.
>
> . . . I am sorry Sir W. Thomson has declined to stand. He has had practical experience in teaching experimental work, and his experimental corps have turned out very good work. I have no experience of this kind, and I have seen very little of the somewhat similar arrangements of a class of real practical chemistry. The class of Physical Investigations, which might be undertaken with the help of men of Cambridge education, and which would be creditable to the University, demand, in general, a considerable amount of dull labour which may or may not be attractive to the pupils.[11]

A few days later, however, Maxwell changed his mind and decided to stand for the chair "on the understanding that he might retire at the end of a year, if he wished to do so."[12] On February 24, therefore, Blore formally announced Maxwell's candidacy. There were no other candidates for the position and, on March 8, Maxwell was elected to the Professorship of Experimental Physics without opposition.

Interestingly, despite Maxwell's current stature in the history of physics, he was not first on the Cambridge list of candidates for its new physics professorship, but third. Unlike Faraday or William Thomson, Maxwell did not enjoy wide public recognition for his achievements during his lifetime. At the time of his appointment to Cambridge, Maxwell's name was relatively unknown outside scientific circles and unfamiliar even to Cambridge students. Horace Lamb, a student during the first few years of Maxwell's professorship, remembered that when Maxwell was appointed "he was little more than a name to many of us, except that on one or two

---

[10] Stokes sent two letters, on February 16 and 18, 1870, urging Maxwell to stand. See CUL MSS ADD 7655/II, 40 & 42 (also printed in Strutt, *Life of Rayleigh,* 48-49). For Blore's letter to Maxwell, see CUL MSS ADD 7655/II, 38A.

[11] CUL MSS ADD 7655/II, 39: Maxwell to E. W. Blore.

[12] This quotation comes from Campbell & Garnett, *Life of Maxwell*, the second edition (London: MacMillan and Co., 1884), 264.

occasions he had been responsible for some highly original questions set in the Mathematical Tripos."[13] Although Maxwell became better known after the publication of his *Treatise* in 1873, it was only after his death that he was ranked among physics' towering figures.

1.2. Physics Education at Cambridge during the 1870s

As Cambridge University's new Professor of Experimental Physics, Maxwell's first priority was to find the right Cambridge men to fill his lecture room and the future laboratory. Those targeted were, naturally enough, candidates for and graduates of Cambridge's Mathematical Tripos (MT) and Natural Sciences Tripos (NST). Nevertheless, enrolling MT students in professorial lectures during the 1870s was no easy task. Even Stokes' very popular lectures on optics and hydrodynamics—both subjects for the MT—attracted an average of only eighteen students a year during this period.[14] The difficulty in enrolling MT students in professorial lecture courses existed because the colleges had traditionally prepared students for the MT through a unique system of tutoring. Competition among tutors and between tutors and professors was intense, and tutors routinely discouraged their students from attending any professorial lectures that did not directly relate to the MT. For the NST, however, professorial lectures were the main sources of instruction, and professors who taught subjects included in the NST were likely to gain students for their lecture courses. In around 1875, for example, the more important University natural science classes each attracted "from twenty to thirty students," with elementary biology drawing a larger number and chemistry attracting "nearly a hundred."[15]

Maxwell attracted only a very disappointing number of students to his lectures. As he recorded in one of his notebooks, nineteen students registered to study heat during his first series of regular lectures for the Michaelmas term of 1871, a number that rose to twenty-six students for the Lent term, but dropped to ten for the Easter term and dwindled to eight for the Michaelmas term of 1872.[16] J. A. Fleming, who came to Cambridge in 1877, "with the object of benefiting by [Maxwell's] lectures and laboratory teaching," was surprised to find "a teacher who was everywhere regarded as the great living authority on his subjects lecturing to a class of two or three students in place of the 100 or more attentive listeners he would have had in any Scottish or German universities." "In the last year of his life," Fleming recalled,

---

[13] J.J. Thomson, et al. *James Clerk Maxwell: A Commemoration Volume, 1831-1879* (Cambridge: Cambridge University Press, 1931), 142.

[14] David B. Wilson, *Kelvin and Stokes* (Bristol: Adam Hilger, 1987), 48.

[15] *CUR* (17 March 1876): 299-358 on 353. In a Syndicate Report of 1876 on "the requirement of the University in different departments of study," the Board of Mathematical Studies asked for no additional teachers, whereas the Board of Natural Science Studies asked for seven additional teachers, among whom at least one was for physics.

[16] CUL MSS ADD 7655/Vn., 2.

"the only two attendants at his lectures were an American gentleman, Mr. Middleton, who was a resident at St. John's, and myself."[17]

The chief reason for this poor attendance was that Maxwell's lectures were not connected in an essential manner with preparation for the triposes. Judging from Fleming's notebooks on electricity and thermodynamics (1878-79), they also were too difficult and too demanding for most NST students.[18] For those who wanted to study experimental physics for the triposes, more practical ways existed: tutor's lectures and, after the Lent term of 1877, Elementary Experimental Lectures given by the "Demonstrator of Experimental Physics."

These lectures were originated by Maxwell to be given by his demonstrator, William Garnett, who was the 1873 fourth wrangler, and who had "impressed Maxwell [the Additional Examiner of that year] by the knowledge of physics displayed in his papers."[19] These elementary lectures were designed to provide students with a basic understanding of the subject matter and to expose NST and medical students to certain practical experiments. The course covered mechanical physics, heat, electricity, magnetism, and light, and required "no previous knowledge of Mathematics beyond Arithmetic."[20] The lectures were so successful that they "soon became popular and formed some of the largest Science classes in the University."[21] Nevertheless, the Elementary Experimental Lectures were of little help in recruiting future physicists, most of whom still came from the MT.

Cambridge's lack of a concrete and systematic program for physics education remained a real weakness of the University. Although physics had been included in both the MT and the NST, it was not yet treated as an independent subject. Even after the 1873 introduction of questions about heat, electricity and magnetism into the MT, most of the questions remained highly mathematical. In 1874, the Board of Mathematical Studies supervising the MT reported that "the questions on Electrodynamics and Magnetism were hardly attempted by any of the candidates."[22] Worse, college tutors were advising candidates to sacrifice depth for quantity in order to obtain higher marks. This advice ran counter to the intention of the examiners who, in 1876, regretfully observed "that very few of the Candidates appear to confine their reading to two or three of the Divisions." Lord Rayleigh, that

---

[17] Thomson et al., *A Commemoration*, 117-118. Brackets added.

[18] CUL MSS ADD 8082 (Thermodynamics), 8083 (Electricity). The latter contains the following subjects: fundamental experiments, electromotive force, insulator and insulating stands, Gauss' method of charging electrometer so as to avoid vibration, Cavendish's proof of the law of inverse squares, Maxwell's theory of the foregoing experiment, coefficients of induction and capacity, Maxwell's method of measuring specific induction capacity, Boltzman's specific induction capacity of gases, how to measure specific induction capacity, electrical induction, expression for energy of field, current distribution in network of conductors, Wheatstone's bridge, thermo-electricity, and potential. Even though Fleming was a candidate for the NST, the level of mathematics used in his notebook was certainly beyond that of the NST.

[19] *A History of the Cavendish Laboratory, 1871-1910*, 18. Brackets added.

[20] *CUR* (12 December 1876): 146.

[21] *A History of the Cavendish Laboratory*, 35.

[22] *CUR* (5 May 1874): 356.

year's additional examiner, felt "strongly that something ought to be done to prevent men getting up a quantity of bookwork without a proper study of the subject."[23] As long as the MT remained a competitive written examination, students would not achieve the desired level of knowledge about contemporary experimental subjects.

The importance of the NST in breeding future physicists increased considerably during the 1870s. When the NST had been introduced in 1851, physics was not considered a separate subject but was taught with chemistry, mineralogy, geology, and physiology. In 1871, the Board of Natural Sciences Studies, the supervising body of the NST, instituted several highly significant reforms which took effect in 1873. First, physics was considered a separate subject included with chemistry in the first division. Second, the NST was divided into two parts, with the first three days reserved for elementary questions, the last three days reserved for more advanced questions, and two additional days reserved for practical examinations. Third, all candidates for honours in the NST were required to show some knowledge of the elementary principles of chemistry and physics.[24]

The year 1876 saw another change that signaled acknowledgment, at last, of the immensity of the body of knowledge the NST covered. To permit serious students adequate time to study selected subjects in depth, the elementary part of the NST was given in June and the advanced and practical parts of the examination were given in December.[25] The results of the first examination, however, showed that the change had backfired. Some students, the Board of Natural Sciences Studies reported, had confined their efforts too narrowly to a single subject, "so that though they have shewn considerable knowledge of the details of their special subjects, they have not shewn sufficient acquaintance with the principles of the cognate and subsidiary subjects."[26] In response to this unforeseen outcome, the Board recommended regrouping the subjects, and physics was grouped with chemistry and mineralogy.[27] The NST now required a more thorough knowledge of every branch of natural sciences and covered more physical subjects than the MT.[28] The MT still aimed to foster the mathematical mind of gentlemen. Thus, the future of physics seemed brighter in the NST.

NST candidates became the chief beneficiaries of the offerings of the new Cavendish Laboratory, which formally opened on June 16, 1874. The Laboratory became the site of the NST practical examinations and, starting with the Lent term of 1877, also became the site of the Elementary Experimental Lectures. Beginning with the Michaelmas term of 1877, NST students were advised to take "Practical

---

[23] *CUR* (9 May 1876): 451. See also *CUR* (23 May 1876): 514-515.
[24] *CUR* (1 March 1871): 212-218.
[25] *CUR* (14 March 1876): 290-291. See also the Report of the Natural Science Studies in *CUR* (9 June 1874): 440-442. The second part of the examination was open only to "those who are declared to have so acquitted themselves as to deserve a B.A. degree, and no others."
[26] *CUR* (6 March 1877): 270.
[27] *CUR* (29 May 1877): 472-474.
[28] For the physics subjects of the NST in the 1870s, see *CUR* (1 March 1871): 213-214.

Physics" courses under the supervision of the professor or the demonstrator.[29] Moreover, the Laboratory remained open during the summer vacation, permitting candidates for the practical part of the NST to prepare for it during this time.

The NST, however, was not yet a conduit for first-rate physicists. Most of the talented students who came to Cambridge to study physics still took the MT. A major defect of the physics part of the NST was its exclusion of advanced mathematics. Whereas the elementary part of the MT required extensive knowledge of mathematics, the elementary part of the NST was limited to definitions, principles, laws, and simple applications and experiments.[30] NST candidates therefore spent a great deal of time memorizing definitions and laws rather than solving problems. Many Cambridge men decried this regrettable "divorce between Mathematics and Experimental Work" so detrimental to a sound physics education.[31]

The lack of a systematic physics program at Cambridge was well illustrated by the frequent overlap of physics lectures and a discontinuity of physics instruction. For example, in the Lent term of 1877, when Maxwell provided lectures on electricity and electrodynamics, so did three tutors. One year later, during the Lent term of 1878, Maxwell again delivered lectures on electricity, but no tutors did. Throughout the 1870s, only Coutts Trotter regularly delivered tutorial lectures on experimental physics, and these lectures were aimed mainly at candidates for the NST and medical students.

The University seemed unaware of this flaw in its physics instruction. An 1875 syndicate, reporting on "the requirement of the University in different departments of study," completely neglected to investigate the problem of overlapping and discontinuous instruction in physics.[32] The Syndicate merely compared the physics lecture lists of Cambridge, Berlin, and Leipzig and concluded that Cambridge was equal to, or perhaps superior to, its German counterparts in physics education. This conclusion no doubt was reached at least partly in opposition to a movement taking place in the 1870s to greatly increase the number of university professors and to place all teaching into their hands. In a series of articles, "Academic Teachers," Henry Sidgwick attempted to counter this so-called "Germanizing the University" movement by suggesting that Cambridge institute a relaxed system of inter-collegiate lectures and appoint more college lecturers or praelectors instead of more professors.[33]

---

[29] According to Newall, "the first organized course of practical experiments in physics was given in Michaelmas Term of 1879" (*A History of the Cavendish Laboratory*, 109). However, when the class actually began is not quite clear.

[30] See *CUR* (15 June 1875): 491, & (14 March 1876): 290.

[31] R. T. Glazebrook, *James Clerk Maxwell and Modern Physics* (New York: MacMillan & Co., 1896), 76.

[32] *CUR* (12 December 1877): 185-195.

[33] *CUR* (22 February 1871): 204-205; (1 March 1871): 221; (8 March 1871): 235-236.

1.3. Three Cavendish Traditions: Maxwell's Legacy as Director of the Cavendish Laboratory

In October 1871, Maxwell delivered his inaugural lecture, not in the Senate House as expected, but in "an out of the way lecture room."[34] The lecture was preceded by only one announcement and only "a score or so of students" attended. In this lecture, Maxwell described the Cavendish Laboratory's future role in the scientific life of Cambridge, as he intended it to be. After stating his optimistic assessment of the future of science, he outlined his views of experimental physics, and then presented his guidelines for the Laboratory:

> Let me say a few words on these two classes of experiments—Experiments of Illustration and Experiments of Research. The aim of an experiment of illustration is to throw light upon some scientific idea so that the student may be enabled to grasp it. The circumstances of the experiment are so arranged that the phenomenon which we wish to observe or to exhibit is brought into prominence, instead of being obscured and entangled among other phenomena, as it is when it occurs in the ordinary course of nature . . .
>
> In an experiment of research, on the other hand, this is not the principal aim. It is true that an experiment, in which the principal aim is to see what happens in the certain conditions, may be regarded as an experiment of research by those who are not yet familiar with the result, but in experimental researches, strictly so called, the ultimate object is to measure something which we have already seen—to obtain numerical estimate of some magnitude.
>
> *Experiments of this class—those in which measurement of some kind is involved, are the proper work of a Physical Laboratory*. In every experiment we have first to make our senses familiar with the phenomenon, but we must not stop here, we must find out which of its features are capable of measurement, and what measurements are required in order to make a complete specification of the phenomenon. We must then make these measurements, and deduce from them the result which we require to find.[35]

Maxwell here clearly expressed his intention that, under his directorship, the Cavendish Laboratory would not be used solely as a teaching laboratory for "Experiments of Illustration," but would also become a serious research center for carrying out precise measurement, something the University lacked. Maxwell was well aware of the tension existing between members of the University who emphasized teaching and those who emphasized research. He saw one of his responsibilities, as director of the Laboratory, as achieving a balance between these two aims that had been the focus of many lively discussions among Cambridge dons.[36] In a letter to his wife just after his appointment, Maxwell noted the existence of the tension and gave his view of how to resolve the problem:

> 20th March 1871
>
> There are two parties about the professorship. One wants popular lectures, and the other cares more for experimental work. I think there should be a graduation—popular

---

[34] J.J. Thomson et al., *A Commemoration*, 143.

[35] Maxwell, "Introductory Lecture," 242-244. Emphasis added.

[36] *A History of the Cavendish Laboratory*, 14.

> lectures and rough experiments for the masses; real experiments for real students; and laborious experiments for first rate men like Trotter and Stuart and Strutt.[37]

What kind of "real" and "laborious" experiments did Maxwell have in mind? Although Maxwell agreed that the main task of the future Cavendish Laboratory would be to arrive at precise measurements, in his inaugural address he envisioned a broader goal:

> This characteristic of modern experiments—that they consist principally of measurements—is so prominent, that the opinion seems to have got abroad, that in a few years all the great physical constants will have been approximately estimated, and that the only occupation which will then be left to men of science will be to carry on these measurements to another place of decimals.
>
> If this is really the state of things to which we are approaching, our Laboratory may perhaps become celebrated as a place of conscientious labour and consummate skill, *but it will be out of place in the University, and ought rather to be classed with the other great workshops of our country, where equal ability is directed to useful ends.*
>
> . . . But the history of science shew that even during that phase of her progress in which she devotes herself to improving the accuracy of the numerical measurement of quantities with which she has long been familiar, she is preparing the materials for the subjugation of new regions, which would have remained unknown if she had been contented with the rough methods of her early pioneers. I might bring forward instances gathered from every branch of science, shewing *how the labour of careful measurement has been rewarded by the discovery of new fields of research, and by the development of new scientific ideas*. But the history of the science of terrestrial magnetism affords us a sufficient example of what may be done by Experiments in Concert, such as we hope someday to perform in our Laboratory.[38]

Thus, for Maxwell, measurement was not the ultimate goal of experimentation. Rather, precise measurement should serve as a vehicle for opening new domains of inquiry. As he observed, "The increase in the accuracy and completeness of magnetic observations which was obtained by the new method, opened up fields of research which were hardly suspected to exist by those whose observations of the magnetic needle had been conducted in a more primitive manner."[39] For Maxwell, the proper work of the future Laboratory would not be merely to carry experiments "to another place of decimals." The "real" and "laborious" experiments Maxwell had in mind would be aimed at more ambitious understanding based on more complicated approaches that employed theoretical study:

> There is no more powerful method for introducing knowledge into the mind than that of presenting it in as many different ways as we can . . . It is therefore natural to expect that *the knowledge of physical science obtained by the combined use of mathematical analysis and experimental research will be of a more solid, available, and enduring kind than that possessed by the mere mathematician or the mere experimenter* . . .
>
> We shall therefore arrange our lectures according to the classification of the principal natural phenomena, such as heat, electricity, magnetism and so on.

---

[37] Campbell & Garnett, *Life of Maxwell*, 381.
[38] Maxwell, "Introductory Lecture," 244. Emphasis added.
[39] Ibid., 245.

> In the laboratory, on the other hand, the place of the different instruments will be determined by a classification according to methods, such as weighing and measuring, observations of time, optical and electrical methods of observation, and so on.[40]

Maxwell's view of experimental physics and his vision of the future Cavendish Laboratory certainly reached far beyond the expectation of the two parties. He believed firmly that experimental physics could flourish in the University only through the successful synthesis of mathematical and experimental sciences. "The combined use of mathematical analysis and experimental research" was not a new slogan: it had been the ideal of scientists since Francis Bacon. What was new in Maxwell's lecture was his aim to institutionalize this ideal in Cambridge University that had recognized only mathematical physics.

Several German-speaking physicists of the mid-nineteenth century had been advancing the harmonization of mathematics and the experimental sciences under the new title, theoretical physics. In an impressive study, *Intellectual Mastery of Nature: Theoretical Physics from Ohm to Einstein,* Christa Jungnickel and Russell McCormmach thoroughly recounted these developments.[41] As these authors pointed out, whereas mathematical physicists aimed to abstract natural phenomena and experimental physicists aimed to measure to "another place of decimals," theoretical physicists were doing both with different perspectives in mind: they often did experiments with their own hands, as was illustrated by Gauss, Weber, Helmholtz and, later, Hertz; and they were quite competent in mathematics and would "consult or even collaborate with a mathematician."[42] The strength of German theoretical physics rested not in its novelty or in the large number of physicists who practiced it, but in its successful institutionalization in the universities.

As Cambridge's sole professor of physics and director of the University's only physics laboratory, Maxwell had the opportunity to establish close links between lectures, laboratory training, and research by introducing a German-style physics curriculum. "If we succeed too well," he warned Strutt on March 15, 1871, just after his appointment, "and corrupt the minds of youth till they observe vibrations and deflections and become Senior Optimes instead of Wranglers, we may bring the whole University and all parents about our ears."[43] However, Maxwell believed, it was only through this new method of inquiry that the Laboratory could breed competent researchers to carry out "real" and "laborious" experiments as the Germans did, and thus realize Maxwell's ultimate dream:

> *Our principal work, however, in the Laboratory must be to acquaint ourselves with all kinds of scientific methods, to compare them, and to estimate their value.* It will, I think, be a result worthy of our University, and more likely to be accomplished here than in any private laboratory, if, by the free and full discussion of the relative value of

---

[40] Ibid., 247-249. Emphasis added.

[41] Christa Jungnickel & Russell McCormmach, *Intellectual Mastery of Nature: Theoretical Physics from Ohm to Einstein*, 2 vols. (Chicago: University of Chicago Press, 1986).

[42] Ibid., vol. 1, 185.

[43] Strutt, *Life of Rayleigh*, 49-50. Maxwell also said in the letter, "To wrench the mind from symbols and even from experiments on paper to concrete apparatus is very trying at first, though it is quite possible to get fascinated with a course of observation as soon as we have forgotten all about the scientific part of it."

> different scientific procedures, we succeed in *forming a school of scientific criticism, and in assisting the development of the doctrine of method.*[44]

Maxwell's emphasis on "criticism" and "method" foreshadowed the characteristics of the Laboratory-to-be: it would not be the school of one man.

In short, Maxwell's inaugural lecture of 1871 unveiled the three basic guidelines of the future Cavendish Laboratory. First, the new laboratory would be a serious research center as well as a teaching laboratory. Second, a major job of the new laboratory would be precise measurement with theory in mind. Third, the new laboratory would be a "school of criticism" that would embrace "all kinds of scientific methods," fostered by Maxwell's laissez-faire policy.

Maxwell's first job as Director of the Laboratory was to design and supervise the erection of the Laboratory building. He visited William Thomson's laboratory in Glasgow and R. B. Clifton's in Oxford to see what features of their arrangements might be adopted.[45] The new facility's architect, W. M. Fawcett of Cambridge, meticulously carried out all of Maxwell's suggestions. The work of arranging and furnishing the Laboratory took up a considerable amount of Maxwell's time, and he found himself lecturing wherever he could discover suitable space. "Lectures begin 24th" he wrote on October 19, 1872. "Laboratory rising, I hear, but I have no place to erect my chair, but move about like the cuckoo, depositing my notions in the Chemical Lecture-room 1st term; in the Botanical in Lent, and in Comparative Anatomy in Easter."[46] The building was finally completed in June of 1874. On the 16th of that month, Chancellor William Cavendish, the Duke of Devonshire, formally presented the University with his gift of the Cavendish Laboratory.

The Laboratory's structure attracted considerable interest among scientists, and nine days after its opening *Nature* published a detailed description of it (Figure 1. 2).[47] The ground floor held a magnetic room, a clock and pendulum room, a room for balances, a room devoted to calorimetry and other heat experiments, a battery room, and a storage room. On the first floor was a large laboratory room intended for the general use of students. The professor's private room was connected to the laboratory room by two hatches, through which he could glance, from time to time, to observe what was going on in the laboratory room. A large apparatus room and a large lecture room with connecting preparation room also were located on the first floor. The second floor was devoted to more advanced experiments. Here were rooms for studying acoustics, radiant heat, optics, electricity, and photography, and for calculating and drawing. The building was so well-designed that, on days when the air in the lecture room (on the first floor) was too damp for good electrical experiments, "an insulated wire connected with the prime conductor of the electrical machine [on the second floor] . . . [could be used to] supply electricity on the lecture

---

[44] Maxwell, "Introductory Lecture," 250. Emphasis added.

[45] A blueprint of the Clarendon Laboratory at Oxford is included in Maxwell's manuscripts. See CUL MSS ADD 7655, Vj., 1.

[46] Glazebrook, *Maxwell and Modern Physics*, 73.

[47] Anonymous Editorial. "The New Physical Laboratory of the University of Cambridge," *Nature 10* (1874): 139-142.

table."[48] A comparison of this facility with those of German physics institutes (for example, that of the University of Strasbourg's Physical Institute, completed in 1883) shows one big difference.[49] German physics institutes featured many *small rooms* for use by individual researchers, but the Cavendish, intended primarily for teaching and not research, had no such private space.

After three years as Cavendish Professor, Maxwell realized that the link between his lectures and laboratory training (or research) would not be cemented in the near future, and that his endeavor to do so would inevitably result in some friction between him and other physics teachers in the University. The traditional system was too firmly entrenched to permit radical change in only a few years. Maxwell's alternative was to seek students who had already passed the examinations and therefore were free to follow their own interests. Such students had been the first to enter the Cavendish Laboratory, and it was through such students that a research tradition could be implanted and nurtured in the University. Maxwell concentrated his efforts on training such students, safely leaving the burden of teaching elementary classes to Garnett.

Maxwell's technique for training research students was unique. Instead of providing know-how, he left it to students to find out what apparatus to use and how to use them. He allowed students to choose their own experiments. He provided students with ideas for research subjects only when they asked him for suggestions, as did G. Chrystal and Fleming. Maxwell always encouraged students, "even when he thought a student was on a wrong track."[50] He once told Arthur Schuster, "I never try to dissuade a man from trying an experiment. If he does not find what he wants, he may find out something else." This "laissez-faire" approach did not mean that Maxwell neglected his duty as director. Rather, "Maxwell took an active part in every research that was carried out in the Laboratory. He used to come round daily, enter the various rooms, and say a few words, discussing results or making suggestions." [51] A recollection by Glazebrook provides one example of the seriousness with which Maxwell viewed his role as director:

> When difficulties occurred Maxwell was always ready to listen. Often the answer did not come at once, but it always did come after a little time. I remember one day, when I was in a serious dilemma, I told him my long tale, and he said:—"Well, Chrystal has been talking to me, and Garnett and Schuster have been asking questions, and all this has formed a good thick crust round my brain. What you have said will take some time to soak through, but we will see about it." In a few days he came back with—"I have been thinking over what you said the other day, and if you do so-and-so it will be all right."[52]

---

[48] Ibid., 141. Brackets added.
[49] See David Cahan, "The Institutional Revolution in German Physics, 1865-1914," *HSPS 15:2* (1985): 1-65 on 28-29.
[50] *A History of the Cavendish Laboratory*, 38-39.
[51] Ibid., 31.
[52] Glazebrook, *Maxwell and Modern Physics*, 78-79.

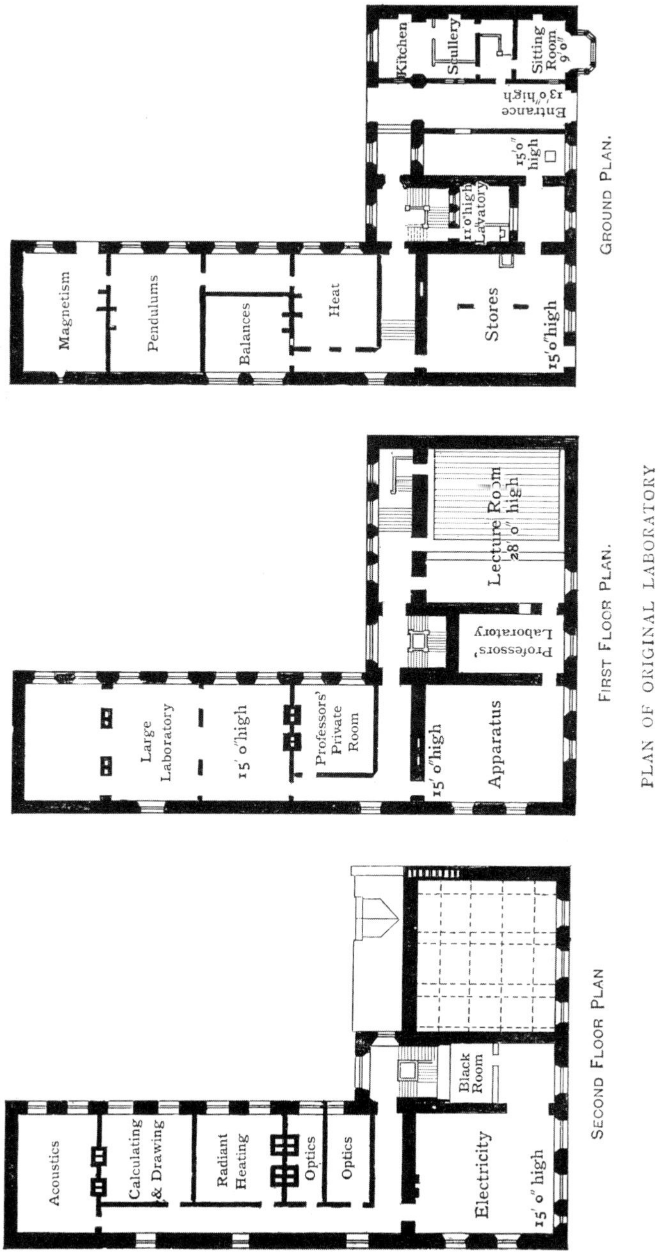

Figure 1.2. Floor Plan of the Cavendish Laboratory [From A History of the Cavendish Laboratory, 16].

Figure 1.3. Lecture Room in the Cavendish Laboratory [Courtesy of the Cavendish Laboratory].

Maxwell strove to encourage students to think creatively. He was firmly committed to the conviction that research should not be determined by the director's opinion or wish.

On the whole, Maxwell played the role of a great teacher rather than that of a bureaucratic director. Because the research group was quite small, Maxwell was easily approached by his students. He was a generous counselor who permitted his students or former students to contact him whether he was at Cambridge or at Glenlair. The correspondence between Maxwell and Chrystal shows that Maxwell continued to act as Chrystal's adviser after the latter left the Laboratory in 1877. Maxwell answered Chrystal's questions about thermo-electricity, provided him with "the most ingenious if not the best method of comparing capacities," and suggested that Chrystal pursue an interesting research topic of "testing the resistance of batteries by the telephone."[53] Schuster, who was deeply influenced by Maxwell, summarized the impact of Maxwell's encouragement on the highly gifted and self-confident students who came to study at the Cavendish:

> . . . Having originated new and fertile ideas in all branches of Physics, Maxwell might easily have found students eager to work out in detail some problem arising out his theoretical investigations. This would have been the recognized method of a teacher anxious to found a "school"; but it was not Maxwell's method. *He considered it best*

[53] CUL MSS ADD 8375. 15 (1 January 1878), 18 (17 July 1878), 19 (8 August 1878): Maxwll to G. Chrystal.

> *both for the advance of science and for the training of the student's mind, that everyone should follow his own path.* His sympathy with all scientific inquiries, whether they touched points of fundamental importance or minor details, seemed inexhaustible . . . It was the seriousness with which he discussed all ideas put before him by his students that, from the beginning, gave the Laboratory its atmosphere of pure and unselfish research.[54]

Schuster's comments carry special weight because he was familiar with physics instruction in both England and Germany, having received his Ph.D. under G. Kirchhoff at Heidelberg and having performed research under W. Weber at Göttingen and Helmholtz at Berlin.[55]

The philosophy that Maxwell attempted to implant in the Cavendish Laboratory was that of laissez-faire. He carefully separated his researches from those of his students, carrying out his own researches essentially without assistance from his students. Significantly, during his tenure at the Cavendish, Maxwell never attempted to carry out the experiment that might directly prove the existence of electromagnetic waves. This was "always a matter of surprise" to Fleming and to many other readers of his works.[56] Maxwell's famous *Treatise* had been published in 1873, only one year before the Cavendish laboratory opened in 1874. The Cavendish might have seemed the right place in which to attempt to prove the existence of electromagnetic waves under the guidance of Maxwell, now the director of the Laboratory, but this did not happen. Nor did Maxwell push any Cavendish researcher in the direction of a theoretical examination of the problem. This behavior might be explained, at least in part, by Maxwell's desire to foster a laissez-faire policy in the Laboratory.

---

[54] *A History of the Cavendish Laboratory*, 38-39. Emphasis added.

[55] For the life and work of A. Schuster, see A. Schuster, *The Progress of Physics during 33 Years (1875-1908)*, reprint (New York: Arno, 1975) and his *Biographical Fragments* (London: MacMillan, 1932), both of which include many vivid descriptions of scientific figures of his time. For more information about Schuster's laboratory in Manchester, see *The Physical Laboratories of the University of Manchester: A Record of 25 Years' Work* (Manchester: Manchest University Press, 1906), 1-38. This book was prepared by Schuster's former students and assistants in commemoration of the 25th anniversary of his election to the professorship.

[56] J.J. Thomson et al., *A Commemoration*, 123. For more on this interesting question, see Thomas K. Simpson, "Maxwell and the Direct Experimental Test of His Electromagnetic Theory," *Isis 57* (1966): 411-432. Simpson suggested that Maxwell's main concern lay in "the nature of light and of light-bearing medium," not in electromagnetic phenomena. To support this argument, Simpson pointed out that the topic of electrical oscillations—the key to proving the theory experimentally—was not fully investigated by Maxwell but by William Thomson. Furthermore, in Maxwell's investigation of electromagnetic radiation from an oscillating circuit, "the point at which Maxwell solves the basic equations to yield the wave equation—falls in Chapter 20 of Part 4, on the 'Electromagnetic Theory of Light,' the interest of which is explicitly with optics." However, Simpson overlooked Maxwell's life-long interest in experiments, which is easily corroborated by the *Treatise* as well as by Maxwell's manuscripts. Simpson's argument seems weak because, whether Maxwell's emphasis lay in optics or electromagnetism, the possibility of testing the theory experimentally nevertheless existed.

Instead, and surprisingly, Maxwell spent a great deal of time on a seemingly much less important task: editing Henry Cavendish's manuscripts since 1874.[57] *An Account of the Electrical Researches of the Honourable Henry Cavendish, F.R.S., between 1771 and 1781* was published in 1879, and it contributed significantly to Cavendish's reputation. Immediately after finishing this project, Maxwell started to revise the *Treatise*: "I am preparing a new edition of Elec. & Mag.," he wrote to Chrystal in July of 1878, "Have you any errata on improvements in Electrostatics? I am revising the Chapter on System of Conductors."[58] He was prevented from completing this task by his failing health, and died the following year. Although Maxwell succeeded in revising only the first nine chapters of the *Treatise*, these chapters clearly indicate "how extensive were the changes intended by Professor Maxwell both in the substance and in the treatment of the subject."[59] We are left to wonder what Maxwell might have achieved if he had concentrated his valuable last few years on improving his electromagnetic theory or attempting to prove it in the Laboratory.

The real importance of Maxwell's decision not to push this specific line of research was that the Cavendish lost its chance to become the preeminent center of electromagnetic research. In Britain, most of the leading "Maxwellians" were to be found outside Cambridge: J. H. Poynting in Birmingham, S. Thompson and Oliver Heaviside in London, G. F. FitzGerald in Dublin, and Oliver Lodge in Liverpool. Except for Poynting, who had carried out some research on the gravitational constant under Maxwell's direction, the Maxwellians had learned Maxwell's theory from his books and had little connection with the Cavendish Laboratory.[60] In Cambridge, Maxwell's theory was developed in two directions. Studies in the theoretical and mathematical aspects of Maxwell's work were carried out in various College rooms by Niven, Glazebrook and, later, by J.J. Thomson. In the Laboratory, some attempts were made to experimentally confirm the theory by measuring the ratio, known as "v," between the electrostatic and the electromagnetic units. These attempts were made by W. M. Hicks, J. E. H. Gordon, and G. Crystal in the 1870s

---

[57] In his role as a historian of science, Maxwell "copied out [the manuscripts that he obtained from the Duke of Devonshire] with his own hand; he saturated his mind with the scientific literature of Cavendish's period, he repeated his experiments, being especially attracted by his device to use himself as a galvanometer to measure (Cavendish measured everything he came near) currents of electricity by the physiological effects they produced when they passed through him." See J.J. Thomson et al., *A Commemoration*, 22.

[58] CUL MSS ADD 8375. 16 (9 July 1878): Maxwell to Chrystal.

[59] James Clerk Maxwell, *A Treatise on Electricity and Magnetism*, 3rd edition, 2 vols. (New York: Dover, 1954), vol. 1, xiii (preface to the second edition by Niven). The remaining work was carried by W. D. Niven, with assistance from Professor Charles Niven and J.J. Thomson, then a fellow of Trinity College. Except for the first nine chapters, the second and third editions were essentially reprints of the first, with some additional footnotes and revised mathematics.

[60] See Jed Z. Buchwald, *From Maxwell to Microphysics* (Chicago: University of Chicago Press, 1985), & Bruce J. Hunt, *The Maxwellians* (Ithaca: Cornell University Press, 1991). Both studies indicate that it was these Maxwellians who developed and disseminated Maxwell's electromagnetic theory.

and by J.J. Thomson, L. R. Wilberforce, W. Cassie, and H. F. Newall in the 1880s.[61] These researchers, however, attempted to prove Maxwell's electromagnetic theory of light indirectly, and their efforts often were poorly organized. During these two decades, all of the Cambridge men who explored Maxwell's electromagnetic theory, with the sole exception of J.J. Thomson, were less enthusiastic about it than the "Maxwellians" outside Cambridge.

1.4. Researchers and Researches

Who were the Cavendish Laboratory researchers of the 1870s? This question is difficult to answer because no official records were kept of the students who registered to study there, and the Laboratory was open to every member of the University. One valuable source of names of such researchers is the "List of Those Who have carried out Researches at the Cavendish Laboratory" in *A History of the Cavendish Laboratory 1871-1910,* which was published to celebrate J.J. Thomson's twenty-five years of tenure of the Cavendish Professorship. This source will be used to identify the Cavendish researchers, with some modifications whenever further evidence is found.[62]

The number of research workers in the Laboratory was very small, and never exceeded ten per year. The majority of these researchers had already passed triposes, meaning that they were Cambridge graduates. Although the Laboratory was open to undergraduates, Cambridge tutors often discouraged their students from studying at the Laboratory by characterizing the Cavendish as "intended for the research work of graduates, . . . [with] no accommodation for young students."[63] This characterization was inaccurate. When undergraduate H. F. Newall courageously applied to Garnett, he was welcomed and the demonstrator "took much trouble in finding simple apparatus for [him] to work with."[64] The only other undergraduates who worked in the Laboratory were A. P. Trotter and Fleming. Fleming, however, was not a typical undergraduate. When he went to Cambridge in 1877 specifically to study and do research under Maxwell, Fleming was a thirty-two-year-old scholar who had received a Bachelor of Science from the University of London in 1870 and had taught science at Rossall School and at Cheltenham College.[65] In 1879, one year before Fleming took the NST, he received a Doctor of Science degree from the University of London.

[61] *A History of Cavendish Laboratory*, 19-20 & 132-135. For the British effort to measure "v," see Simon Schaffer, "Accurate Measurement is an English Science," in M. Norton Wise (ed.), *The Values of Precision* (Princeton: Princeton University Press, 1995), 135-172.

[62] M. Price, J. A. Hughes & S. Schaffer, *The Cavendish Laboratory: Introduction to Prosopography,* unpublished typescript (Cambridge: Whipple Museum for the History of Science, 1991) is also consulted but this work covers only the period between 1874 and 1895.

[63] *A History of the Cavendish Laboratory*, 104. Brackets added.

[64] Ibid.

[65] "Fleming, Sir (John) Ambrose 1849-1945," *DNB* (1941-1950), 258-260.

Cambridge University had no formal graduate programs or systems that might have supported Maxwell's ambitious programs. By the nineteenth century, residence requirements for the M.A. had virtually disappeared, and at both Cambridge and Oxford the degree was awarded to graduates with three or four years' standing.[66] After graduation, students could remain at Cambridge for more study, to prepare for contests such as the Smith or Adams Prize, or to write dissertations for College fellowships. However, only highly talented and highly motivated graduates could or would remain at Cambridge after graduation with the hope of being elected as college fellows. With the number of students from the middle or lower classes increasing, the problem of providing financial support for them was becoming more serious; scholarships for graduates were few in number, and the teaching opportunities on which most graduates depended for financial support were generally given first to elected fellows.

Table 1.1. Researchers at the Cavendish Laboratory, 1874-1879

Name	*Years at the Cavendish*	*Cambridge Years*	*MT/NST*
W. M. Hicks *	1873-74	M. 1869, A.B. 1873, A.M. 1876	7th Wrangler (1873)
W. Garnett	1874-80	M. 1869, A.B. 1873, A.M. 1876	5th Wrangler (1873)
W. D. Niven *	1875	M. 1862, A.B. 1866, A.M. 1869	3rd Wrangler (1866)
G. Chrystal ***	1875-77	M. 1871, A.B. 1875, A.M. 1878	2nd Wrangler (1875)
J. E. H. Gordon	1875-78	E. 1872, A.B. 1875	Junior Optime 12th (1875)
S. A. Saunder	1875-76	M. 1871, A.B. 1875, A.M. 1878	14th Wrangler (1875)
A. W. Clayden	1876-78	M. 1873, A.B. 1877, A.M. 1880	NST Second Class (1876)
C. T. Heycock	(1876?)	M. 1877, A.B. 1881, A.M. 1884	NST First Class (1880)
A. Schuster *	1876-81		
R. T. Glazebrook *	1876-97	M. 1872, A.B. 1876, A.M. 1879	5th Wrangler (1876)
A. P. Trotter	1876-79	M. 1876, A.B. 1880	NST Second Class (1879)
J. A. Fleming *	1877-82	M. 1877, A.B. 1881, A.M. 1884	NST First Class (1880)†
D. MacAlister **	1977-78	M. 1873, A.B. 1877, A.M. 1880	1st Wrangler (1877)
A. W. Sunderland	1877	M. 1872, A.B. 1876, A.M. 1879	7th Wrangler (1876)
J. H. Poynting *	1878-80	M. 1872, A.B. 1876, A.M. 1879	3rd Wrangler (1876)
W. N. Shaw *	1879-1900	M. 1872, A.B. 1876, A.M. 1879	16th Wrangler & NST First Class (1876)†

*: *F.R.S.*
**: *1st Smith Prize Winner*
***: *2nd Smith Prize Winner*
†: *Distinguished in Physics*

M: entered in Michaelmas Term
E: entered in Easter Term

The Cavendish Laboratory thus played the role of an *informal graduate school* filled with highly talented and motivated graduate students. The first group of research students was a very select one. As Table 1.1 indicates, ten out of the total of fifteen were wranglers (with eight out of these ten having placed within the first

[66] For more infromation, see E. Rudd & R. Simpson, *The High Education: A Study of Graduate Education in Britain* (London: Routledge & Kegan, 1975), 6-15.

tenth), two were Smith Prize men, two obtained First Class in the NST with distinction in Physics, nine were Fellows of Colleges, and seven eventually became Fellows of the Royal Society. It was a remarkable group. More surprising is the fact that these students came to the Laboratory of their own will; neither outside pressure nor academic requirements propelled them to the Cavendish. Schuster, Fleming, Poynting, and Shaw came to study under Maxwell: Schuster, the Laboratory's only non-Cambridge man, was recruited by Maxwell; Fleming entered St. John's College with the specific aim of working with Maxwell; Poynting resigned a demonstratorship at Owens College to return to Cambridge to study under Maxwell; Shaw was recalled from Berlin at Maxwell's request. Most of the remaining students came to the Cavendish because it provided rooms, apparatus, and an eminent teacher. Hicks, Chrystal, Gordon, Saunder, Glazebrook, and MacAlister entered the Laboratory the same year they received their B.A. degrees; Clayden entered after taking the NST but before receiving his B.A.; Sunderland entered after one year of independent study. Niven was a tutor and, after 1875, an examiner for the MT. Garnett was Maxwell's demonstrator.

As a graduate school, the Cavendish Laboratory faced many problems. To begin with, there were no regulations to aid in the supervision of the graduate researchers. Only Maxwell and Garnett were available to direct them. Maxwell did not provide these students with solid experimental training programs, an omission that was a serious defect because the students desperately needed such guidance: they were principally mathematicians with little experience in experiments (see Table 1.1). For example, Glazebrook had come to the Laboratory to find a dissertation topic for his Trinity Fellowship. On this occasion, Maxwell took the opportunity to explain to him the principle of the Wheatstone bridge. Glazebrook was a fifth wrangler who a few weeks earlier had failed to successfully answer the question posed by Rayleigh to "explain the Wheatston's [sic] Bridge method of measuring resistance," a question many research students could not have answered when they first came to the Cavendish.[67] Although these students were talented and although Maxwell's laissez-faire policy had great merits, these students needed the basic training which Maxwell failed to provide in any systematic fashion. Maxwell also lacked any means by which to provide financial aid to the research students who, understandably, often were lured away by paying jobs offering security. This unfortunate circumstance led to relatively short stays at the Laboratory and to numerous abandoned projects.

These defects of the Cavendish are illustrated in the number of research papers published from 1874 to 1879 (Table 1.2). The only productive researcher in the Laboratory was Maxwell, who produced forty articles.[68] Schuster, who had earned a German Ph.D. before coming to Cambridge, was the second most prolific writer,

---

[67] J.J. Thomson et al., *A Commemoration*, 133.

[68] Since 1871 Maxwell had written several popular articles relating to physics, among them entries for the *Encyclopaedia Britannica*, articles for the *Kensington Museum Handbook*, several book reviews, and even a few brief biographical notes. For these articles, see Niven, *Scientific Papers of Maxwell*, vol. 2.

producing eight papers. The true Cambridge graduates failed to match the productivity of Maxwell and Schuster because they spent a good deal of time learning how to carry out experiments: J. E. H. Gordon and Chrystal each produced two papers; S. A. Saunder, Niven, Glazebrook, W. N. Shaw, and D. MacAlister each published one paper. There were also two collaborative works (Chrystal-Saunder and Clayden-Heycock). Garnett, the demonstrator, was the only Cambridge researcher beside Maxwell who did not need to worry about financial security, but he was unable to perform his own research without interruption.[69] Later, when Rayleigh reduced the demonstrator's burden by creating a second demonstratorship, both demonstrators became the Laboratory's chief researchers.

Table 1.2. Number of Research Papers Published at the Cavendish, 1874-1879

Name	*Individual Work*	*Co-Work*
J. C. Maxwell	40	
A. Schuster	9	
J. E. H. Gordon	2	
G. Crystal	2	1/2 x 1
S. A. Saunder	1	1/2 x 1
W. D. Niven	1	
R. T. Glazebrook	1	
D. MacAlister	1	
W. N. Shaw	1	
A. W. Clayden		1/2 x 1
C. T. Heycock		1/2 x 1
Total: 11	58	2

* *Each co-work is divided by the number of participants.*

Although the number of research papers produced at the Cavendish was not impressive, the contents of these papers were. They often reflected their authors' considerable talents and presaged brilliant future careers. Electricity and magnetism were the most popular subjects of these papers and represented roughly half the studies performed during the first two decades of the Laboratory. Among them were three unsuccessful attempts to prove Maxwell's electromagnetic theory of light. W. M. Hicks, who was "fired with the desire of measuring experimentally the velocity of propagation of electromagnetic waves," designed "a piece of apparatus" that he hoped would do so.[70] Gordon carried out a series of experiments to determine Verdet's constant, the ratio of the amount of rotation of plane polarized light and the intensity of the magnetic field. He intended his experiment to produce "the number

[69] *A History of the Cavendish Laboratory*, 35.
[70] Ibid., 19.

in optical measure which is equivalent to unity in electromagnetic measure," which would make it possible to express quantities in electromagnetic measure "in optical measure by multiplying them by this number."[71] Gordon also attempted to test Maxwell's electromagnetic theory of light by working "on the measurement of the specific inductive capacities of transparent dielectrics."[72] However, neither of these experiments produced the satisfactory results Gordon anticipated.

The Laboratory's stock of apparatus was still poor during the 1870s, and the transfer of a number of British Association (B. A.) resistance coils to the Laboratory from the Kew Observatory provided inspiration for several studies. Maxwell initiated an experiment to measure the B. A. units of electricity at frequent intervals, and he even planned a redetermination of the units.[73] Around 1875-76, Chrystal launched a series of experiments in the Cavendish to test Ohm's law. The project was originally conceived by a B.A. committee comprised by Maxwell, Schuster, and J. D. Everett, but Chrystal carried out all the experiments. Under Maxwell's supervision, Chrystal devised three different kinds of experiments and virtually confirmed the law.[74] The first results of the use of the resistance coils at the Laboratory appeared in 1876, when Chrystal and Saunder reported having carefully measured "the difference between the resistances of the several coils at some standard temperature, and also the coefficients of variation of resistance with temperature in the neighborhood of the standard temperature."[75] Chrystal also investigated the influence of induced magnetism on the galvanometer that caused "an apparent departure from Ohm's law."[76] After Chrystal left Cambridge, Glazebrook and Fleming continued the project, and Fleming designed a very convenient bridge, known as "Fleming's banjo," for quick comparison of resistances. Saunder also worked on a new form of Leclanché's cell and illustrated "how the electromotive force of this cell varies when it works for a considerable time through circuits of various resistances."[77]

---

[71] J. E. H. Gordon, "On the Determination of Verdet's Constants in Absolute Units," *Phil. Trans. 167* (1877): 1-34 on 33.

[72] J. E. H. Gordon, "Measurements of Electrical Constants. No. II. On the Specific Inductive Capacities of Certain Dielectrics. Part I," *Phil. Trans. 170* (1879): 417-446.

[73] For this project, see Schaffer, "Accurate Measurement is an English Science." Schaffer argued that Maxwell carried out the project of determination of ohm with concrete ideas and determination.

[74] Report of the Committee, consisting of Professor Clerk Maxwell, Professor J. D. Everett, and Dr. A. Schuster, "For Testing Experimentally Ohm's Law," *B. A. Report* (1876): 36-63. The first experiment adopted a bridge which consisted of five resistance coils, but it had some defects. To remedy this, the second experiment employed a method in which resistances were compared by strong and weak currents. The currents passed so fast alternatively in a second that "the temperature of the wire could not sensibly alter during the interval between one current and the next." Ohm's law was confirmed by this second method. The third experiment employed an induction coil, and its peculiar results led Chrystal to his next paper, "On Bi- and Unilateral Galvanometer Deflection".

[75] G. Chrystal & S. A. Saunder, "Results of a Comparison of the British-Association Units of Electrical Resistance," *B. A. Report* (1876): 13-19 on 13.

[76] G. Chrystal, "On Bi- and Unilateral Galvanometer Deflection," *Phil. Mag. 2* (1876): 401-414 on 402.

[77] S. A. Saunder, "On the Variations of the Electromotive Force of a New Form of Leclanché's Cell," *Nature 12* (1875): 564-565 on 564.

Beside electricity and magnetism, other areas of investigation flourished in the Laboratory. Niven worked to find a simpler way to calculate trajectories of projectiles; he employed three different methods to reduce the number of calculations required.[78] After A. W. Clayden and C. T. Heycock, during a course of chemistry lectures on spectrum-analysis, initiated an experiment on the spectrum of indium in which they found sixteen lines "instead of the three lines expected," they moved their research to the Cavendish and spent a summer there producing a new table.[79] Glazebrook confirmed Fresnel's theory of double refraction by measuring the velocity of the normal propagation of plane waves in a biaxal crystal.[80] Sunderland worked on the anomalous dispersion in fuchsin to determine the refractive indices as close as possible to the absorption band, but never finished this work.[81] MacAlister worked with Maxwell's improved version of the set of the sphere and hemispheres to test the inverse square law, the subject of Henry Cavendish's research.[82] Poynting's research on the gravitational constant and Shaw's meteorological experiments both were initiated during Maxwell's tenure, but they were completed only after his death.

Schuster proved himself an exceptional figure among the early Cavendish researchers by producing as many papers as all the other Cambridge men combined. Initially Maxwell had recruited him because of his interest in Ohm's law but, as soon as Schuster arrived at Cambridge, his interest changed to spectroscopy and he became a pioneer of that field in Britain.[83] His studies of the presence of oxygen in the sun and on the spectra of metalloids are particularly important. Schuster also became one of the first physicists to examine the conductivity and discharge of electricity through gases.[84]

As can be seen from this survey, a variety of research subjects characterized the Cavendish under Maxwell. This variety was due partly to students' "definite purpose" for coming to the Laboratory "of working out either an idea of their own or one which had been suggested to them."[85] This variety also was due to the wide latitude in choosing dissertation topics that Maxwell allowed students, like Poynting and Glazebrook, who came to the Laboratory to complete their fellowship dissertations. However, most of the credit for the extensive range of research conducted in the early Cavendish belongs to Maxwell and his policy of laissez-faire.

---

[78] W. D. Niven, "On the Calculation of the Trajectories of Shot," *Proc. Roy. Soc. 26* (1876): 268-287.
[79] A. W. Clayden & C. T. Heycock, "The Spectra of Indium," *Phil. Mag. 2* (1876): 387-389.
[80] R. T. Glazebrook, "An Experimental Determination of the Values of the Velocities of Normal Propagation of Plane Waves in different directions in a Biaxal Crystal, and a Comparison of the Results with Theory," *Phil. Trans. 170* (1878): 287-377.
[81] *A History of the Cavendish Laboratory*, 28.
[82] Ibid., 33.
[83] A. Schuster, "Researches on the Spectra of Metalloids," *Nature 15* (1877); 401-402; "Spectra of Metalloids," *Nature 15* (1877): 447-448; "On the Presence of Oxygen in the Sun," *Nature 17* (1877): 148-149; "On the Spectra of Metalloids. Spectrum of Oxygen," *Pro. Roy. Soc. 27* (1878): 383-388, & *Phil. Trans. 170* (1879): 37-54; "An easy Method for Adjusting the Collimator of a Spectroscope," *Phil. Mag. 7* (1879): 95-98; "On Spectra of Lightning," *Phil. Mag. 7* (1879): 316-321.
[84] A. Schuster, "On the Passage of Electricity through Gases," *Proc. Camb. Phil. Soc. 3* (1877): 57-61.
[85] *A History of the Cavendish Laboratory*, 29.

The Cavendish researchers were, in general, very ambitious. Despite their limited experience and lack of training, they aimed high. Due to lack of training, Hicks and Gordon failed; Clayden, Heycock, Saunder had to settle for simple experiments; and Glazebrook required about two years to produce results. However, once they had gained experience, their productivity rose rapidly. For example, in the four years from 1880 to 1884, Glazebrook produced more than fifteen research papers.

As a graduate school, the Cavendish Laboratory desperately needed organizational support for providing training and guidance to its talented but inexperienced students. Maxwell failed to create this structure. Under Maxwell, the Cavendish had no mechanism for coordinating students' activities: each student was related to the Laboratory or to other students only through Maxwell. The relationship between the director and the students existed on a one-to-one basis only. The Cavendish researchers had in common that they were all Maxwell's students, but they were not yet a community. Maxwell was a man of ideas, not of organization. By 1879, he had managed to turn the Cavendish Laboratory into a place for doing research but he had not motivated its students to become a research community or a school. This task would fall to Maxwell's successors, Lord Rayleigh and J.J. Thomson.

CHAPTER 2

RAYLEIGH'S DIRECTORSHIP, 1880-1884

To myself they were perhaps the happiest I ever spent.

R. T. Glazebrook[1]

The remarkable credibility of this landed aristocrat, wrangler, administrator and physicist first emerged during a brief but decisive period of work on the standard unit and value of electrical resistance between 1879 and 1884.

Simon Schaffer[2]

2.1. The Election of Lord Raleigh

Maxwell's health began to fail early in 1879, and he died in Cambridge on November 5, 1879 at the age of 48. For the Cambridge scientific community, his death was a double calamity: not only had they lost a great scientist, but his passing also threatened the future of the Cavendish Laboratory. According to the regulations of his professorship, the chair was to "terminate with the tenure of office of the Professor first elected, unless the University by Grace of the Senate shall decide that the Professorship shall be continued." On November 20, the Senate passed a resolution that "the Professorship of Experimental Physics, established by Grace of the Senate Feb. 9, 1871, be continued, subject to the regulations then enacted so far they are now applicable."[3] Cambridge leaders concerned with physical science were eager to attract to the position a worthy successor to Maxwell.

William Thomson led the list of potential candidates, but it soon became clear that he could not be induced to reconsider his refusal of 1871. Thereafter, John William Strutt, who in 1873 had succeeded his father to become the third Lord Rayleigh, emerged as the preferred candidate. In 1865, he had been both senior wrangler and the first winner of the Smith Prize. Since 1869, he had published more than sixty papers on acoustics, wave propagation, jet propulsion, optics, spectroscopy, and electromagnetism.[4] His monumental two-volume treatise, *Theory*

---

[1] *A History of the Cavendish Laboratory*, 74.
[2] Simon Schaffer, "Rayleigh and the Establishment of Electrical Standards," *European Journal of Physics 15* (1994): 227-285 on 277.
[3] *CUR* (18 November 1879): 121.
[4] See *Scientific Papers by John William Strutt, Baron Rayleigh*, 6 vols., reprinted (New York: Dover, 1964), vol. 1.

Figure 2.1. Lord Rayleigh (John William Strutt), Second Cavendish Professor (1879-1884) [Courtesy of the Cavendish Laboratory].

of Sound (1877-1878), had made him one of the most respected scientists of his generation. His scientific talents were such that "his mathematical analysis seemed to flow naturally into the most concise and elegant form, and, whatever might be the difficulty of the subject, it was never increased by any obscurity or ambiguity as to the meaning of the writer."[5] Rayleigh also was an excellent experimentalist who "needed nothing for his experiments but some glass tubing and a few pieces of sealing wax," and who could see "distinctly the essential points of an experiment or a measurement" and keep "that in view throughout."[6] With these talents, Rayleigh was easily regarded as Maxwell's proper successor.

Rayleigh also had special ties to the Cavendish Laboratory. He had helped overcome Maxwell's reluctance to stand for the Professorship and had been considered the most suitable alternative should Maxwell reject the post.[7] It is likely, therefore, that he felt some responsibility for guaranteeing the continuity of the Professorship. After Maxwell's death, he was urged by many, including William Thomson and Stokes, to stand for the office.[8] The Chancellor of the University himself attempted to persuade Rayleigh to accept the chair:

> Though it is perhaps somewhat unreasonable to ask you to undertake duties the discharge of which would involve heavy demands on your time, and might very probably be attended with no small personal inconvenience, I feel so strongly the advantage the University would derive from your acceptance of the office, that I hope you will allow me as Chancellor of the university, and also as taking a special interest in this Professorship, to support the appeal which I am told is about to be made to you, and to express a hope that you will consent to take the proposal into your favourable consideration.[9]

To his letter the Chancellor attached a memorial expressing the widespread view at Cambridge that Rayleigh's election "would tend greatly to the advance of Physical Science and the advantage of the University."

At first, Rayleigh was reluctant to stand for the chair. He was a hereditary peer, and at that time it seemed unthinkable to many that a member of the nobility might become a professor (at the annual salary of a mere £500). Eventually he did accept the candidacy, despite his mother's objection, and it has been suggested that economic considerations played a role in his decision. The agricultural depression of 1879 had made it difficult for Rayleigh to maintain his private laboratory at Terling, and the salary attached to the chair would provide him with "means of experimenting and zealous and duly instructed assistants and volunteers." Rayleigh

---

[5] J.J. Thomson et al., "Lord Rayleigh. O. M., F. R. S. (a collective obituary)," *Nature* 103 (1919): 365-369 on 365 (J.J. Thomson).

[6] Ibid., 366 & 368 (R. T. Glazebrook). Rayleigh's interest in experimentation went back to his Cambridge years of the 1860s. He attended Stokes' lectures and tried to learn experimental skills from him but was soon disappointed. He became a self-trained experimentalist.

[7] Strutt, *Life of Rayleigh,* 48-49. In his letter from Cambridge, Rayleigh said, "Some people thought that if [Maxwell] would not, I was the proper person. It is now I believe nearly certain that he will come, and so I am relieved of having to make a difficult decision."

[8] Ibid., 99-100.

[9] Ibid., 100-101.

agreed to stand for the chair on the condition that he would take the post for a short period only, perhaps "3 or 4 years, if they can get no one else fit for the post."[10] He was elected to the chair on December 12, 1879.

2.2. Organizational Changes

Rayleigh's first task as director of the Cavendish Laboratory was to consolidate its organization. Over the next five years, he would create another demonstratorship and two new assistant demonstratorships, improve the Laboratory's teaching system, raise new funds, and expand the stock of instruments. As a result of these improvements, the Cavendish Laboratory would emerge as the University's center for the teaching of experimental physics to undergraduates as well as a center for graduate research.

During Maxwell's stewardship of the Cavendish, the number of students attending the Elementary Lectures given by demonstrator Garnett had increased beyond the capacity of a single teacher. Garnett's lecture courses drew from 18 to 42 undergraduate students and "from 2 or 3 to about 20" of the more advanced students who conducted private studies at the Laboratory.[11] When Garnett resigned his post for a college job, Rayleigh asked the Vice-Chancellor to divide Garnett's duties and stipend into two demonstratorships at salaries of £75 each per year. The Vice-Chancellor obliged, and in the fall of 1880 Rayleigh appointed R. T. Glazebrook and W. N. Shaw, two able researchers who had worked with Maxwell, to the new positions.[12]

In creating the additional demonstratorship, Rayleigh had two goals. The first was to strengthen the Cavendish's teaching system. The second was to ease the burden on the individual demonstrator so he could pursue his own research. The demonstratorships indeed offered "the best possible training for independent teaching posts both in Cambridge and elsewhere."[13] Each demonstrator was expected to fulfill his teaching responsibility by attending the Laboratory from 10:00 a.m. to 5:00 p.m. three days a week. "On the other three days [they] were free for research work, College duties, or private teaching, as the case might be."[14] In time, both new demonstrators were appointed to additional University teaching posts: in 1884 Glazebrook became Mathematical Lecturer and in 1887 Shaw

---

[10] Ibid., 99-100.

[11] *CUR* (3 June 1879): 676. Maxwell's annual report of 1879 gives an idea of the number of students attending the Demonstrator's elementary lectures: for the 1878 Easter term, lectures on Electrostatics and the Elements of Thermo-electricity and Voltaic Electricity, a class of 33; for the 1878 Michaelmas term, lectures on Voltaic Electricity and Electro-magnetism, including the principal electrical measurements required in telegraphic engineering, & on Experimental Mechanics, two classes, together numbering 18; for the 1879 Lent term, lectures on Heat, & on Hydrostatics, two classes, together numbering 42.

[12] *CUR* (15 March 1881): 407; (29 March 1881): 438. It is interesting to note that the official recognition of the establishment of the second demonstratorship by the Senate was long delayed.

[13] *CUR* (2 March 1880): 329.

[14] *A History of the Cavendish Laboratory*, 47-48.

became Natural Science Lecturer. Both men also received remuneration as examiners and their stipends from the University would soon be increased. Thus, although the initial demonstrator stipend had been reduced by half, the real income of the new demonstrators eventually became larger than that of the former demonstrator, Garnett. The Cavendish demonstratorships (and later the assistant demonstratorships) were to become posts highly sought after by Cavendish researchers who wanted to remain at the Laboratory to conduct their own studies.

The first challenge for Rayleigh and his two new demonstrators was designing an efficient teaching system to meet the University's increasing demand for physics education. One source of the growing demand for the Laboratory's services was a flow of students preparing for medical examinations. Since 1875, the Board of Medical Studies (which became the Special Board for Medicine in 1882) had regularly published a list of physics lectures required to be attended by candidates for the first examination for the medical degree (M. B.).[15] These requirements made medical students a large majority of the students attending elementary physics lectures. Another source of the demand was precipitated by changes in the NST regulations made two months before Rayleigh's election, in October of 1879, when the Board of the NST published new regulations to take effect in 1881.[16] In the reformed NST, the elementary and advanced parts of the examination would be taken a year apart, and physics would become a completely separate subject. A third source of demand for physics education stemmed from a highly encouraging recommendation made by the Board for Biology and Geology and the Board for Physics and Chemistry. In December of 1882, the two Boards issued a report in which they deemed it desirable "to encourage Mathematical Students to study Physics not only in books but in the Laboratory, and not to confine their reading to those parts of the subject which admit of Mathematical treatment."[17] To this end, the Boards recommended that the advanced part of the NST be opened to honours graduates of the Mathematical Tripos, a recommendation that was instituted at the University in 1883. Meanwhile, from 1882 on, the MT was divided into three parts. The first two parts were almost the same as those of the 1870s, but the newly introduced third part included higher subjects of mathematics and was open only to wranglers.[18] All these changes in the MT and the NST brought more students to the Laboratory.

To meet the diverse needs of the different groups served by the Laboratory and

---

[15] *CUR* (5 October 1875): 15-17. To meet both oral and practical examinations, hydrostatics, heat, and electricity were recommended subjects.

[16] *CUR* (28 October 1879): 64-65; a slightly amended report was published in the next year (*CUR* (10 February 1880): 280-282).

[17] *CUR* (12 December 1882): 225.

[18] See *CUR* (27 February 1883): 423-424, & Wilson "Experimentalists among Mathematicians," 369-371. The more mathematical the MT became, the more central the NST was to physics education. The combination of the MT and the NST then became the major way to train Cambridge physicists: Wilson pointed, "From 1871 to 1881 half of the students emphasizing physics in the NST had already taken the MT, and from 1882 to 1889 about two-thirds taking physics in part II of the NST had already taken the MT (p. 352)."

its staff, Rayleigh and his two demonstrators organized two levels of courses, elementary and advanced. Courses at both levels included lecture classes and practical classes. The object of the elementary courses was to provide basic knowledge and practical experience to students who aimed to take the first part of the NST, the first examination for the M. B, or the special examination for natural sciences. In these elementary courses, the demonstrators instructed large numbers of students using "Experiments of Illustration." The demonstrator would set up a selected experiment in such a way that the students would observe intended effects. Glazebrook and Shaw later compiled these experiments in textbooks.[19]

The Laboratory's advanced courses were designed to address the needs of candidates for the advanced part of the NST, and these courses differed significantly from the elementary courses. In the advanced courses, the demonstrators lectured on more difficult subjects, used higher levels of mathematics, and allowed students

Table 2.1. Courses Taught by Glazebrook and Shaw (1883-84)

[From CUR (17 October 1883): 74-75; (27 November 1883): 213; (16 January 1884): 354-355; (21 April 1884): 611-612]

Michaelmas Term

*Demonstrations on Heat (advanced) by Demonstrator
*Demonstrations on Electricity and Magnetism (Advanced) by Demonstrator
*Practical Physics by Professor or Demonstrators
*Geometrical Optics by Mr. Glazebrook (Trinity)
*Advanced Physics by Mr. Shaw (Emmanuel)
*Elementary Physics for the NST (Part I) by Mr. Shaw (Emmanuel)

Lent Term

*Demonstrations on Mechanics and Heat (elementary) by Demonstrator
*Demonstrations on Electricity and Magnetism (advanced) by Demonstrator
*Practical Physics by Professor or Demonstrators
*Advanced Physics (continued) by Mr. Shaw (Emmanuel)
*Elementary Physics (continued) by Mr. Shaw (Emmanuel)
*Hydrostatics and Heat by Mr. Shaw (Emmanuel)
*Elementary Physics by Mr. Glazebrook (Trinity)
*Wave Theory of Light by Mr. Glazebrook (Trinity)

Easter Term

*Demonstrations on Optics and Electricity (elementary) by Demonstrator
*Demonstrations on Light and Sound (advanced) by Demonstrator
*Elementary Physics (continued) by Mr. Glazebrook (Trinity)
*Elementary Physics (continued) by Mr. Shaw (Emmanuel)

[19] *A History of the Cavendish Laboratory*, 46-47.

much greater freedom in experimentation. Students in the advanced courses were expected to attempt to perform "original work" or to endeavor to replicate some "important researches." There, "practically nothing in the way of description of the experiments or explanation of their theory or instruction as to their details was written out by the Demonstrator."[20]

An important feature of Rayleigh's teaching innovations was that they were aimed at Cambridge undergraduates and thus helped fulfill the intent of the Laboratory's original planners. Glazebrook and Shaw, acting as demonstrators or college lecturers, carefully organized their elementary and advanced classes so that the lectures and experiments would complement each other. The two demonstrators would first lecture on heat, electricity, magnetism, optics, or another subject and then move the class to the laboratory room to supervise student experiments, as can be seen in the list of lectures they gave during the academic year of 1883-84 (See Table 2.1).

While Glazebrook and Shaw concentrated their energies on undergraduate teaching, Rayleigh gave his lectures to the relatively few most advanced students

Figure 2.2. An experimental class at the Cavendish [Courtesy of the Cavendish Laboratory].

[20] Ibid., 47 & 253.

who used the Laboratory for their own studies. These students were allowed to perform research in the Laboratory during the Long Vacation.[21] Whereas Maxwell had offered regular lectures each term on heat, electricity, and magnetism, Rayleigh lectured not only on traditional subjects, such as "Electricity and Magnetism" or "Acoustics," but also on topics related directly to his own experimental work, such as "Unit of Electrical Resistance," "Electrical Measurement," or "Current Electricity and Its Practical Application." Rayleigh's lectures drew smaller audiences than those of his demonstrators, but much larger audiences than the lectures of his predecessor, Maxwell. For example, Rayleigh's lectures on current electricity and on acoustics were attended by "from 20 to 30" students, not by three, as had been true for Maxwell's lectures.[22]

Rayleigh's consolidation of the Cavendish teaching system had the desired outcome of providing the University with unified and organized physics instruction. Rising enrollments marked Raleigh's approach as a clear success. Under Maxwell, Garnett's Easter-term elementary course of 1878 had been attended by thirty-three students whereas, under Rayleigh, Glazebrook's Easter-term elementary course of 1884 drew forty-seven students and Shaw's drew twenty-three.[23] During Rayleigh's tenure, the number of students attending practical classes rose dramatically, from thirteen in the Easter term of 1880 to thirty-two in the Michaelmas term of 1881, seventy-two in the Lent term of 1882, and fifty-eight in the Easter term of 1882.[24] The number of demonstrator-run courses rose even more impressively, from three (excluding Practical Physics) in 1879-80, to eight in 1880-81, nine in 1881-82, and fifteen in 1882-83 and in 1883-84.[25]

In light of the rising number of physics students and courses, Rayleigh could justifiably ask the University for better salaries for his demonstrators and for a larger staff. In the spring of 1881, Rayleigh asked the Vice-Chancellor to raise the demonstrators' stipends from £75 to £100, and his request was granted within ten days.[26] In the fall of 1883, the General Board of Studies recommended that the demonstrators' stipends be increased by £50, and that two future assistant demonstrators each receive stipends of £60.[27] In January of 1884, those two new positions were filled by J. H. Randell and J. C. McConnel.[28]

When Rayleigh announced his resignation in the fall of 1884, the teaching program of the Cavendish Laboratory was far better organized than the one he had inherited only five years earlier. Three levels of lectures in various experimental subjects now were being offered, undergraduate enrollment had dramatically

---

[21] *CUR* 7 June 1881): 638. The Laboratory was kept open for these advanced students in July and August.
[22] *CUR* (16 May 1884): 733.
[23] *CUR* (3 June 1879): 676, and *A History of the Cavendish Laboratory*, 41.
[24] *CUR* (13 June 1883): 880.
[25] The data are based on the lecture lists in the *CUR* from 1879 to 1884.
[26] *CUR* (15 March 1881): 407.
[27] *CUR* (13 November 1883): 153.
[28] *CUR* (8 January 1884): 330.

increased, graduate enrollment had increased slowly but steadily, and the staff size had increased by one demonstrator and two assistant demonstrators. The change in the Laboratory was really "striking," according to Newall, who had been the Laboratory's first undergraduate student during the 1870s and who had returned to the Laboratory after a four-year absence.[29]

Rayleigh made another important change at the Cavendish, a change in the Laboratory's financial footing. Finance is a crucial component of every institution, but Maxwell's contribution to this important aspect of the Laboratory was poor. After the laboratory building was completed, Maxwell asked almost nothing from the University. To meet the needs for ordinary maintenance and laboratory attendant wages, the Museums and the Lecture Rooms Funds supplied the Cavendish with the minimal amount of approximately £250 per year. Further necessary funds were supplied by Maxwell from his own purse. As Glazebrook remembered, "During the last few years of his life Maxwell had himself provided what was wanted, expending many hundreds of pounds in this manner, and the demands made on University funds had been but small."[30] Despite Maxwell's 1876 report that "the Chancellor ha[d] now completed his munificent gift to the University by furnishing the Cavendish Laboratory with apparatus suited to the present state of science,"[31] the reality was quite different. Cambridge University, unlike its German counterparts, did not give financial support to research.

After Rayleigh accepted the Cavendish directorship, he quickly moved to set up a new fund, the so-called "Apparatus Fund," by circulating a letter for this purpose in the spring of 1880. A slightly modified version of this letter was published later in the annual report of the Museums and Lecture Rooms Syndicate:

> On visiting the Cavendish Laboratory in December last after my appointment to the Professorship of Experimental Physics, I was at once struck by the great deficiency of apparatus. . . . To illustrate my meaning I may mention that there is no steam engine or other prime mover; nor among the acoustical apparatus is there any musical instrument. Even with an adequate outfit, a considerable annual expenditure is necessary for renewals, and to meet the wants of students engaged in original research.
>
> Knowing that the University is not likely for several years to be in a position to meet the want, and feeling that Cambridge ought not to remain in this respect behind several Continental and American Universities, I have been endeavouring to raise an Apparatus Fund, to be spent in 8 or 10 years at the discretion of the Professor, by inviting contributions from persons interested in Cambridge and in Science. . . . I should suppose that £2500 will be required during the next ten years to put the institution upon a proper footing . . .[32]

Rayleigh's proposal met with a warm response from various circles, including nobles and scientists, and within two months £1,825 was pledged. By 1881

---

[29] *A History of the Cavendish Laboratory*, 110.
[30] Ibid., 43.
[31] *CUR* (15 May 1877): 433, in Maxwell's annual report. According to Maxwell's account book, the Chancellor paid about £943 for the apparatus in June, 1876. Brackets added.
[32] CUL MSS ADD 7655. III (c). 1: Circulating letter of Lord Rayleigh for the Apparatus Fund for the Cavendish Laboratory, dated 10 March 1880. This letter was published in *CUR* (18 May 1880): 557.

Rayleigh was able to report that twenty-four contributors had donated £2,025. About £500 of that amount had been spent, and the remaining £1,500 had been invested for future needs.[33] At last, the Cavendish Laboratory had secured an independent source of funding. Although the funds were not large enough to meet the entire need, they could be spent entirely for research purposes.

Rayleigh raised more money by levying fees for the use of certain laboratory facilities. His first official notice as Professor announced that students taking the practical physics course would pay two guineas per term for instruction by the professor or demonstrator.[34] The number of students in the practical physics course grew rapidly, and so did the amount collected in fees. For the 1881-82 academic year, these fees totaled £343, which was used to defray "the larger part of the current expenditure on materials and small apparatus."[35] Rayleigh assumed responsibility for collecting these fees and making the necessary purchases.

As the result of Rayleigh's efforts, the Laboratory's stock of instruments increased substantially. Maxwell had relied heavily on instruments which had been privately donated or constructed within the Laboratory itself. In 1875, according to an annual report by Maxwell, a number of instruments had been presented to the Cavendish and three instruments had been constructed on site by Garnett and others. The chief donors of the apparatus were the Chancellor and Maxwell; others included the late C. Babbage, Mrs. Faraday, Rayleigh, Prof. James Thomson, H. M. Tresca; and the British Association of Electrical Standards.[36] Under Rayleigh, donations of equipment declined sharply while direct purchases increased steadily, if slowly. Rayleigh first purchased instruments needed for undergraduate instruction, and then purchased more sophisticated instruments for research use.[37] In 1882, for example, he purchased the instruments he used in his celebrated experiments to determine the ohm.

Rayleigh's annual reports revealed that the Cavendish was purchasing a greater number of instruments manufactured by instrument companies. This development was aided by the opening of new instrument manufacturing firms in Cambridge. During Maxwell's tenure at the Cavendish, Cambridge had no scientific instrument-maker of note. By comparison, William Thomson in Glasgow was fully supported in his experimentation by the local optician and instrument-maker, James White.[38]

---

[33] *CUR* (17 May 1881): 562.

[34] *CUR* (3 February 1880): 272. During the Long Vacation (July and August), the fee was £2. 2s. See *CUR* (7 June 1881): 638.

[35] *CUR* (13 June 1883): 880 & 901. Rayleigh thought that large fraction of these fees "should go to the Demonstrators, whose work is very laborious and inadequately remunerated by the official salaries."

[36] *CUR* (27 April 1875): 352-354.

[37] Students in the elementary practical classes required basic instruments, "such as galvanometers." In his annual report in 1881, Rayleigh stated that a large portion of the total expenditure had been spent on such miscellaneous apparatus (*CUR* (17 May 1881): 562). In his last annual report, in 1884, Rayleigh stated that "a large part of the current expenditure is upon materials and small goods which it would be useless to enumerate" (*CUR* (16 May 1884): 733).

[38] M. J. G. Cattermole and A. F. Wolfe, *Horace Darwin's Shop: A History of the Cambridge Scientific Instrument Company, 1878 to 1968* (Bristol: Adam Hilger, 1987), 11.

At Cambridge, the considerable distance between experimenter and instrument-maker had made for severe drawbacks, as Maxwell noted:

> It has been felt that experimental investigations were carried on at a disadvantage in Cambridge, because the apparatus had to be constructed in London. The experimenter had only occasional opportunities of seeing the instrument maker, and was perhaps not fully acquainted with the resources of the workshop, so that his instructions were imperfectly understood by the workman. On the other hand the workman had no opportunity of seeing the apparatus at work, so that any improvements in construction which his practical skill might suggest were either lost or misdirected.[39]

According to Maxwell's account book, one of the Cavendish's chief instrument suppliers, between 1873 and 1878, had been the London firm, Elliot Brothers, from which Maxwell had ordered resistance coils and a magnetometer.[40]

However, beginning in the spring of 1878, James Stuart, the University's Professor of Mechanism, began to construct instruments for University members, and the Cavendish Laboratory soon became one of his main clients.[41] According to Rayleigh's annual report in 1881, Stuart's workshop, during the academic year of 1880-81, supplied the Cavendish with reflecting galvanometers, electromagnetic regulators, an optical bank, and a double-refraction apparatus for combining prismatic colors.[42]

Less than a year after Stuart began this enterprise, in January of 1879, Robert Fulcher and A. G. Dew-Smith opened a mechanical shop in Cambridge "to supply apparatus of all kinds used in Physiological Research, including Recording and Electrical Apparatus."[43] Fulcher had worked under Maxwell beginning in 1876. He had "constructed in a satisfactory manner several pieces of new apparatus" for the Cavendish, and then moved to Stuart's workshop where he continued to construct instruments for the Cavendish.[44] Dew-Smith had worked with Michael Foster, Professor of Physiology, and had been providing the physiological laboratory with financial support for laboratory equipment since the 1870s.[45] Fulcher and Dew-Smith's business flourished, but their partnership lasted less than a year. On December 8, 1880, Fulcher left the company and Dew-Smith formed a partnership

---

[39] *CUR* (15 May 1877): 434.

[40] CUL MSS ADD 7655. V. (j). 3 (Maxwell's account book); II 144 (31 October 1877) & 165 (28 October 1878), both from the Elliot Brothers to Maxwell.

[41] For Stuart's workshop, see T. J. N. Hilken, *Engineering at Cambridge University, 1783-1965* (Cambridge: Cambridge University Press, 1967), 63-74.

[42] *CUR* (17 May 1881): 563. In 1882, Fresnel's interference apparatus and, in 1883, galvanometers were delivered from Stuart's workshop to the Laboratory.

[43] From an advertisement dated January 10, 1879. See Cattermole & Wolfe, *Horace Darwin's Shop,* 14.

[44] Quotations from *CUR* (2 April 1978): 420. Maxwell continued, "He has also made improvements in the working of instruments already in the Laboratory. He has also done work for other departments of the University . . . It now seems likely that his time will continue to be fully employed, and it is expected that much of the apparatus to be used in Cambridge, which has hitherto been ordered from London, may, in future, be constructed in Cambridge, and tested while in the maker's hands by those who are to use it."

[45] See Gerald L. Geison, *Michael Foster and the Cambridge School of Physiology* (Princeton: Princeton University Press, 1978), 182.

with the engineering genius, Horace Darwin. On January 1, 1881, the Cambridge Scientific Instrument Company opened its doors.

From its beginning, the Cambridge Scientific Instrument Company, popularly known as "Horace's shop," enjoyed a special relationship with the Cavendish. Darwin had some personal connections with the Laboratory: he had once worked there with his brother, George Darwin, on explorations of the gravitational effects of the moon; he also assisted Rayleigh's experiment on the determination of the ohm by testing and mounting the water engine.[46] By March of 1881, three months after its opening, the company was billing the Laboratory for a newly constructed commutator and for repair work on apparatus donated by the British Association.[47] During the company's first year, it also sold the Laboratory a fine wire galvanometer and a circular iron table.[48] In subsequent years it supplied the Laboratory with other instruments, including divided circles for polarimeters and a reading microscope.[49] Although Rayleigh still ordered many instruments from Elliot Brothers in London—for example, most of the coils and apparatus employed in his famous ohm-measurement experiments—,[50] he now was able to enjoy the benefits of having a scientific instrument-maker close at hand. Horace's shop also took advantage of the proximity of the Laboratory by seeking advice from the Cavendish staff; Glazebrook once helped Darwin design a cathetometer.[51]

The Cavendish also was developing its own workshop. Under Maxwell, a small workshop had been completed on the ground floor, and "two instrument makers [one of them very probably Fulcher] were frequently employed."[52] However, when Stuart opened his workshop in 1878, both of these instrument-makers moved to his workshop and were not replaced until Rayleigh took charge of the Laboratory. After Rayleigh and his two demonstrators sought an able mechanic to lead the workshop, George Gordon was appointed to the position of Laboratory instrument-maker. Gordon was an impressive man. "Much of the apparatus used in the Laboratory was his handiwork, and, though he was far from being a skilled instrument-maker, he soon acquired from the Professor some of his own capacity for attending closely to essentials, and the results were admirable."[53] Later, Rayleigh took Gordon with him to Terling to work in his private laboratory. As the Cavendish researchers became increasingly active, the development of the workshop became an increasingly important concern. The most economical way to acquire needed instruments was to

---

[46] H. Darwin & G. H. Darwin, "First Report of B. A. Committee on Lunar Disturbance of Gravity," *B. A. Report* (1880): 25-26; G. H. Darwin, H. Darwin, et al., "Second Report of B. A. Committee on Lunar Distarbance of Gravity," *B. A. Report* (1882): 95; Rayleigh & A. Schuster, "On the Determination of the Ohm [B. A. Units] in Absolute Measure," in *Scientific Papers of Rayleigh*, vol. 2, 5-6.
[47] Cattermole & Wolfe, *Horace Darwin's Shop*, 27.
[48] *CUR* (22 May 1882): 582.
[49] *CUR* (16 May 1884): 753. See also Cattermole & Wolfe, *Horace Darwin's Shop*, 27-31.
[50] *Scientific Papers of Rayleigh*, vol. 2, 38, 40, 88, and 290.
[51] Cattermole & Wolfe, *Horace Darwin's Shop*, 31.
[52] *A History of the Cavendish Laboratory*, 35. Brackets added. See also J.J. Thomson et al., *A Commemoration*, 111.
[53] *A History of the Cavendish Laboratory*, 46. See also Strutt, *Life of Rayleigh*, 104.

make them in the Laboratory where, importantly, a researcher could exert control on an instrument's manufacture and also learn from the manufacturing process.

2.3. Rayleigh's Determination of the Ohm

As discussed in the previous section, Rayleigh managed to improve the Cavendish's organization within a very short period of time. During that time, he also was conducting the important experiments that resulted in his determination of the ohm. Rayleigh had not come to Cambridge with a specific research program in mind for himself and the Laboratory's research students. However, he soon decided to begin a new project in order to "identify the laboratory with some research planned on an extensive scale so that a common interest might unite a number of men sharing in the work."[54]

Finding subject for the intended research project was not difficult. Cambridge was "so saturated with the subject [of electricity]," Rayleigh later remembered, "that I quickly came to the conclusion that it would be best to make it the subject of my own research."[55] In the spring of 1880, Rayleigh made up his mind to resume work on the determination of the unit of electric resistance in the hope of producing a measurement of greater accuracy. Maxwell had once called the determination of the value of electrical resistance "the cardinal operation in electricity, in the same sense that the determination of weight is the cardinal operation in chemistry."[56] Rayleigh's choice was influenced by the availability of the necessary apparatus. It was also prompted by industry's urgent need for standardized electrical units. With the advent of the telegraph, the measurement of electrical resistance had acquired commercial importance. No standard value of resistance had been accepted, and scientists in different countries adopted different sizes of copper or iron wire as their standards.[57]

To remedy the confusion, in 1861 the British Association organized the Committee on Standards of Electrical Resistance. In 1862, in its first report, this committee recommended the development of a system which would "bear a definite relation to units which may be adopted for the measurement of electrical quantity, current, and electromotive force," and which also would "bear a definite relation to

[54] Strutt, *Life of Rayleigh*, 109.
[55] *CUR* (2 February 1882): 283. Brackets added. Rayleigh said that he had no particular interest in electricity before coming to Cambridge. During the 1870's, Rayleigh published more than sixty papers on various subjects, but only two of these were more or less related to electricity. See *Scientific Papers of Rayleigh*, vol. 1.
[56] Maxwell, *Treatise*, vol. 1, 465 [article 335].
[57] See *A History of the Cavendish Laboratory*, 59. Glazebrook described the situation as follows: "Thus Lenz in 1838 used the resistance of one foot of No. 11 copper wire. Wheatstone in 1843 proposed one foot of copper wire weighing 100 grains. In 1846 Hankel used a length of iron wire, while in 1848 Jacobi issued as a standard a certain length of copper wire known since as Jacobi's standard. When telegraphs were introduced . . . in England a mile of No. 16 copper wire was adopted as a unit, in Germany the German mile of No. 8 iron wire was used, and in France the kilometre of iron wire 4 millimetres in diameter."

the unit of work, the great connecting link between all physical measurement."[58]

To serve these purposes, the Committee chose for a standard Weber's approach, in which the unit of resistance was defined by the quantity of energy dissipated in unit time when unit current was passed through a length of wire. For the experiment, it adopted a method devised by William Thomson in which a circular coil of insulated wire revolved about a vertical axis with uniform velocity, and a small magnet was suspended at the center of the coil. When current flowed through the revolving coil, the deflection of the magnet was easily observed. This deflection also was influenced by Earth's magnetism but was independent of its intensity. The amount of deflection, therefore, depended only on the resistance of the coil and some constants of the apparatus. The resistance of the coil then was obtained from the measurement of the constants of the apparatus and from the magnetic deflection on the angular velocity of the coil.

The Committee also decided that, for practical purposes, material units 10^9 times larger than the absolute unit would be issued and that, from time to time, correction would be made. The experimental determination of the resistance had been entrusted to: Maxwell, when he was Professor of Natural Philosophy at King's College, London; Balfour Stewart, who was Director of the Kew Observatory where the B. A. units would be stored; and Fleeming Jenkin, an excellent electric engineer. A detailed account of the experiment was published in the British Association Report of 1863.[59]

Although great efforts were made to eliminate every possible error in the experiment, subsequent investigations revealed that the measurement was inaccurate by more than 1 per cent, a value too great to accept.[60] Thereafter, scientists in Germany, the United States, and Britain attempted to discover the sources of error and to design more suitable methodologies. When Maxwell became Professor of Experimental Physics at Cambridge in 1871, he was in an ideal situation to carry out such work. That year, the British Association General Committee decided to transfer all the electrical apparatus of the Committee of Electrical Standards, including the original apparatus used in the 1863 experiment, to the future Cavendish Laboratory

---

[58] "Report of the Committee appointed by the British Association on Standards of Electrical Resistance," *B. A. Report* (1862): 126.

[59] For more details, see "Report of the Committee appointed by the British Association on Standards of Electrical Resistance," *B. A. Report (1863)*: 111-176. The 1863 committee consisted of "Professor Wheatstone, Professor Williamson, Mr. C. F. Varley, Professor Thomson, Mr. Balfour Stewart, Mr. C. W. Simens, Dr. A. Mattiessen, Professor Maxwell, Professor Miller, Dr. Joule, Mr. Fleeming Jenkin, Dr. Esselbach, and Sir C. Bright." For the theory of the experiment, see also Maxwell, *Treatise*, vol. 2, 408-411 [article 763-766]. The early reports were later complied by Jenkin. See F. Jenkin ed., *Reports of the Committee on Electrical Standards* (London: Taylor and Francis, 1873). For a short but plausible description of the project, see I. B. Hopley, "Maxwell's Work on Electrical Resistance: I. The Determination of the Absolute Unit of Resistance," *Annals of Science*, 13 (1957): 265-272; "Maxwell's Work on Electrical Resistance: II. Proposals for the Re-Determination of the B.A. Unit of 1863," *Annals of Science*, 14 (1958): 197-210.

[60] Hopley, "Maxwell's Work on Electrical Resistance I," 269.

to be placed at the disposal "of the Professor of Experimental Physics."[61] Moreover, Maxwell now supervised a group of research students able to carry out experiments under his direction.

Nevertheless, Maxwell never finished the project.[62] He devoted insufficient attention to the efforts of the Cavendish research students involved in this effort, including Chrystal, who devoted himself entirely to comparing the British Association units of resistance and to finding errors in the original experiment. Glazebrook, Saunder, Garnett, and later Fleming also became involved in some aspects of this work while at the Laboratory. Maxwell, the man of ideas, provided these students—and Chrystal, in particular—with suggestions and criticisms, but he failed to organize a team effort to solve the problem or to urge other students to join the project. As always, his priority was to encourage students to follow their own interests. The laboratory itself no doubt offered its own institutional barriers to completion of the ohm-measurement project. It was a new institution, and everything there was still in a state of change.

This was the situation in the spring of 1880 when Rayleigh decided to resume determination of the unit of electric resistance. His experimental efforts can be roughly divided into four phases. The first phase was repetition of the original British Association method and adoption of a modified experiment. The second phase was determining the value of the ohm in terms of the length of a column of mercury. The third phase was employing L. Lorenz's method for determining the ohm. The final phase was determining the values of the other two units of electricity, the volt and the ampere.

In June of 1880, Rayleigh began the first phase of experimentation, repetition of the original British Association experiment. For the most part, this repetition was carried out by Rayleigh himself, Schuster, and Mrs. Sidgwick, Rayleigh's assistant and sister-in-law.[63] Horace Darwin and Professor Stuart also provided some help. Rayleigh made a few improvements to the original experiment but, more important, he made theoretical corrections of errors. Through re-calculation, Rayleigh showed that the Committee had "seriously under-estimated" the coefficient of self-induction, which was significant because the self-inductance of the coil greatly influenced the result, and he achieved a more accurate value.[64] As a result of the first phase of experimentation, Rayleigh obtained the following value of the ohm: 1 B. A. unit =

---

[61] "Report of the Committee appointed by the British Association on Standards of Electrical Resistance," *B. A. Report* (*1871*): ixix. However, the apparatus remained the property of the British Association and at the disposal of the Committee.

[62] See Simon Schaffer, "Late Victorian Metrology and its Instrumentation: A Manufactory of Ohm," in Robert Bud & Susan E. Cozzens (eds.), *Invisible Connections: Instruments, Institutions, and Science* (Bellingham, WA: Spie Optical Engineering Press, 1992), 23-56, and his "Accurate Measurement is an English Science."

[63] See Rayleigh and A. Schuster, "On the Determination of the Ohm [B. A. Unit] in Absolute Measure," *Proc Roy. Soc.* 37 (1881): 104-141: ". . . Dr. Schuster took all the readings of the principal magnetometer. Mrs. Sidgwick observed the auxiliary magnetometer; while the regulation of the speed by stroboscopic observation fell to me."

[64] Ibid., 116. The Committee's estimate was L = 437,000, while Rayleigh's was 451,000.

Figure 2.3. Early apparatus to measure the ohm at the Cavendish [Courtesy of the Cavendish Laboratory].

0.9893 x 10^9 C.G.S. units. This result (0.9893) agreed closely with that of H. Rowland (0.9911) and confirmed Rowland's finding that the original British Association unit was about 1 per cent too small.

As soon as he finished the first experiment, in the spring of 1881, Rayleigh began the next phase of experimentation. He constructed a new apparatus based on the design used in the previous phase but with a linear dimension enlarged at a ratio of about 3:2.[65] Again, the principal experimenters were Rayleigh, Schuster, and Mrs. Sidgwick, but after Schuster left for Manchester, only Rayleigh and Mrs. Sidgwick were regularly involved in the project. Most of the measurements were carried out during the Michaelmas term of 1881, and the result was: 1 B. A. unit = 0.98651 x 10^9 C.G.S. units. This value was very accurate and was in satisfactory agreement with other independent results, particularly Joule's value of the dynamical equivalent of heat as determined electrically and mechanically.[66]

This second phase of Rayleigh's project was influenced by a recommendation of the Paris Electric Conference of 1882 to express the value of the ohm in terms of the length of a column of mercury of 1 square millimeter at 0°C. This method first had been proposed by Werner Siemens, and it had become widely accepted on the Continent as the most suitable method. Eventually, despite opposition from the English representatives headed by William Thomson, the Electrical Conference of 1884 decided to adopt, as the legal ohm, the resistance of a column of mercury 106 cm in length and 1 square millimeter in cross-sectional area.[67] The necessity of expressing the ohm in terms of the length of a column of mercury became clear to Rayleigh in early 1882. The experiment of the second phase of the project was relatively simple, and Rayleigh and Mrs. Sidgwick carried it out during February and March of 1882. The result was: 1 mercury unit = 0.95418 B. A. unit = 0.94130 ohm, or 1 ohm = 106.24 centimeter of column of mercury.[68]

In the third phase of the project, Rayleigh employed a method for determining the value of the ohm introduced by L. Lorenz in 1873.[69] This method, Rayleigh found, possessed the important advantage of producing results that were free of the influence of terrestrial magnetism and of thermoelectric effects. Rayleigh modified Lorenz's method to make it more efficient, but without altering its essential character. The result was: 1 B.A. unit = 0.98677 x 10^9 C.G.S. units.[70]

Having obtained a satisfactory value of the ohm (or B. A. unit), Rayleigh launched the fourth and last phase of the project: determination of the ampere and the volt. To define the ampere, he measured the quantity of silver deposited by electrolysis, which reflected the magnitude (in absolute units) of the current that passed through the solution of silver nitrate or silver chloride causing the initial

---

[65] The new apparatus was constructed by Messrs. Elliot's shop under the superintendence of Rayleigh and Schuster. For details of the new apparatus, with a diagram, see Rayleigh, "Experiments to Determine the Value of the British Association Unit of Resistance in Absolute Measure," *Phil. Trans.* 173 (1882): 661-697. By resolution of the Paris Electric Congress (1882), the ohm became the standard unit (10^9 C.G.S.) . Therefore the title of the paper was intentionally changed before publishing.

[66] Strutt, *Life of Rayleigh*, 117-118.

[67] Ibid., 123-127.

[68] Rayleigh and Mrs. Sidgwick, "On the Specific Resistance of Mercury," *Phil. Trans. 174* (1882): 173-185.

[69] Rayleigh, "Comparison of Methods for the Determination of Resistances in Absolute Measure," *Phil. Mag. 14* (1882): 329-346.

[70] Mrs. Sidgwick and Rayleigh, "Experiments, by the Method of Lorenz, for the Further Determination of the Absolute Value of the British Association Unit of Resistance, with an Appendix on the Determination of the Pitch of a Standard Turning-Fork," *Phil. Trans. 174* (1883): 295-322.

deposit of silver. The value obtained by Rayleigh and Mrs. Sidgwick was: quantity of silver deposited by 1 ampere per second = 0.00111794 grams. This value (0.00111794) was fairly close to that found by Kohlrausch (0.0011183), and it corrected Mascart's finding (0.00118).[71]

For convenience, Rayleigh determined the volt by measuring the value of the electromotive force of a Clark Cell. If a measured current traverses a known standard resistance, the electromotive force at the extremities of the resistance can be measured and compared with the output of a Clark Cell. With the values of both the ohm and the ampere known, determining the volt was theoretically an easy task, but the preparation of the cells and the maintenance of a constant voltage across them were experimentally difficult tasks. The necessary experiments often were carried out simultaneously with that for determining the ampere. The final result was: Electromotive Force of a Clark Cell at 15°C = 1.4345 volt.[72]

With the completion of this series of experiments, Rayleigh had accomplished the difficult task of standardizing the electrical units, and he had done so within the short span of four years. The results he achieved were so accurate that his son, Robert John Strutt, could boast that "they practically left nothing to be desired."[73]

Rayleigh's achievement was much more than a personal success. From the point of view of the Cavendish Laboratory, the work he completed during his tenure at the Cavendish left important tracks. First, although the idea of teamwork had been somewhat foreign at the Cavendish under Maxwell, Rayleigh did not undertake the ohm determination project by himself but with a group of other researchers. Schuster later remembered :

> On his own initiative Rayleigh adopted a definite and novel policy, its essential point being the fostering of a spirit of community among the advanced students. To make a beginning he desired to identify the laboratory as a work with some research in which a combination of workers was necessary . . . Rayleigh was disappointed in his original desire, no volunteers besides myself offering to help [in the measurement of the ohm project], until Mrs. Sidgwick joined us.[74]

By enlisting the assistance of various volunteers for his project—measurement of the ohm, Rayleigh proved that a research team could be a powerful and effective means of conducting experiments. The legacy of this action was an increasing number of collaborative efforts at the Cavendish during his tenure and later. During his tenure at the Cavendish from 1800 to 1884, six joint works were conducted there, including four in which Rayleigh directly participated. During Maxwell's tenure from 1874 to 1879, only two joint works were conducted, and Maxwell did not participate in either of them.

Another of Rayleigh's legacies was better utilization of the Laboratory's

---

[71] *A History of the Cavendish Laboratory*, 69.

[72] Rayleigh and Mrs. Sidgwick, "On the Electro-Chemical Equivalent of Silver, and on the Absolute Electromotive Force of Clark Cells," *Phil. Trans. 175* (1884): 411-460. Additional experiments were conducted by Rayleigh and Mrs. Sidgwick in June, November, and December.

[73] Strutt, *Life of Rayleigh*, 120.

[74] A. Schuster, "Lord Rayleigh," *Supplement to Nature 118* (1926): 47-49 on 48.

facilities. The apparatus made by research students during the 1870s, such as Fleming's resistance balance and Chrystal's coils, the original British Association apparatus, and newly constructed instruments such as the water engine all were put to use.

One reason for Rayleigh's research success was that he always endeavored to introduce into the methodology as many improvements as possible. He conducted theoretical investigations and carefully analyzed all potentially fruitful methods. Rayleigh's introduction of a second auxiliary magnetometer into the first phase of his experiment proved so successful that it became standard in his following experiments. In his papers related to the measurement of the ohm, Rayleigh devoted a great many pages to explaining how the apparatus should be arranged and why it should be arranged in that specific manner. He aimed to provide more accurate calculations. Even as he performed the measurement experiment according to the British Association method, he examined every other known method in his search for a better result. The fruit of this labor was his paper, "Comparison of Methods for the Determination of Resistances in Absolute Measure," in which he analyzed six different ways of measuring the ohm. Only after he conducted this comparative assessment did he perform the experiment that he considered "the best of all," the one in which he followed the Lorenz method. Rayleigh's aim was not merely to improve experimental know-how in order to reach "another place of decimals." He wanted to demonstrate that theoretical research was a necessary component of experimental work. His comparison of different methodologies was therefore an important aspect of the Cavendish, not only in theory but in practice. For example, Glazebrook, J. M. Dodd, and E. B. Sargant carried out an ohm-measurement experiment employing a method different from that used by Rayleigh.[75] With his measurement of the ohm project, Rayleigh not only realized Maxwell's dream of a "school of scientific criticism" but also helped to consolidate the research tradition of the Cavendish Laboratory. The Cavendish soon emerged as "a centre of Victorian electrochemical metrology."[76]

2.4. Researchers and Researches

The evolution of the Cavendish into a research school also was stimulated by the researchers who worked there. The Laboratory was still fortunate in attracting talented research students. Although the number of Cavendish research students did not rise sharply during Rayleigh's tenure, these students constituted a select group. Thirteen were wranglers, six had obtained the First Class in the elementary part of the NST (both new and old versions), five had obtained the First Class in the

---

[75] R. T. Glazebrook, J. M. Dodds, & E. B. Sargant, "Experiments on the Value of the British Association Unit of Resistance," *Phil. Trans. 174* (1883): 223-268. The experiment consisted of two parts: Part I, which was regarded as preliminary, contained three series of experiments that were carried out by Glazebrook and Dodds; in Part II another three experiments were conducted by Glazebrook and Sargant.

[76] Schaffer, "Late Victorian Metrology and its Instrumentation," 24.

advanced part of the NST, two had won the Smith Prize, and nine later became Fellows of the Royal Society.

Under Rayleigh, however, the demography of the Laboratory had begun to change. Entering students tended to be a few years younger than their counterparts under Maxwell's directorship. Under Maxwell, almost all students came to the Cavendish after having passed triposes, but the success of the practical physics classes under Rayleigh attracted undergraduates, who came to the Laboratory not only to prepare for examinations but also to pursue their own interests. These undergraduates included such notable figures as L. R. Wilberforce, R. Threlfall, T. C. Fitzpatrick, and S. Skinner. Although undergraduate students did not produce any research papers at the Cavendish, they learned how to perform research. The first group of undergraduates under Rayleigh became prolific researchers during the next decade. Their productivity demonstrated that Rayleigh's innovations had worked. A link between undergraduate and graduate study had begun to be realized.

Another demographic change under Rayleigh was a sharp rise in the number of Cavendish researchers who took and passed the NST. During the 1870s, only four of fifteen Cavendish researchers took the NST. However, among Rayleigh's fifteen newcomers (not counting those from outside Cambridge), seven took either one or both parts of the NST. Three of those who took the advanced part of the NST had previously taken both the elementary and advanced parts of the MT, as permitted by the new NST regulations.[77] This combination of MT and NST studies would eventually replace sole MT study as the accepted route for Cambridge's future physicists.

A third change under Raleigh was the introduction of women to the Laboratory. Early in the 1880s, Cambridge at last allowed women to study alongside men in lecture rooms and laboratories.[78] When the Cavendish Laboratory opened its doors to women in 1882, it was one of the University's first institutions to do so, largely because Rayleigh was more liberal in this regard than Maxwell. The Laboratory's first woman student was I. Freund, who took both parts of the NST and was included in the First Class. The other two women present in the Laboratory during Rayleigh's tenure were not students; Mrs. Sidgwick worked as Rayleigh's indispensable assistant in his experiments on ohm-measurement and other subjects; and Mrs. Shaw assisted her husband in an experiment to determine the atomic weight of silver and copper.[79]

The most important change at the Cavendish under Rayleigh, however, was in the number of studies carried out by research students. The number of published research papers increased from fifty-eight under Maxwell to eighty-six under Rayleigh. The number of authors increased from nine to sixteen, and the number of researchers who wrote more than two papers increased from three to eight.

---

[77] I omit Wilberforce because he first took the NST (part I only) and then the MT (parts I & II).

[78] From 1882 on, men and women took the MT and the NST separately.

[79] See John N. Howard, "Elanor Mildred Sidgwick and the Rayleighs," *Applied Optics 3* (1964): 1120-1122. For Mrs. Shaw, see *A History of the Cavendish Laboratory*, 73.

Table 2.2. Number of Research Papers Published at the Cavendish, 1880-1884

Name	*Individual Work*	*Co-Work*
Lord Rayleigh	29	1/2 x 4
R. T. Glazebrook	17	1/3 x 1
J.J. Thomson	13	
A. Schuster	5	1/2 x 1
W. N. Shaw	5	
G. H. Darwin	2	1/2 x 1
J. H. Poynting	2	
J. C. McConnel	2	
L. R. Wilberforce	2	
Mrs. Sidgwick		1/2 x 3
J. A. Fleming	1	
R. Threlfall	1	
C. Spurge	1	
H. Darwin		1/2 x 1
J. M. Dodds		1/3 x 1
E. B. Sargant		1/3 x 1
Total: 16	80	6

The increasing maturity of the Cavendish Laboratory was reflected in the content and level of these research papers. Next to Rayleigh himself, Glazebrook was the most prolific researcher during Rayleigh's tenure. Most of his papers were related to his own experimental work at the Laboratory, but some were purely theoretical and highly mathematical. His experimental activities can be divided into two categories, optics and spectroscopy—his favorite subjects—and electromagnetism. He also wrote four mathematical/theoretical papers on electromagnetism: one concerning the electromagnetic theory of light, one on the molecular vortex theory of electromagnetic action, one comparing Maxwell's equations of the electromagnetic field with those of Helmholtz and Lorenz, and one examining general equations of electromagnetism.[80] On the whole, Glazebrook's experimental papers were more impressive than his theoretical papers.

After Glazebrook, J.J. Thomson was the next most productive writer in the Laboratory during Rayleigh's tenure. Most of his papers were highly mathematical and theoretical, and only two related to experiments. One of these two, "On the Determination of the Number of Electrostatic Units in the Electromagnetic Unit of

[80] Some examples of Glazebrook's work as follows: "On the Measurement of Small Resistance," *Phil. Mag. 11* (1881): 291-295; "On Molecular Vortex Theory of Electromagnetic Action," *Phil. Mag. 11* (1881): 397-413; "On some Equations connected with Electromagnetic Theory of Light," *Pro. Camb. Phil. Soc. 4* (1881): 155-167; "On the Refraction of Plane polarized Light at the Surface of a Uniaxal Crystal," *Phil. Trans. 173* (1882): 595-620; "On Spectrophotometer," *Proc. Camb. Phil. Soc. 4* (1883): 304-308; "On Curved Diffraction of Gratings," *Phil. Mag. 15* (1883): 414-423.

Electricity," was closely related to Maxwell's electromagnetic theory of light.[81] J.J.'s experimental result was v = 2.963 x 10^{10} C.G.S. units, a fair value. Quite remarkably, this was only his third experiment. J.J.'s research interests were diverse. He wrote two papers on mathematical functions, seven papers on electromagnetism, a lengthy thesis and a summary on the vortex motion which earned him the Adams Prize, a paper on the theory of electric discharge in gases, and another pioneering paper concerning a problem in physical chemistry. J.J.'s works will be discussed more thoroughly in Chapter 3.

Shaw and Schuster also were prolific writers during this period, producing five papers each. Two of Shaw's papers centered on meteorology. Two others examined electricity: one was about dimensional equations and change of units, and the other was comparison of resistances. His fifth paper was a short investigation of the focal lines of lenses. Schuster's four papers resulted mainly from spectroscopy studies he had carried out during the 1870s when he had examined the spectra of metalloids, the influence of temperature and pressure on the spectra of gases, and the harmonic ratios in the spectra of gases. His fifth paper focused on the dynamical theory of radiation in support of the molecular theory of heat and light.

The next group of researchers produced two papers each. G. H. Darwin, with his brother H. Darwin, worked on the measurement of the gravitational effects of the moon. During their experiments, the Darwins discovered "the slower oscillations of the soil," which became the subject of G. H.'s paper, "On the Formation of Ripple-Marks in Sand."[82] Poynting's studies made use of two practical instruments, the sonometer and the saccharimeter. J. C. McConnel worked on the measurement of the dark rings in quartz—using two Nicol's prisms—and on the effects of the self-induction of the galvanometer in the determination of the capacity of a condenser. In his first paper, L. R. Wilberforce reported on his attempt to measure the capacity of a condenser, in which he adopted the methodology and apparatus used by J.J. in his measurement of "v." [83] Wilberforce's other paper addressed a problem in electrostatics.

The others produced one paper each. Fleming reported on a new design of a resistance-balance "recently constructed for the Cavendish Laboratory, and which experience shows to have several decided advantages over the old form."[84] R. Threlfall reported on the India-rubber method of mounting sections. C. Spurge conducted a mathematical analysis of the curves of constant intensity in the passage

---

[81] J.J. Thomson, "On the Determination of the Number of Electrostatic Units in the Electromagnetic Unit of Electricity," *Phil. Trans. 174* (1883): 707-721. His first experimental paper at the Cavendish was "On Some Electromagnetic Experiments with Open Circuits," *Phil. Mag. 12* (1881): 49-60.

[82] G. H. Darwin et al., "Second Report of B.A. Committee on Lunar Disturbance of Gravity"; G. H. Darwin, "On the Formation of Ripple-Mark in Sand," *Proc. Roy. Soc. 36* (1883): 18-43.

[83] L. R. Wilberforce, "On Some Experiments on the Measurement of the Capacity of a Condenser," *Proc. Camb. Phil. Soc. 5* (1884): 175-182.

[84] J. A. Fleming, "On a New Form of Resistance-Balance adapted for comparing Standard Coils," *Phil. Mag. 9* (1880): 109-117 on 110. Fleming said this resistance-balance had been constructed in Stuart's workshop under his direction.

of homogeneous polarized light through an uniaxal crystal cut.

In conclusion, by the end of Rayleigh's tenure, the Cavendish Laboratory had evolved into a bona fide research institution. The number of research papers and researchers producing them had increased sharply. Rayleigh had served not only as administrator but also as an exemplary model for research students. His two demonstrators, as his deputies, also had become model researchers and the teachers of and assistants to a new crop of research students. Also during Rayleigh's tenure, there emerged one exceptional student who had carried out as much research as his teachers—J.J. Thomson.

A hallmark of Rayleigh's leadership was the exchange of information and apparatus among Cavendish researchers. Under Maxwell, each researcher had communicated primarily with the director and collaborative efforts were rare, but Rayleigh openly encouraged researchers to share information and efforts. To help establish a sense of unity and a spirit of cooperation among the Cavendish researchers, Rayleigh introduced a daily afternoon "tea time" to provide opportunities for informal discussions of common topics.[85]

Although Rayleigh fostered a "meeting of the minds" at the Cavendish, Maxwell's philosophy remained a guiding force. Maxwell's vision of a research school for graduate students, his laissez-fair spirit and willingness to explore different methods, and his respect for precise measurement could be detected in the works of the Cavendish researchers, who frequently referred to Maxwell's books in their theoretical analyses. They also turned to Maxwell's works in their search for experimental subjects.

2.5. Rayleigh and the Continuation of Maxwell's Guidelines for the Cavendish Laboratory

Rayleigh's five-year tenure at the Cavendish Laboratory often has been neglected by historians of science primarily because of its brevity. However, this period was pivotal in the development of the Laboratory. The basic organization and management directions established by Rayleigh at the Laboratory remained almost unchanged for several decades. More importantly, during his tenure Rayleigh correctly understood and implemented Maxwell's guidelines for the Laboratory. Under Rayleigh's directorship, Maxwell's dream of linking lectures with laboratory training and practical and theoretical research became a reality. Rayleigh also worked to realize Maxwell's dream of breaking down the barrier between experimental and mathematical/theoretical physics. Rayleigh's 1882 address to the British Association at Southampton was reminiscent of Maxwell's inaugural lecture of 1871:

> Even within the limits of those departments whose foundation is evidently experimental, there is room, and indeed necessity, for great variety of treatment. One

---

[85] John N. Howard, "John William Strutt, Third Baron Rayleigh," *Applied Optics 3* (1964): 1091-1101 on 1097.

> class of investigators relies mainly upon reiterated appeals to experiment to resolve the questions which appear still to be open, while another prefers, with Thomas Young, to base its decisions as far as possible upon deductions from experiments already made by others. It is scarcely necessary to say that in the present state of science both methods are indispensable . . . In many others, where the interest is mainly theoretical, we cannot afford to neglect the confirmation which our view may derive from the comparison of measurements made in different fields and in face of different experimental difficulties . . .
>
> Examples such as this, . . . show *how difficult it often is for an experimenter rightly to interpret his results without the aid of mathematics. It is eminently desirable that the experimenter himself should be in position to make the calculations, to which his work gives occasion, and from which in return he would often receive valuable hints for further experiment. I should like to see a course of mathematical instruction arranged with especial reference to physics, within which those whose bent was plainly towards experiment might, more or less completely, confine themselves.*[86]

Rayleigh was faithful to Maxwell's vision as director of the Cavendish Laboratory. Despite his unique social status and frequent illness, Rayleigh took charge of the Cavendish and accomplished his job successfully.[87] As D. J. de Solla Price indicated, in serving the Cavendish for his five-year tenure, Rayleigh made a genuine sacrifice in order to continue Maxwell's ideas and to advance his nation's standing.[88] Rayleigh said of Maxwell:

> To estimate rightly his influence upon the present state of science, we must regard not only the work that he executed himself, important as that was, but also the ideas and the spirit which he communicated to others. Speaking for myself as one who in a special sense entered into his labours, I should find it difficult to express adequately *my feeling of obligation.*[89]

In the history of the development of the Cavendish Laboratory, Rayleigh's tenure represented a firm bridge between Maxwell and J.J. Thomson, the next professor.

However, his contemporaries at Cambridge seemed unappreciative of Rayleigh's maintenance of Maxwell's tradition. When Rayleigh announced his resignation in the fall of 1884, to most Cambridge dons the election of the next professor meant choosing a successor to Maxwell, not Rayleigh. Although Rayleigh was in every sense the proper interpreter of Maxwell, in the eyes of the more conservative dons at the University, the Cavendish Laboratory under his leadership seemed far too concerned with practical matters. Why did the *only physics laboratory* of the

---

[86] "Address to the Mathematical and Physical Science Section of the British Association," in *Scientific Papers by Rayleigh*, vol. 2, 118-124 on 118-119 & 121. Emphasis added. The goal of establishing a mathematical class for physics students was realized under J.J. See section 3.3.3 of this book.

[87] When Rayleigh was appointed, *Punch* (13 December 1879, 273) ran an article, "Lord and Professor," satirizing his election as Professor. However, his peerage contributed to his success as Director of the Cavendish. The Senate House made exceptional speed in approving nearly all his requests. For example, it approved the appointment of his two demonstrators in only ten days. Rayleigh's organizational reform therefore advanced very smoothly. His rheumatic fever returned in October, 1882, and he spent several months in France, Italy, and Germany during 1882-83. See Strutt, *Life of Rayleigh*, 133-136.

[88] See Strutt, *Life of Rayleigh*, 413. Rayleigh later recalled that "I could not have gone on working as hard as I was doing then." (Ibid., 148.)

[89] "Presidential Address," in *Scientific Papers by Rayleigh*, vol. 2, 333-354 on 350. Emphasis added.

University carry out such down-to-earth work as the measurement of the ohm? Should not that project have been carried out by the Professor of Engineering? Or, as Schuster indicated, was the idea of the "fostering of a spirit of community among the advanced students" too "noble" for the Cambridge men? J.J. Thomson's doubts about the character of the Laboratory must have been shared by many Cambridge dons:

> In starting the work on electrical units [Rayleigh] had in mind a scheme of identifying the laboratory with a research planned on an extensive scale in which a large number of workers in the laboratory might take part. I think it is doubtful whether, considered from the point of view of training for young physicists, such a scheme is advantageous in a university laboratory. Most men are much more interested in, and more enthusiastic about, one branch of physics than another, and a young man develops his powers more rapidly and puts more independent thought into his work if tackles a problem which he has, to some extent, selected for himself, than if he joins a scheme of work already planned.[90]

Such opinions would significantly influence the election of the next professor.

[90] J.J. Thomson, "The Life of Lord Rayleigh (book review)," *Nature 114* (1924): 814-816 on 816.

CHAPTER 3

J.J. THOMSON'S FIRST TEN YEARS AT THE CAVENDISH, 1885-1894

> My doubt was whether Thomson should be professor of experimental physics. He had done very little experimenting at that time, though enough to show that he could do it. But he has shown since that it was right to appoint him.
>
> Lord Rayleigh[1]

> I had looked on you as a mathematician, not an experimental physicist, and could not at first bring myself to regard you in that light.
>
> R. T. Glazebrook[2]

3.1. The Election of J.J. Thomson

Rayleigh was elected President of the British Association for the Advancement of Science meeting in Montreal in 1884, that association's first meeting outside the British Isles. When he returned in early November, he notified the Vice Chancellor of his resignation from the professorship of experimental physics. Rayleigh's resignation was formally announced on November 17, and election of his successor was scheduled for December 22.[3] Rayleigh left Cambridge on December 13, thus clearly indicating that he would not influence the election.

Expressing interest in the post were a number of ambitious young physicists from both inside and outside Cambridge. For the first time, a true competition for the professorship was about to take place. The "official candidates" registered by the University were Arthur Schuster (Professor of Applied Mathematics at Manchester), Osborne Reynolds (Professor of Engineering at the Victoria University of Manchester), Richard T. Glazebrook (Demonstrator and Lecturer at Cambridge), Joseph Larmor (Professor of Natural Philosophy at Queen's College, Galway), and Joseph John Thomson (Lecturer at Cambridge).[4] Also desiring the post, it was rumored, were William Garnett (former demonstrator at the Cavendish Laboratory) and George F. Fitzgerald (Professor of Natural and Experimental

---

[1] Strutt, *Life of J.J. Thomson*, 20.
[2] Ibid., Glazebrook's letter to J.J. Thomson after the election.
[3] *CUR* (18 November 1884): 165 & (25 November 1884): 186.
[4] CUL MSS O. XIX 52, *Elections of Professors from 1826*, vol. 1, 196.

Figure 3.1. Joseph John Thomson, the third Cavendish Professor (1884-1919) [Courtesy of the Cavendish Laboratory].

Philosophy at the University of Dublin). Although all these candidates were well qualified, none had achieved recognition as a first-class physicist. Because the professional reputations of the candidates were well matched, competition for the professorship became fierce. Many suspected that the electors would prefer a senior scholar to a junior one. As Fitzgerald noted in a congratulatory letter to J.J. Thomson after the election:

> I was very much afraid they might appoint one of the senior candidates such as W. G. Adams, or Garnett. . . . I was afraid they might have thought you too junior but I must now express my hopefulness for Cambridge when it does not consider the most important of all qualifications, namely the energy of youth, as a disqualification.[5]

The competition soon narrowed to Glazebrook versus Thomson. A plausible guess was that Glazebrook, the demonstrator of the Cavendish Laboratory who had acted as Rayleigh's deputy, would succeed his former master. Rayleigh in fact had recommended Glazebrook as a safe choice to one of the electors, G. H. Darwin. Electing J.J. Thomson, Rayleigh believed, would be "rash" because of his inexperience in teaching and management.[6]

Nevertheless, on December 22 the electors chose Thomson, a mathematician, the youngest and least experienced of the candidates, as third professor of experimental physics.[7] His election surprised everyone, including himself:

> . . . in December 1884 I was, to my great surprise and I think to that of everyone else, chosen as [Rayleigh's] successor. I remember hearing at the time that a well-known College tutor had expressed the opinion that things had come to a pretty pass in the University when mere boys were made Professors. I had sent in my name as a candidate without dreaming that I should be elected, and without serious consideration of the work and responsibility involved. When after my election I went into these, I was dismayed. I felt like a fisherman who with light tackle had casually cast a line in an unlikely spot and hooked a fish much too heavy for him to land. I felt the difficulty of following a man of Lord Rayleigh's eminence.[8]

Why did the electors make such an unexpected choice? Of the two leading candidates, J.J. Thomson [from now on J.J.] more closely fit the Cambridge ideal. In 1880, he had been a second wrangler (and a second Smith Prize winner), whereas Glazebrook had stood fifth on the list. In 1881, just one year after his graduation, he had been elected a fellow of Trinity College, whereas Glazebrook spent two years at the Cavendish Laboratory preparing his fellowship dissertation. In 1884, just four years after his graduation, J.J. was made a Fellow of the Royal Society, whereas Glazebrook had to wait six years for his election in 1882. J.J.'s outstanding talent also was indicated by his 1883 receipt of the Adams Prize for his essay on "a general investigation of the action upon each other of two closed vortices in a

---

[5] See Strutt, *Life of J.J. Thomson*, 21-22.

[6] Strutt, *Life of Rayleigh*, 415. See also Strutt, *Life of J.J. Thomson*, 20.

[7] The electors were Vice-Chancellor N. M. Ferrers, Prof. R. B. Clifton, Prof. G. H. Darwin, Sir W. R. Grove, Prof. G. D. Liveing, Prof. W. D. Niven, Prof. G. G. Stokes, Prof. J. Stuart, Prof. W. Thomson.

[8] J.J. Thomson, *Recollections*, 98.

perfect incompressible fluid."[9] Not only was the subject of vortices in fluid motion a suitable choice of topic for a Cambridge mathematician, but J.J.'s receipt of the prestigious Adams Prize at the young age of twenty-six was an impressive accomplishment. Only Maxwell had won this coveted honor at the same young age, twenty-six; Poynting and Larmor would be considerably older when so distinguished.[10] Also in 1883, J.J. had been appointed as one of the University's first lecturers in mathematics, together with four men who were senior to him: A. R. Forsyth, E. W. Hobson, W. H. Macaulay, and Glazebrook.

J.J.'s talent was recognized by his seniors, including Maxwell, Rayleigh, G. H. Darwin, and Niven. Garnett remembered that after J.J. had been working at the Cavendish "for a very short time," Maxwell who "very quickly estimated the capability of his students" said to him, "Thomson will never know the difference between things that are hard and things that are easy for they all come alike to him."[11] Rayleigh also had highly praised J.J.'s talents. Rayleigh had abandoned his effort to measure the ratio of electrostatic units to electromagnetic units (an experiment crucial to proving Maxwell's theory of the electromagnetic nature of light) when he learned that J.J. was working on it, later remembering that "Thomson rather ran away with it." Although Rayleigh wrote a letter of recommendation on behalf of Glazebrook, it was "not in comparison with Thomson or anyone else." When asked by J.J.'s former teacher, Henry Roscoe of Owens College, to support Schuster as the best candidate for the professorship, Rayleigh answered, "I am not sure that he is the best."[12]

The opinions of the electors Darwin and Niven also carried special weight. Darwin, as an examiner for the Adams Prize, had quickly recognized J.J.'s exceptional talent in mathematics. "The problems you have solved," he informed J.J. in his congratulatory letter, "are of amazing difficulty, and the results of the greatest interest. May you go on and discover a true dynamical theory of chemistry."[13] Niven, who had edited the second edition of Maxwell's *Treatise*, could be expected to back the candidate who could best continue Maxwell's research on electromagnetism, and he was among "one of those who pressed his claims most strongly" in favor of J.J. [14] A special connection existed between Niven and J.J. In the academic year of 1877-78, J.J. had attended Niven's lectures on

---

[9] *CUR* (5 April 1881): 446. This topic was announced in 1881, with the suggestion that "the case of two linked vortices should be fully discussed, with the view of determining (1) whether any steady motion is possible, and (2) whether any motion can occur in which there are periodical changes in the forms and dimensions of the vortices." Thomson's essay, a purely mathematical one, won the coveted prize and was published as a separate volume under the title, *A Treatise on the Motion of Vortex Rings* (Cambridge, 1883).

[10] The Adams Prize, established in 1848 to honor John Couch Adams, was awarded every two years "for an Essay on some subject of Pure Mathematics, Astronomy, or other branch of Natural Philosophy." Maxwell won the Prize in 1857, Poynting in 1893, and Larmor in 1899. See *The Historical Register of the University of Cambridge to the Year 1910*, 321.

[11] CUL MSS, ADD 7655, III (d) 5, 41, & also ADD 8385.10, 9.

[12] Strutt, *Life of Rayleigh*, 127, 415 & 413 respectively.

[13] CUL MSS ADD 7654 D4 (25 January 1883): G. H. Darwin to J.J.

[14] Strutt, *Life of J.J. Thomson*, 19.

electricity and magnetism, in which Niven had given "special reference to Maxwell's work."[15] Later, J.J. had assisted Niven in editing the *Treatise*. Niven later "confessed that he was rather afraid of J.J. as a pupil."[16]

The research papers published by J.J. between 1880 and 1884 demonstrated that J.J. fit the mold of the Cambridge-style physicist better than Glazebrook. Of these thirteen papers, eleven were theoretical or concerned mathematical subjects; only two related directly to experiments. Glazebrook's eighteen papers during that period were based largely on his own experiments, and many concerned such practical problems as measurement of resistances. Although Glazebrook also was a competent "Cambridge Mathematician," in mathematical abilities he was overshadowed by J.J.

In their selection of a worthy successor to Maxwell and Rayleigh, the electors chose to give more weight to scientific talent than to experience. Experience—whether in teaching or management (both of which J.J. lacked)—could be acquired, but talent certainly could not. Moreover, in the opinion of some Cambridge dons, Rayleigh's projects had deviated too far from academic subjects; measurement of the ohm easily could be misinterpreted as a practical measurement to the next place of decimals. In the view of some members of the Cambridge elect, the Cavendish Laboratory should leave practical projects to the Engineering Department and encourage the Cavendish researchers to choose more fundamental or theoretical topics, such as experimentally proving the electromagnetic nature of light. Of the two leading candidates, J.J. would be the one more likely to provide research students with such fruitful ideas. The choice of J.J., though it surprised many at the University, was logical and sound. Indeed, as J. G. Crowther noted, that choice was "a manifestation of the quality and health of the Cambridge scientific environment in the 1880s."[17] During the last quarter of the nineteenth century, however, this kind of sound judgment was still novel. In academia, candidate selection often turned more on the job-seeker's patronage and influence than on his talent. The election of J.J. revealed how seriously the electors viewed their responsibility to assure the continuity of the new research tradition at the Cavendish Laboratory.

Joseph John Thomson, now Professor of Experimental Physics at Cambridge University, had been born only twenty-eight years earlier, on December 18, 1856, at Cheetham Hill, near Manchester. His father, a bookseller, had planned an engineering career for J.J. and had sent him to Owens College at age fourteen in hope that he would learn something useful while awaiting a vacant engineering apprenticeship.[18] J.J. later regarded his entrance into Owens College as "the most critical event" of his life. [19] It was there that he first seriously studied science, and he did so under a "brilliant staff of Professors": Osborne Reynolds, Balfour Stewart,

---

[15] Strutt, *Life of J.J. Thomson*, 8, & J.J. Thomson, *Recollections*, 42-43.
[16] Strutt, *Life of J.J. Thomson*, 9.
[17] Crowther, *The Cavendish Laboratory*, 102.
[18] For his early life, see J.J. Thomson, *Recollections*, chapter 1.
[19] Ibid, 2.

Henry Roscoe, and Thomas Barker. Engineering Professor Reynolds used Rankine's textbook in "the most original and independent" manner. Physics Professor Stewart paid "special attention to the principle of the Conservation of Energy." Roscoe's lectures, although J.J. was not charmed by them, "were carefully arranged, were clear, and illustrated by plenty of experiments." Barker, who had been senior wrangler of 1862, taught mathematics in such a clear manner that J.J. claimed he had never known a better teacher of mathematics "and in some respects no one so good." At Owens College, J.J. also come in contact with two junior scientists who had worked in the Cavendish Laboratory, Poynting and Schuster, and it was Schuster who introduced J.J. to Maxwell's *Treatise*.[20]

J.J.'s intended engineering career came into jeopardy when his father died; his mother "could not afford the very large premium required to become an apprentice."[21] Barker suggested an alternative, that J.J. stay at Owens College "for another year and go on with mathematics and physics, and then try for an entrance scholarship at Trinity College, Cambridge." While preparing for the scholarship, he published his first paper about the experiment on "Constant Electricity between Non-Conductors" under the supervision of Stewart and Schuster.[22] J.J. tried twice for an entrance scholarship and on his second attempt won a minor one. In October of 1876, he left for Cambridge, "where he would spend the rest of his life."[23]

In many ways, J.J.'s academic life while a student at Cambridge was typical of talented and ambitious Mathematical Tripos candidates. His coach was the legendary private tutor, E. J. Routh, and he attended college lectures given by T. Dale, J. W. L. Glaisher, and Niven. However, unlike most other undergraduates, he also attended professorial lectures such as Stokes' lectures on optics and Adams' and Cayley's lectures on mathematics.[24] Interestingly, J.J. attended none of Maxwell's lectures, and he also kept his distance from chemistry courses. During his undergraduate years, as his notebooks show, he concentrated on mathematics and mathematical physics, studying calculus, differential equations, theory of chances, hydrodynamics, elasticity, sound, waves, vibrations of rods, capillary theory, heat, geometrical optics, planetary and lunar theory, terrestrial magnetism, electrostatics, and electricity and magnetism.[25]

His talents soon were apparent to his teachers. With the encouragement of Glaisher, he published a series of mathematical papers for the *Messenger of Mathematics*.[26] Such publication was very unusual for a student who had not yet

---

[20] Ibid., 12-30.

[21] Ibid., 30.

[22] J.J. Thomson, "Experiments on Contact Electricity between Non-Conductors," *Proc. Roy. Soc. 25* (1876): 169-171.

[23] John L. Heilbron, "Thomson, Joseph John," *DSB, 13*, 362.

[24] For more about Thomson's undergraduate study, see J.J. Thomson, *Recollections*, chapter 2.

[25] CUL MSS ADD 7654, NB 18-32.

[26] J.J. Thomson, "On the Resolution of the Product of Two Sums of Eight Squares into the Sum of Eight Squares," *Messenger of Mathematics 7* (1877): 73-74; "An Extension of Arbogast's Method of Derivations," *Messenger of Mathematics 7* (1878): 142-143; "Vortex Motion in a Viscous Incompressible Fluid," *Messenger of Mathematics 8* (1879): 174-175.

taken the MT. Only undergraduates of the intellectual stature of William Thomson or J.J. had the ability to publish. In fact, J.J. proved himself a better researcher than examination candidate, ranking behind Larmor as second wrangler, a rank with which he was "very well satisfied."[27]

J.J.'s real scientific career began just after he took his degree, when he entered the Cavendish Laboratory. Like for other Cambridge seniors who had preceded him to the Cavendish, for J.J. the Laboratory played the role of graduate school. Interestingly, at the Cavendish, J.J continued his mathematical and theoretical studies instead of pursuing experimentation.[28] The topic of his first major work, his Trinity College fellowship thesis, was the transformation of energy. In this dissertation, he suggested in highly mathematical and theoretical terms that all energy might be kinetic. Parts of this work were published in 1885 and 1887 in *Philosophical Transactions,* and the entire work was republished in 1888 in an expanded version as a book, *Application of Dynamics to Physics and Chemistry*.[29]

The generality of this dissertation topic demonstrated young J.J.'s self-confidence and ambition, as did his examination of vortex motion in the essay that won him the Adams Prize. In that essay, J.J. adopted William Thomson's theory of the vortex atom and mathematically described its motions.[30] J.J. was a young mathematician who could skillfully handle difficult equations in their proper Lagrangian forms to arrive at interesting conclusions. John Heilbron argued that J.J., like other Cambridge mathematicians, "took it for granted that an appropriate Lagrangian could always be found, or in other words, that in principle all physical phenomena could be explained mechanically."[31]

From the very beginning of his work at the Cavendish, one of J.J.'s favorite research topics was electromagnetism, an interest which yielded him seven papers. The first of these concerned Maxwell's electromagnetic theory of light and J.J.'s attempt to "obtain equations a little more general than those used by Prof. Maxwell, and to develop some of the consequences of the theory."[32] His second paper, "On the Electric and Magnetic Effects produced by the Motion of Electrified Bodies," was an important contribution to the study of electromagnetism. In this paper he introduced the concept of the electromagnetic mass or extra inertia of a charged particle, and demonstrated that a moving charged body would experience an

---

[27] J.J. Thomson, *Recollections*, 63.

[28] He published two papers on elliptic functions and integration in *Messenger of Mathematics*.

[29] J.J. Thomson, "On Some Applications of Dynamical Principles to Physical Phenomena," *Phil. Trans. 176* (1885): 307-342; "Some Applications of Dynamical Principles to Physical Phenomena-part II," *Phil. Trans. 178* (1887): 471-526; *Applications of Dynamics to Physics and Chemistry* (London: MacMillan, 1888).

[30] J.J. Thomson, "On the Vibrations of a Vortex Ring and the Actions of Two Vortex Rings upon each other," *Proc. Roy. Soc. 33* (1881): 145-147; "On the Vibration of a Vortex Ring, and the Action upon each other of Two Vortices in a Perfect Fluid," *Phil. Trans. 173* (1882): 493-521.

[31] Heilbron, *DSB 13*, 364.

[32] J.J. Thomson, "On Maxwell's Theory of Light," *Phil. Mag. 9* (1880): 284-291on 284.

apparent increase in mass and, in a magnetic field, would behave like an element of a conductor carrying a current.[33] In his third paper, he reported the results of his first experiment at the Cavendish Laboratory, an experiment that concerned the electromagnetic behavior of an open circuit containing a coil that acted like a condenser and possessed some electrostatic capacity.[34]

In two 1882 papers of the same title, "On the Dimensions of a Magnetic Pole in the Electrostatic System of Units," J.J. supported Maxwell against Clausius.[35] In 1883, in an experimental paper that he wrote on the determination of the ratio between the electrostatic and the electromagnetic unit, "v," J.J. adopted "a very slight modification of the method described in §776 of Maxwell's '*Electricity and Magnetism*'" to produce a result that was not entirely convincing.[36] The experiments on which he reported in this paper employed several pieces of apparatus designed by Rayleigh, and in the paper J.J. acknowledged Rayleigh's "valuable advice . . . throughout the investigation." J.J.'s 1884 paper, "On Electric Oscillations and the Effects produced by the Motion of an Electrified Sphere," was another mathematical exercise on electromagnetism.[37]

During J.J.'s student days at the Cavendish he produced two exceptional works: one on electrical discharge through gases and one on physical chemistry. His 1883 paper on electrical discharge through gases signaled the start of his life-long pursuit of the subject, but it did not report on an experiment. It was a theoretical paper in which J.J. attempted to explain electrical discharge through gases as a disturbance "in the electric field of stresses consisting of tension along the lines of force combined with pressures at right angles to them."[38] In an equally pioneering work on physical chemistry, J.J. supported the view of Clausius that "the atoms which form the molecules of a compound gas are continually changing partners" and argued that this viewpoint had many advantages.[39] He also attempted to provide Clausius's idea with some dynamical foundations. He investigated the role of time, mass, and temperature on chemical combination and suggested that the same approach could be applied to liquids. J.J.'s exceptional works on physical chemistry

---

[33] J.J. Thomson, "On the Electric and Magnetic Effects produced by the Motion of Electrified Bodies," *Phil. Mag. 11* (1881): 229-249. For more about this work, see also David R. Topper, *J.J. Thomson and Maxwell's Electromagnetic Theory*, unpublished Ph.D. dissertation (Cleveland: Case Western Reserve University, 1970), chapter v.

[34] J.J. Thomson, "On Some Electromagnetic Experiments with Open Circuits," *Phil. Mag. 12* (1881): 49-60. He acknowledged that "the above experiments were made in the Cavendish Laboratory, Cambridge."

[35] J.J. Thomson, "On the Dimensions of a Magnetic Pole in the Electrostatic System of Units," *Phil. Mag. 13* (1882): 427-429, & same title, *Phil. Mag. 14* (1882): 225-226. The latter was a reply to criticism by Clausius.

[36] J.J. Thomson, "On the Determination of the Number of Electrostatic Units in the Electromagnetic Unit of Electricity," *Phil. Trans. 174* (1883): 707-721.

[37] J.J. Thomson, "On Electrical Oscillations and the Effects produced by the Motion of an Electrified Sphere," *Proceedings of the London Mathematical Society 15* (1884): 197-218.

[38] J.J. Thomson, "On a Theory of the Electric Discharge in Gases," *Phil. Mag. 15* (1883): 427-434 on 427.

[39] J.J. Thomson, "On the Chemical Combination of Gases," *Phil. Mag. 18* (1884): 233-267.

and electrical discharge through gases demonstrated not only his enormous talent but also his wide interest in the physical sciences.

By the end of 1884, J.J. had displayed his talents as a researcher, and he had done so at Cambridge's own Cavendish Laboratory. His election as the University's third "Professor of Experimental Physics" therefore was not a matter of chance. The Laboratory truly had served as his graduate school by providing him with an excellent advisor in the person of Rayleigh, necessary space and apparatus, and, most important, stimulating surroundings. Through his effective use of the Cavendish Laboratory, J.J. undoubtedly had provided a good example to other researchers.

3.2. J.J. Thomson as a Researcher

When J.J. accepted his election as Cavendish Professor of Experimental Physics, he undertook to discharge a special responsibility, one that his predecessors, Maxwell and Rayleigh, had never borne. This responsibility was to prove himself a first-rate physicist.

J.J.'s reputation as a physicist would be very important to the development of the Cavendish Laboratory. To grow as a research institution, the Cavendish needed at its helm a leading figure, someone who had proven himself to be the very best in some areas of the physical sciences. For J.J., this unstated burden was more demanding than managing the Laboratory or providing lectures to accompany experiments. Yet, within ten years of accepting his professorship, J.J. managed to fulfill this serious responsibility. He did so by authoring important research papers and books. His publication of books would help consolidate his authority as professor. His publication of research papers would change his public image from that of mathematician to that of an experimental physicist.

3.2.1. Books

For scientists seeking recognition, authoring books was a traditional method of gaining a broader audience. Between 1885 and 1895, J.J. published three books, *Applications of Dynamics to Physics and Chemistry* (1888), *Notes on Recent Researches in Electricity and Magnetism* (1893), and *Elements of Mathematical Theory of Electricity and Magnetism* (1895). These publications revealed changes in J.J.'s style and his growing self-confidence.

J.J. based his first book, *Applications of Dynamics*, on his Trinity fellowship thesis concerning the transformation of energy. The contents of this dissertation had been greatly augmented by "the substance of a course of lectures delivered at the Cavendish Laboratory in the Michaelmas Term of 1886" (Application of Dynamical Principles to Physical Phenomena).[40] Parts of this book had been published previously in article form in 1885 and 1887 in *Philosophical Transactions*, but J.J.

---

[40] J.J. Thomson, *Applications of Dynamics to Physics and Chemistry* (London: MacMillan, 1888), v.

thought that a book would be "of service to students of Physics."[41] He devoted about a third of this book to the applications of dynamics in their Lagrangian formulation to chemical subjects, such as "Evaporation," "Properties of Dilute Solutions," "Dissociation," "General Case of Chemical Equilibrium," "Effects Produced by Alterations in the Physical Conditions on the Coefficient of Chemical Combination," "Change of State from Solid to Liquid," and "The Connexion between Electromotive Force and Chemical Change." [42] By emphasizing the Lagrangian approach, J.J. revealed a proclivity toward unification of all natural phenomena. On the whole, *Applications of Dynamics* was a collection of J.J.'s theoretical researches in physical chemistry, which was an emerging field in the late nineteenth century. It did not, however, reveal any new facets of J.J. as "Professor of Experimental Physics."

J.J. intended that his second book, *Notes on Recent Researches in Electricity and Magnetism*, serve as a sequel to Maxwell's *Treatise on Electricity and Magnetism*. "I have endeavoured," he wrote, "to give an account of some recent electrical researches, experimental as well as theoretical, in the hope that it may assist students to gain some acquaintance with the recent progress of Electricity and yet retain Maxwell's *Treatise* as the source from which they learn the great principles of the science." [43] Like Maxwell's *Treatise*, this book was not a monograph but a collection of investigations. Chapter 1 concerned electric displacement and Faraday's tubes of forces. Chapter 2 was devoted to the discharge of electricity through gases. Chapter 3 replaced Maxwell's "somewhat indirect" method of applying Schwarz's method of transformations of two-dimensional problems in electrostatics. Chapter 4 presented information about electrical waves and oscillations. Chapter 5 contributed an examination of electromagnetic waves, with special emphasis on Heinrich Hertz's experiment and modifications of that experiment. Chapter 6 was devoted to rapidly alternating currents. Chapter 7 provided an investigation of electromotive intensity in moving bodies.

Notes on Recent Researches was an important work and had special implications for J.J.' s later work. By updating Maxwell's *Treatise*, J.J. relieved himself of a burden he must have felt since his inauguration as professor, namely, the obligation to advance the cause of Maxwellian electromagnetism. J.J. had developed a special relationship to the *Treatise* while assisting Niven in editing its second edition in 1881 and, later, while himself acting as editor to the third edition of the *Treatise* in 1891. In the preface to the third edition, J.J. revealed a plan:

> When I began to revise this Edition it was my intention to give in foot-notes some account of the advances made since the publication of the first edition, not only because I thought it might be of service to the students of Electricity, but also because all recent investigations have tended to confirm in the most remarkable way the view advanced by Maxwell. I soon found, however, that the progress made in the science had been so

---

[41] Ibid.

[42] These are the titles of chapters 11 to 17.

[43] J.J. Thomson, *Notes on Recent Researches in Electricity and Magnetism* (Oxford: Clarendon Press, 1893), v.

> great that it was impossible to carry out this intention without disfiguring the book by a disproportionate quantity of foot-notes. I therefore decided to throw these notes into a slightly more consecutive form and to publish them separately.[44]

J.J.'s title for this book, *Notes on Recent Researches in Electricity and Magnetism*, reflects both his intention and his modesty.

This was the first book in which J.J. examined experimental researches on a large scale. By emphasizing "the physical method," he stressed the value of experimentation as vigorously as the value of mathematics and theory. J.J. straightforwardly summed up his outlook in the preface:

> In the first place, though no instrument of research is more powerful than Mathematical Analysis, which indeed is indispensable in many departments of Electricity, yet analysis works to the best advantage when employed in developing the suggestions afforded by other and more physical methods. One example of such a method, and one which is very closely connected with the initiation and development of Maxwell's theory, is that of the 'tubes of force' used by Faraday. . . .
>
> The physical method has all the advantages in vividness which arise from the use of concrete quantities instead of abstract symbols to represent the state of the electric field; it is more easily wielded, and is thus more suitable for obtaining rapidly the main feature of any problem; when, however, the problem has to be worked out in all its details, the analytical method is necessary.
>
> In a research in any of the various fields of electricity we shall be acting in accordance with Bacon's dictum that the best results are obtained when a research begins with Physics and ends with Mathematics.[45]

J.J.'s argument was not novel, but it was surprising, coming as it did from a scholar with a staunchly Cambridge mathematical background. Chapter 5, entitled "Electromagnetic Waves," most clearly demonstrated J.J.'s transformation from mathematician to physicist. In it, J.J. described and analyzed *in detail* several "Experiments to determine the velocity of electromagnetic waves through various dielectrics," particularly Hertz's experiments. Similarly, he discussed, in both theoretical and experimental terms, J. Kerr's experiments on the effects produced by a magnetic field on plane polarized light, A. Kundt's experiments on thin metallic films, and E. Hall's experiments on the effect of magnetic field in various conductors. Whereas his previous book, *Applications of Dynamics*, contained no figures, no description of apparatus, and few tables, *Notes on Recent Researches* had these aplenty: it contained 144 figures alone.

Importantly, *Notes on Recent Researches* indicated J.J.'s growing interest in a new field of activity. J.J. devoted one quarter of the book to a single topic, the passage of electricity through gases, and he included vivid descriptions of experiments. Maxwell had previously noted the importance of this topic in his *Treatise*: "When [electric discharges through gases] are better understood, they will probably throw great light on the nature of electricity as well as on the nature of

---

[44] Maxwell, *Treatise*, vol. 1, xv-xvi.
[45] *Notes on Recent Researches*, vi.

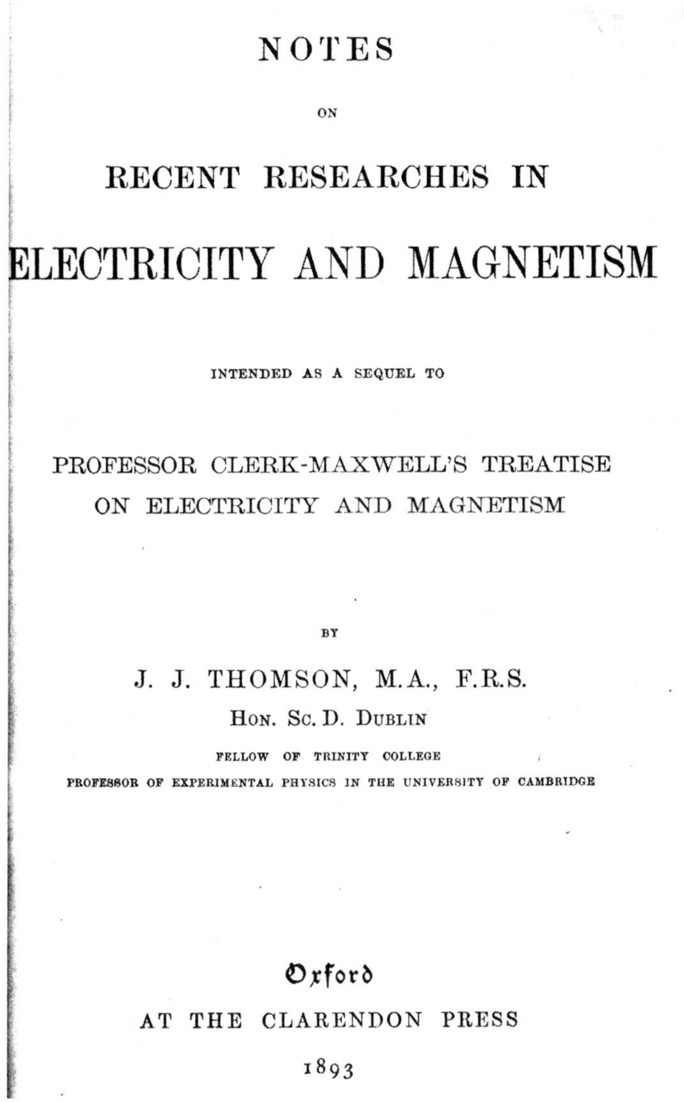

NOTES

ON

RECENT RESEARCHES IN

ELECTRICITY AND MAGNETISM

INTENDED AS A SEQUEL TO

PROFESSOR CLERK-MAXWELL'S TREATISE
ON ELECTRICITY AND MAGNETISM

BY

J. J. THOMSON, M.A., F.R.S.
HON. SC. D. DUBLIN
FELLOW OF TRINITY COLLEGE
PROFESSOR OF EXPERIMENTAL PHYSICS IN THE UNIVERSITY OF CAMBRIDGE

Oxford
AT THE CLARENDON PRESS
1893

Figure 3.2. Title page of Notes on Recent Researches in Electricity and Magnetism.

gases and of the medium pervading space."[46] But, among British scientists, only William Crookes and J.J. had paid much attention to this prediction. As J.J.

---

[46] Maxwell, *Treatise*, vol 1., 60-61 [article 57]. Brackets added. For some early history of electric discharge through gases, see Erwin N. Hiebert, "Electric Discharge in Rarefied Gases: The Dominion of Experiment. Faraday, Plücker, Hittorf," in A. J. Fox (ed.), *No Truth Except in the Details: Essays in Hounour of Martin J. Klein* (Dodrecht: Kluwer Academic Publishers, 1995), 95-134.

indicated at the beginning of Chapter 2, "There is no summary in English text books of the very extensive literature on this subject."[47] With his capacity for theoretical analysis and respect for experimentation, J.J. was poised to become the forerunner in this field.

The publication of *Notes on Recent Researches in Electricity and Magnetism* thus revealed a mathematical physicist in the process of becoming the "Professor of Experimental Physics." Although *Notes on Recent Researches* never became as popular as Maxwell's *Treatise* or J.J.'s later books on electric discharge and conduction, during the 1890s and the first decade of the 1900s it was thoroughly studied by Cavendish researchers and by serious physicists around the world. In letters to J.J., other researchers mentioned the book's influence. In 1893, Hertz wrote that he had determined to "study thoroughly a great deal of it."[48] In 1906, Peter Zeeman wrote that he had "been applying just new theories of yours about the effect of placing a vibrating electrical system near other such systems (§ 432 *Recent Research*)."[49]

Whereas *Notes on Recent Researches* displayed J.J.'s mastery of the details of experimentation, his third book, *Elements of the Mathematical Theory of Electricity and Magnetism,* demonstrated that he also was an impressive teacher and expositor. This was his first textbook, and it was based on the lectures on "Electricity and Magnetism" and "Mathematics for Students of Physics" that J.J. had given during the previous ten years.[50] In it—like his predecessor Rayleigh—J.J. emphasized the advantages of some mathematical training for physics students:

> It is not at all necessary to make use of advanced analysis to establish the existence of some of the most important electromagnetic phenomena. There are always some cases which will yield to very simple mathematical treatment and yet which establish and illustrate the physical phenomena as well as the solution by the most elaborate analysis of the most general cases which could be given.
>
> The study of these simple cases would, I think, often be of advantage even to students whose mathematical attainments are sufficient to enable them to follow the solution of the more general cases. For in these simple cases the absence of analytical difficulties allows attention to be more easily concentrated on the physical aspects of the question, and thus gives the student a more vivid idea and a more manageable grasp of the subject than he would be likely to attain if he merely regarded electrical phenomena through a cloud of analytical symbols.[51]

These comments reaffirmed J.J.'s conviction that physical approaches are more valuable than purely mathematical approaches, as he had previously revealed in *Notes on Recent Researches.*

---

[47] J.J. Thomson, *Notes on Recent Researches*, 53.

[48] CUL MSS ADD 7654 H49 (19 May 1893): H. Hertz to J.J.

[49] CUL MSS ADD 7654 Z4 (22 March 1912): P. Zeeman to J.J.

[50] Beginning in 1885, J.J. lectured regularly on electricity and magnetism. He introduced his lecture, "Mathematics for Students of Physics," in the Michaelmas term of 1888.

[51] J.J. Thomson, *Elements of the Mathematical Theory of Electricity and Magnetism* (Cambridge: Cambridge University Press, 1895), v-vi.

Elements of the Mathematical Theory of Electricity and Magnetism was very readable and covered a wide range of subjects, from "General Principles of Electrostatics" and "Lines of Force" to "Dielectric Currents and the Electromagnetic Theory of Light" and "Thermoelectric Currents." It enjoyed wide popularity, and a new edition was demanded within two years. By 1921, it had gone through five editions.[52] Although J.J. lacked self-confidence as a teacher when he was first elected to his professorship, his authorship of this elementary textbook was a manifestation of his new confidence in himself in his role as a teacher.

3.2.2. Research Papers

J.J.'s research papers from 1885 to 1894 also reveal his changing style and growth as a researcher. An examination of the distribution of his research papers during this period, as seen in Table 3.1, reveals some important aspects of J.J. as researcher.[53] His transformation from mathematical physicist to experimental physicist is unmistakable. As Professor of Experimental Physics, J.J. changed year by year. He shifted the focus of his papers from largely mathematical and theoretical studies of electromagnetism to reports of experimental and theoretical studies of electrical discharge. After 1890, most of his papers no longer were

Table 3.1. J.J. Thomson's Research Papers, 1885-1894

Year	*Mathematical/Theoretical*	*Experimental*	*Total*
1885	4	1[A]	5
1886	1	3[B]	4
1887	1	3[C]	4
1888	1	1[D]	2
1889	3	3	6
1890	0	3[E]	3
1891	1	4	5
1892	0	2	2
1893	0	2	2
1894	0	4	4
Total	11	26	37

A: One joint work with H. F. Newall.
B: Two joint works with R. Threlfall.
C: Two joint works with H. F. Newall.
D: One joint work with J. Monckman.
E: One joint work with G. F. C. Searle.

[52] The second edition of *Elements* appeared in 1897; the third in 1904; the fourth in 1909; the fifth in 1921.

[53] The numbers appearing in Table 3.1 slightly differ from those in "A List of Memoirs containing an Account of Work done in the Cavendish Laboratory" in *A History of the Cavendish Laboratory*.

theoretical interpretations of results obtained by other researchers but instead included his own experimental results. These papers were filled with data, figures, and tables. With the help of his assistants and friends at the Cavendish, J.J. gradually overcame his lack of experience in experimental matters, and seven of his papers were the result of collaborative experimentation with various colleagues (Table 3.1). By the end of 1894, the figure of J.J. sitting in front of a piece of apparatus, engaged in an experiment on electrical discharge in gases, was a familiar one at the Cavendish (See Figure 4.3). Even Hertz admired J.J.'s "beautiful experiments on discharge in gases without electrodes," appreciation that had been inconceivable in the early 1880s.[54]

J.J. nevertheless continued to devote much time and energy to electromagnetism and related topics. He compared various electrical theories (1885) and wrote about the rotation of the plane of polarization of light by a moving medium (1885), electrical oscillations on cylindrical conductors (1886, 1888), magnetization of iron (1887), electric leakage (1887), resistance of electrolytes (1889), propagation of electrical disturbances (1899), transmission of alternating currents along wires (1889), specific inductive capacity of dielectrics (1889), various effects and properties of electric fields (1889, 1891), and electrification of a steam jet (1893).[55] His repetition of the experiment to determine the ratio of the electromagnetic to the electrostatic unit, "v," for which J.J. got the fairly good value of 2.9955×10^{10} cm/sec, suggests an almost obsessive commitment to Maxwell's electromagnetism.[56] In 1886, J.J. and Newall attempted to prove Maxwell's electromagnetic theory "by attacking the vector potential through an experimental

[54] CUL MSS ADD 7654 H49 (19 May 1893): H. Hertz to J.J.

[55] J.J. Thomson, "Report on Electrical Theories," *B. A. Report* (1885): 97-155; "Note on the Rotation of the Plane of Polarization of Light by a Moving Medium," *Proc. Camb. Phil. Soc. 5* (1885): 250-254; "Electrical Oscillations on Cylindrical Conductors," *Proceedings of the London Mathematical Society 17* (1886): 310-328; "Electrical Oscillations on Cylindrical Conductors," *Proceedings of the London Mathematical Society 19* (1888): 520-549; -& H. F. Newall, "Experiments on the Magnetization of Iron Rods, especially on the Effect of Narrow Crevasses at Right Angles to their Length," *Proc. Camb. Phil .Soc. 6* (1887): 84-90; —& H. F. Newall, "On the Rate at which Electricity leaks through Liquids which are Bad Conductors of Electricity," *Proc. Roy. Soc. 42* (1887): 410-429; "The Resistance of Electrolytes to the Passage of very rapidly alternating Currents, with some Investigations on the Times of Vibrations of Electrical Systems," *Proc. Roy. Soc. 45* (1889): 269-290; "Note on the Effect produced by Conductors in the Neighborhood of a Wire on the Rate of Propagation of Electrical Disturbances along it, with a Determination of this Rate," *Proc. Roy. Soc. 46* (1889): 1-13; "The Application of the Theory of the Transmission of Alternating Currents along a Wire to the Telephone," *Proc. Camb. Phil. Soc. 6* (1889): 321-325; "Specific Inductive Capacity of Dielectrics when acted on by very rapidly alternating Electric Forces," *Proc. Roy. Soc. 46* (1889): 292-295; "On the Magnetic Effects produced by Motion in the Electric Field," *Phil .Mag. 28* (1889): 1-14; "On the Illustration of the Properties of the Electrical Field by Means of Tubes of Electrostatic Induction," *Phil. Mag. 31* (1891): 149-171; "On the Absorption of Energy by the Secondary of a Transformer," *Proc. Camb. Phil. Soc. 7* (1891): 249; "On the Effect of Electrification and Chemical Action on a Steam jet, and of Water Vapour on the Discharge of Electricity through Gases," *Phil. Mag. 36* (1893): 313-327; "The Electrolysis of Steam," *Proc. Roy. Soc. 53* (1893): 90-110; "On the Electricity of Drop," *Phil. Mag. 37* (1894): 341-358.

[56] J.J. Thomson and G. F. C. Searle, "A Determination of 'v', the Ratio of the Electromagnetic Unit of Electricity to the Electrostatic Unit," *Phil. Trans. 181* (1890): 583-621.

investigation of the question of the continuity of displacement currents in dielectrics."[57] Their experiment was halted by Hertz's celebrated work of 1887. Although J.J. and Newall were disappointed to lose the opportunity to prove the correctness of Maxwell's theory, Hertz's results were welcomed enthusiastically at the Cavendish.[58] After Hertz's successful experiment and his own experiments on the determination of "v," J.J. evidently felt less pressure to remain in the mainstream of research into electromagnetism, and he slowly departed from it.

During this period, electric discharge through gases became J.J.'s main research topic. Early in this research, he aimed to learn the characteristics of electric discharge in various gases at different temperatures and pressures: in pure nitrogen (1886), in halogen gases (1887), in carbonic acid gases (1889), and in hot gases (1890).[59] From around 1890, J.J. accepted the challenge of more fundamental questions. He designed new experiments to test electric discharge without electrodes (1891, 1892). He also measured the velocities of propagated electric discharges (1891) and cathode rays (1894).[60]

From their beginning, J.J.'s experiments on electric discharges through gases were tied closely to theory. For example, in an 1886 paper, J.J. used his vortex ring theory to explain the decomposition of molecules in gaseous discharge.[61] In an 1894 paper, J.J. predicted that research on electric discharge would "throw light on the connection between electricity and matter."[62] As Isobel Falconer has pointed out, J.J.'s mind always was focused on the possibility that modeling the "interaction between the electromagnetic field and matter on an atomic scale" could connect physics and chemistry.[63] The study of electric discharge through gases was a topic

---

[57] *A History of the Cavendish Laboratory*, 135.

[58] Ibid., 87. J.J. recalled that when he repeated the experiment during an undergraduate lecture in 1888, the student's "enthusiasm spread to all the workers in the Laboratory, and soon experiments on electric waves were going on all over the building."

[59] J.J. Thomson & R. Threlfall, "On an Effect produced by the Passage of an Electric Discharge through Pure Nitrogen," *Proc. Roy. Soc. 40* (1886): 329-340; J.J. Thomson, "On the Dissociation of Some Gases by the Electric Discharge," (Bakerian Lecture) *Proc. Roy. Soc. 42* (1887): 343-345; "On the Effect of Pressure and Temperature on the Electric Strength of Gases," *Proc. Camb. Phil. Soc. 6* (1889): 325-329; "On the Passage of Electricity through Hot Gases," *Phil. Mag. 29* (1890): 358-366 & 441-449.

[60] J.J. Thomson, "On the Discharge of Electricity through Exhausted Tubes without Electrodes," *Phil. Mag. 32* (1891): 321-336 & 445-464; "On the Electric Discharge through Rarefied Gases without Electrodes," *Proc. Camb. Phil. Soc. 7* (1891): 131; "Some Experiments on Electric Discharge," *Proc. Camb. Phil. Soc. 7* (1892): 314; "On the Pressure at which the Electric Strength of a Gas is a Minimum," *Proc. Camb. Phil. Soc. 7* (1892): 330; "On the Rate of Propagation of the Luminous Discharge of Electricity through a Rarefied Gas," *Proc. Roy. Soc. 49* (1891): 84-100; "On the Velocity of the Cathode Rays," *Phil. Mag. 38* (1894): 358-365.

[61] J.J. Thomson, "Some Experiments on the Electric Discharge in a Uniform Electric Field, with some Theoretical Considerations about the Passage of Electricity through Gases," *Proc. Camb. Phil. Soc. 5* (1886): 391-409.

[62] J.J. Thomson, "Electric Discharge through Gases," *Not. Proc. Roy. Inst. 14* (1894): 239-247 on 239.

[63] Isobel Falconer, "Corpuscles, Electrons and Cathode Rays: J.J. Thomson and the 'Discovery of the Electron,'" *BJHS 20* (1987): 241-276 on 252-253. The author describes J.J. not only as an "experimentalist" but also as a serious theoretician whose work "seems to have been motivated by a quest for unity in science."

quite suitable for association with the Cavendish because it was related to a wide variety of themes in physics and even chemistry.

Why did J.J. evolve from mathematical physicist to experimental physicist? The answer is provided by his acceptance of the Professorship of Experimental Physics. Acceptance of this post compelled J.J. to spend more time and energy on experimental physics, at the expense of mathematical physics. The post also gave him access to the Laboratory's experimental equipment and help with its use. Another important by-product of the professorship was financial stability, which accelerated J.J.'s transformation from mathematician to experimentalist. Prior to accepting the post, J.J., unlike Maxwell and Rayleigh, had depended on income from his University salary and private-teaching tuitions. During the years from 1880 to 1884, J.J.'s annual income was far less than the £150 stipend paid to a Cavendish demonstrator.[64] In 1884, for example, he had earned a mere £50 as University lecturer in mathematics plus a few pounds from private students. As Professor of Experimental Physics, however, he received £650 per year from the University and thus could (and did) spend a considerable part of his salary on his experiments.[65]

The post also gave the professor with middle-class roots a greatly improved social status. Financial security combined with his new social standing enabled him to marry (in January of 1890) Miss Rose Paget, who was the daughter of an influential and widely respected Cambridge physician, George E. Paget. She was one of his research students at the Cavendish.[66]

By the end of 1894, J.J. had succeeded in becoming a prominent—though not yet world-class—physicist with credentials in both theoretical and experimental physics, and he no longer worried about his authority as a scientist. His growing confidence as a researcher became a critical factor in his future development as the leader of the Cavendish school. An important element of this self-confidence was his highly satisfactory performance of his duties as professor and as director of the Cavendish Laboratory.

3.3. Consolidating the Organization of the Cavendish Laboratory

J.J.'s lack of experience in teaching and management had been an obstacle to his election as Professor of Experimental Physics. J.J. overcame this handicap with great energy and good cheer, and with the generous assistance of Glazebrook and Shaw. Under J.J.'s direction, the organization of the Cavendish Laboratory was successfully streamlined. As a whole, J.J.'s changes to the Cavendish seemed less dramatic than Rayleigh's, but J.J.'s slow and steady organizational consolidation of the Laboratory indicated that the Cavendish was entering maturity.

---

[64] In addition, Glazebrook and Shaw had earnings from teaching several courses in the colleges and the Cavendish Laboratory.

[65] The salary attached to the professorship of experimental physics had been £850 since 1885, but by regulation this sum was to be reduced to £650 "if the Professor hold a Headship or Fellowship" [see *The Historical Register of the University of Cambridge to 1910*, 103].

[66] Strutt, *Life of J.J. Thomson*, 34-35.

3.3.1. Glazebrook, Shaw, and J.J. Thomson

The Cavendish's founding syndicate had emphasized that one of the Laboratory's main functions would be to provide undergraduates with proper instruction in experimental physics. Under Rayleigh, these responsibilities were partially met: by 1884, the Cavendish's teaching staff consisted of two demonstrators and two assistant demonstrators. J.J., who "had never taught any classes in practical physics," was responsible for directing this larger department, and to do so he desperately needed Glazebrook and Shaw's valuable experience.[67] Their decision to stay at the Laboratory was a great relief to him.

This arrangement was not without difficulties, however. Glazebrook had difficulty accepting his defeat for the professorship. Just after the election, on 28 December 1884, he wrote to Rayleigh, "Of course, the election is a very great blow to me and I can not help feeling it very illogical and unfair . . . Yet I do not like to give [the demonstratorship] up partly for the sake of the Laboratory and partly for the advantage for my own work."[68]

Glazebrook and Shaw's grave concern for the Cavendish saved the Laboratory as an institution, and J.J. as its leader, from the débâcle that their departure would have created. Because Glazebrook and Shaw were J.J.'s seniors and former teachers, he was not in a position to direct them at will. This uncomfortable situation could not last. Ultimately, Glazebrook and Shaw would have to relinquish their demonstratorships. Respectable ways would have to be found to secure other positions for these two men so their experience and contributions would not be lost by the University.

First, a post was arranged for Shaw. In February 1887, the Special Board for Physics and Chemistry recommended an organizational change because "one of the Demonstrators [Shaw] who is at present delivering lectures on Physics for his College as well as demonstrating finds that he cannot continue to undertake the duties of both parts." The Board proposed that "a University Lecturer in Physics be appointed who shall lecture on Experimental Physics, and occasionally assist in the general work of the Laboratory at the Cavendish Laboratory."[69] This scheme was a shrewd one because it did not require raising money for the new post: the £50 stipend paid to Shaw from the Common University Fund simply would be reallocated to his new physics lectureship, without disturbing the £100 official salary of the demonstratorship. The proposal was approved without opposition in May, and Shaw resigned his demonstratorship in June. The lectureship in physics was formally announced in October and, as planned, Shaw was elected.[70]

---

[67] J.J. Thomson, *Recollections*, 98. Both Glazebrook and Shaw were re-appointed by J.J. (*CUR* (24 March 1885): 546).

[68] London Imperial College, MSS Rayleigh (28 December 1884): Glazebrook to Rayleigh. Brackets added.

[69] CUL MSS, vol. 39.33, 67. Brackets added. The report was slightly amended and published in the *CUR* (22 March 1887): 567.

[70] *CUR* (31 May 1887): 784; (14 June 1887): 850; (4 October 1887): 1; (18 October 1887): 65. The tenure of the post was five years.

Arranging a new position for Glazebrook was a bit more complicated because he was not only a Cavendish demonstrator but also a University lecturer in mathematics. In December of 1890, the Special Board for Physics and Chemistry proposed a new post for him, that of "Assistant Director of the Cavendish Laboratory," responsible for "general supervision of Demonstrations, and the selection and custody of the apparatus required for such demonstrations and other similar purposes."[71] This proposal was highly unusual because the post was to be created for a specific person and could have easily provoked opposition in the Senate, even though it did not "involve any increase in expenditure" because the stipend of £50 was to be reallocated from the Common University Fund. Interestingly, although the report on the proposal was dated April 19, 1890, it was not published by the General Board of Studies until December 3 of that year, about eight months later; those involved in this proposal needed time to negotiate and to reach an agreement before its publication.

Figure 3.3. Richard Tetley Glazebrook and William Napier Shaw, two influential demonstrators under Rayleigh and J.J. Thomson [Courtesy of the Cavendish Laboratory].

[71] *CUR* (9 December 1890): 292.

Some complaints could not be avoided. At the January (1891) meeting to discuss the proposal, James B. D. Mayo, an ardent debater in almost every discussion of the Senate, expressed his dissatisfaction with this novel proposal, arguing that "it was only a personal compliment, and this was not a time for personal compliments at the cost of the University."[72] Mayo's opposition was overcome by Henry Sidgwick and J.J., both of whom emphasized the "benefit by Glazebrook's experience and ability." In February, Glazebrook was appointed Assistant Director of the Cavendish Laboratory, a post which he held until he left the Laboratory in 1898. At that time, Shaw replaced Glazebrook in the post, but the post was discontinued when Shaw left the Cavendish in 1900.

3.3.2. Teaching Staff

Glazebrook and Shaw's semi-departures launched J.J.'s new regime, and new members of the teaching staff were appointed. Until 1894, frequent changes were made in the teaching staff. Four different men served as Cavendish demonstrators (H. F. Newall, L. R. Wilberforce, G. F. C. Searle, and S. Skinner) and eleven different men served as assistant demonstrators (J. H. Randell, J. C. McConnel, R. Threlfall, H. F. Newall, H. L. Callendar, L. R. Wilberforce, T. C. Fitzpatrick, R. S. Cole, C. E. Ashford, W. C. D. Whetham, and J. W. Capstick). Table 3.2 shows the staff transition at the Cavendish Laboratory during these years.

A major change in the teaching organization was the addition of a new demonstratorship (see Table 3.2). The timing of this addition is noteworthy. When Glazebrook resigned his demonstratorship to become assistant director, J.J. shrewdly exploited the opportunity to increase the size of the teaching staff. In November 1890, just before publishing its recommendation for the establishment of Glazebrook's new post, the General Board of Studies approved and recommended a proposal by J.J. for "the appointment of another Demonstrator in Experimental Physics . . . on the understanding that the salary of the new Demonstrator shall be paid out of the Laboratory fees."[73] J.J. justified his proposal on the ground that the number of Cavendish students had increased so rapidly that an unofficial staff member, whose name "does not appear in the list of Demonstrators in the Calendar," was giving assistance and should be rewarded. Largely because J.J. intended to provide the new salary from the Laboratory's purse ("fees received for the Demonstration") rather than the University's purse, his proposal was deemed acceptable. By proposing another demonstratorship, J.J. was easing the way for establishment of Glazebrook's assistant directorship. If both proposals were passed, Glazebrook's relinquishment of the demonstratorship would seem more honorable because it would make "two" openings for the younger generation. J.J. succeeded in his scheme. Four days after Glazebrook was named Assistant Director, Searle and Skinner were appointed Demonstrators.[74]

---

[72] *CUR* (27 January 1891): 463.
[73] *CUR* (25 November 1890): 247.
[74] *CUR* (10 February 1891): 483 & 503.

The frequent changes in the assistant demonstratorships are shown in Table 3.2. These positions were by nature relatively unstable because they were not considered regular employment. They had originally been intended to be temporary jobs, as Rayleigh had pointed out when recommending the establishment of assistant demonstratorships:

Table 3.2. Demonstrators and Assistant Demonstrators, 1885-1919

Demonstrators			
W. N. Shaw (1880 M - 87 E)	→ H. F. Newall (1887 M - 90 E)	→ L. R. Wilberforce (1890 M - 1900 E)	→ C. T. R. Wilson (1900 E - 1919 E)
R. T. Glazebrook (1880 M - 91 M)	→ G. F. C. Searle (1891 L - 1935 E)		
	→ S. Skinner (1891 L - 1904 L)	→ P. V. Bevan (1904 M - 08 M)	→ A. Wood (1909 L - 11 L)
	→ T. G. Bedford (1911 M -1927 E)		
Assistant Demonstrators			
J. H. Randell (1884 L - 87 E)	→ H. L. Callendar (1887 M - 88 E)	→ T. C. Fitzpatrick (1888 M - 1906 M)	→ T. G. Bedford (1907 L - 1911 E)
	→ J. A. Crowther (1911 M - 1921 E)		
J. C. McConnel (1884 L - 85 E)	→ R. Threlfall (1885 M - 86 E)	→ H. F. Newall (1886 M - 87 E)	→ L. R. Wilberforce (1887 M -90 M)
	→ R. S. Cole (1890 M - 92 L)	→ C. E. Ashford (1892 E)	→ W. C. D. Whetham (1892 M - 94 L)
	→ J. W. Capstick (1894 E - 96 L)	→ P. E. Bateman (1896 E -99 E)	→ R. G. K Lempfert (1899 M - 1900 E)
	→ J. S. Townsend (1900 M)	→ P. V. Bevan (1901 L -04 E)	→ C. Chittock (1904 M -1909 L)
	→ G. I. Taylor (1909 E - 1911 E)	→ H. Thirkill (1911 M -1919 E)	

M: Michaelmas Term **L**: Lent Term **E**: Easter Term

> The best arrangement would seem to be two Senior Demonstrators and as many Junior Demonstrators as the size of the classes may from time to time render necessary. The Junior Demonstrators would usually be the men who have just taken their degrees, who would hold the appointment temporarily, perhaps from term to term, while on the lookout for regular employment. Their services would be obtainable at cheaper rate.[75]

Thus, although by regulation Cavendish demonstrators and assistant demonstrators were appointed and removed by the professor with the consent of the Vice-Chancellor, in practice different hiring procedures were followed for the two positions. Under J.J., the Cavendish's need for assistant demonstrators was greater than before, but their services did not gain the assistant demonstrators much official recognition.[76] It was natural, then, for them to seek permanent jobs and move on. Only a few fortunate assistants, like Newall and Wilberforce, were promoted to Cavendish demonstratorships. Nevertheless, the two assistant demonstratorships provided junior researchers with good opportunities to train themselves at the Laboratory; this position was an excellent steppingstone for ambitious young researchers.

Together with Glazebrook and Shaw, J.J. lectured on experimental physics and directed the junior teaching staff and other research students. The three demonstrators and two assistant demonstrators taught subjects not covered by the senior staff. By the early 1890s, the Cavendish Laboratory had secured as large and competent a teaching staff as any Continental universities.

A good deal of the credit for this development rests with J.J., who wisely planned and managed the events which secured the expansion of the teaching staff. However, almost equal credit rests with Glazebrook and Shaw. They not only made their valuable teaching and management experience available to their former student, J.J., but also trained the assistant demonstrators and taught research students how to handle the Laboratory apparatus. J.J.'s appreciation of Glazebrook and Shaw's decision to remain at the Laboratory certainly was deeply felt. Their decision to continue to "take charge of the classes in practical physics" rescued J.J. from a potentially very difficult situation.[77]

Glazebrook and Shaw also helped J.J. succeed in University politics. Young and recently elected, J.J. was ill-equipped to advance his causes or to protect the interests of the Cavendish in the Senate House, where seniority and political dynamics dominated both discussions and votes. For example, when Professor Stuart wanted to strengthen his case for the purchase of a new gas engine for the Engineering Department, he maintained that the engine was also needed "for driving the Dynamo of the Cavendish Laboratory." When J.J. protested that Stuart

---

[75] *CUR* (13 June 1883): 879.

[76] In his statement in support of establishment of another demonstratorship, J.J. expressed his appreciation for the assistant demonstrators' efforts, noting that "[in some cases] the Assistant Demonstrators . . . have proved so efficient that they are now doing work of as much responsibility and importance as is done by those who hold the title of University Demonstrator." See *CUR* (25 November 1890): 247.

[77] J.J. Thomson, *Recollections*, 98.

had advanced this argument without notifying J.J., Stuart apologized, but it seems doubtful that the Engineering Professor would have made such an omission when the Laboratory was headed by Maxwell or Rayleigh.[78]

J.J.'s political vulnerability made Glazebrook and Shaw's experience and support invaluable. Glazebrook and Shaw were influential players in University politics. Both were well connected to their colleges, Trinity and Emmanuel; both had been examiners of the MT, NST, and other examinations for many years; and both were members of several University boards. Glazebrook even had been elected University proctor. J.J. desperately needed their political savvy during his first years at the Cavendish when he lacked authority and recognition, and their cooperation with him in the interests of the Laboratory was crucial. Thus, in May 1896, when neither J.J., the Laboratory's director, nor Glazebrook, its assistant director, were able to attend an important Senate discussion concerning the extension of the Laboratory building, it was Shaw rather than a Cavendish demonstrator who defended J.J.'s argument, even though Shaw held no office at the Cavendish at that time.[79]

3.3.3. Physics Teaching at the Cavendish Laboratory

With an increasing number of teachers at hand, J.J. undertook to reform the curriculum and introduce new physics courses. The new courses fell into two categories, those taught by the professor and those taught by the demonstrators. According to regulation, the Cavendish Professor's first duty was to deliver "one course of Lectures each of two terms at least" on "Heat, Electricity, and Magnetism." Maxwell had offered three such courses each academic year, and Rayleigh two. J.J., not content merely to fulfill that responsibility, gave four to seven lectures each academic year. During the first ten years of his directorship, J.J. taught almost twice as many courses as Maxwell and Rayleigh together had taught during the previous fourteen years. J.J.'s enthusiasm for teaching was very different from the reluctant participation in teaching of many Cambridge professors.

J.J.'s teaching methods and the subjects he taught also differed from those of his predecessors (see Table 3.3). Although J.J. lectured on the required traditional topics, such as electricity, magnetism, properties of matter, and kinetic theory of gases, he also lectured on the dynamical principles that had been his specialty before his election as professor. He added new courses designed specifically for tripos candidates, such as "Papers for Candidates for the Natural Sciences Tripos, Part II," "Mathematics for Students of Physics," and "Demonstration in Physics required for the Mathematical Tripos." J.J. believed that the role of physics courses for candidates of the NST was to teach mathematics. The role of physics courses for candidates for the MT, he thought, was to teach experimentation through demonstration.

---

[78] *CUR* (11 May 1886): 593; (8 June 1886): 737-743. Stuart's request was included in the "Regulations for the management of the Mechanical Workshop."
[79] *CUR* (3 June 1896): 885.

J.J.'s offering of an increasingly varied selection of courses was prompted by his interactions with students, and the order in which he introduced the new courses reflected those interactions (See Table 3.3). The first new course, offered beginning in the Easter term of 1886, targeted advanced students, specifically, candidates for the NST, Part II.[80] Two years later, J.J. introduced "Mathematics for Students of Physics," a course that eventually was recommended by both the Special Board for Physics and Chemistry and the Special Board for Mathematics. Natural sciences students understandably were in need of a mathematics course, but why did mathematics students who spent considerable time and energy involved in advanced mathematics need another mathematics course? The answer is that J.J.'s mathematics course, unlike the others, emphasized physics. J.J. approved of Rayleigh's idea that "a course of mathematical instruction arranged with especial reference to physics" was valuable even for MT students, and J.J.'s approach to this course eventually shaped his first textbook, *Elements of the Mathematical Theory of Electricity and Magnetism* (1895). In 1892, J.J. introduced a demonstration class for MT students. Here again, J.J.'s sagacity was at work: he waited to offer this course until he himself had accumulated considerable experience with experimentation.

J.J.'s introduction of new courses indicated that he had clear ideas and strong opinions about the teaching of physics and that he was ready to promote these ideas. In this respect, he went farther than Maxwell and Rayleigh, who had not developed a systematic physics curriculum. Maxwell's lectures had failed to influence students, and Rayleigh did not pay attention to less advanced subjects. Both had attracted only small numbers of students to their courses. J.J. was different. He was a highly competent lecturer who "always attracted a large class, not only including all the workers at the Laboratory whose heart was in their subject, but also containing a considerable number of students in other branches of learning who were attracted by his personality."[81] In his course during the Michaelmas Term of 1887, for example, J.J. had 52 students, and in the course during the Lent Term of 1888 he had 56 students—large classes at Cambridge.[82]

The expansion of the Cavendish teaching staff meant reallocated duties and an increase in the number of classes. The responsibilities of the assistant demonstrators were increased, and Glazebrook and Shaw no longer were students' only sources of information on practical physics, although they were still the most reliable teachers of this subject. With an increased number of courses, the subject matter of each course became more specific. New audiences were targeted, chief among these being the medical students preparing for their first medical examinations. The implementation of special classes tailored to meet the needs of these medical

---

[80] Starting in 1886, the course was taught by some combination of two from the trio of J.J., Glazebrook, and Shaw.

[81] *A History of the Cavendish Laboratory*, 257. See also Strutt, *Life of J.J. Thomson*, 42-43.

[82] *CUR* (4 June 1888): 759.

Table 3.3. J.J. Thomson's Classes, 1885-1894

Year	*Michaelmas Term*	*Lent Term*	*Easter Term*
1885-86	*Properties of Matter *Electrostatics	*Magnetism *Electrodynamics	*Papers for Candidates for the NST, Pt II
1886-87	*Electricity & Magnetism *Applications of Dynamical Principles to Physical Phenomena	*Properties of Matter	*Papers for Candidates for the NST, Pt II
1887-88	*Electricity & Magnetism *Applications of Dynamical Principles to Physical Phenomena	*Properties of Matter	*(Experimental) Electricity & Magnetism
1888-89	*Properties of Matter *Mathematics for Students of Physics	*Electricity & Magnetism *Kinetic Theory of Gases	*Electricity & Magnetism (cont.) *Papers for Candidates for the NST, Pt II (with Glazebrook)
1889-90	*Properties of Matter *Mathematics for Students of Physics	*Electricity & Magnetism *Physical Dynamics	*Electricity & Magnetism (cont.) *Papers for Candidates for the NST, Pt II (with Glazebrook)
1890-91	*Properties of Matter *Mathematics for Students of Physics	*Electricity & Magnetism *Kinetic Theory of Gases	*Electricity & Magnetism (cont.) *Papers for Candidates for the NST, Pt II (with Shaw)
1891-92	*Properties of Matter *Mathematics for Students of Physics	*Electricity & Magnetism *Physical Dynamics *Demonstration in Physics required for the MT	*Electricity & Magnetism (cont.) *Papers for Candidates for the NST, Pt II (with Shaw)
1892-93	*Properties of Matter *Mathematics for Students of Physics	*Electricity & Magnetism *Kinetic Theory of Gases *Demonstration in Physics required for the MT	*Electricity & Magnetism (cont.)
1893-94	*Properties of Matter *Mathematics for Students of Physics	*Electricity & Magnetism *Demonstration in Physics required for the MT	*Electricity & Magnetism (cont.)

Table 3.4. The Number of Students attending the Demonstration Classes, 1885-1894 (As reported by J.J. Thomson in his annual reports)

Year	*Michaelmas*	*Lent*	*Easter*	*Long Vacation*	*Researchers*
1884-85		90	48	32	10
1885-86	101	99	49	?	12
1886-87	75	85	45	35	10
1887-88	71	153	93	29	10
1888-89	136	144	98	19	12
1889-90	132	141	138	42	10
1890-91	164	173	144	29	9
1891-92	150	164	170	30	10
1892-93	166	175	130	33	?
1893-94	165	189	162	28	?
1894-95	232	202	183	33	?

students was one of the key arguments supporting the appointment of additional teaching staff. From 1891 on, "the temporary dissecting-room" was used to accommodate classes of medical students studying physics, and this "greatly relieved the pressure on our space; if it had not been available it would have been impossible to have accommodated all the students who wished to work in the Laboratory." [83] The growth in the number of students attending several demonstration courses is summarized in Table 3.4.

The increase in the size of the teaching staff also made it possible for the demonstrators to offer few specialized advanced courses. For example, in 1888, L. R. Wilberforce offered a course on "Dynamo-Electric Machines." From 1888 through 1890, Shaw offered advanced courses on "Electrolysis" and "Thermodynamics and Radiation." From 1890 on, Glazebrook gave advanced courses on "Electrical Measurement."

What was the nature of the experimental physics courses taught at the Cavendish Laboratory? Glazebrook and Shaw's text, *Practical Physics* (first edition in 1885, second edition in 1886), which was "intended for the assistance of Students and Teachers in Physical Laboratories," offers valuable insight into the kind of physics experimentation employed by the Cavendish staff in their teaching.[84] In the preface of *Practical Physics*, the authors discussed their teaching aims:

---

[83] *CUR* (17 June 1891): 1064.

[84] R. T. Glazebrook & W. N. Shaw, *Practical Physics*, second ed. (London: Longmans, Greens & Co., 1886), vii.

> Our general aim in the book has been to place before the reader a description of a course of experiments which shall not only enable him to obtain a practical acquaintance with methods of measurement, but also as far as possible illustrate the more important principles of the various subjects. We have not as a rule attempted verbal explanations of the principles, ... In following out this plan we have found it necessary to interpolate a considerable amount of more theoretical information. The theory of balance has been given in a more complete form than is usual in mechanical text-books; the introduction to the measurement of fluid pressure, thermometry, and calorimetry have been inserted in order to accentuate certain important practical points which, as a rule, are only briefly touched upon; while the chapter on hygrometry is intended as a complete elementary account of the subject. We have, moreover, found it necessary to adopt an entirely different style in those chapters which treat of magnetism and electricity. We have selected a few—in our judgment the most typical—experiments in each subject, and our aim has been to enable the student to make use of his practical work to obtain a clearer and more real insight into the principles of the subject.[85]

Following these general guidelines, Glazebrook and Shaw arranged twenty-one chapters on the following topics: physical measurements, units of measurement, physical arithmetic, measurement of the more simple quantities, measurement of mass and determination of specific gravities, mechanics of solids, mechanics of liquids and gases, acoustics, thermometry and expansion, calorimetry, tension of vapour and hygrometry, photometry, mirrors and lenses, spectra, refractive indices and wave-lengths, polarized light, colour vision, magnetism, electricity (definitions and explanations of electrical terms), experiments in the fundamental properties of electric currents (measurements of electric current and electromotive force), Ohm's law (comparison of electrical resistances and electromotive forces), and galvanometric measurement of a quantity of electricity. As a whole, *Practical Physics* was a well-balanced textbook in experimental physics and certainly more modern and complete than F. Kohlrausch's *An Introduction to Physical Measurements* or E. C. Pickering's *Elements of Physical Manipulation*, the texts employed by Glazebrook and Shaw in the early 1880s.[86]

The book's content was not its only merit. *Practical Physics* was unique among physics textbooks in that its use was applicable to large classes. A reviewer of the first edition of the book indicated that "The authors have done a real service to all whose business it is to conduct classes in a physical laboratory by supplying them with a most excellent guide. Not only teachers, but also students, will find this book invaluable."[87] Glazebrook and Shaw had made every effort to arrange "the most typical experiments in each subject" into such a well-ordered scheme that instructors and students could easily manage the experiments in class:

---

[85] Ibid., viii-ix.

[86] F. Kohlrausch, *An Introduction to Physical Measurements with Appendices on Absolute Electrical Measurement, etc.* translated from the second German edition by T. H. Waller & H. R. Proctor (London: J & A Churchill, 1873); Edward C. Pickering, *Elements of Physical Manipulation* 2 vols. (New York: Hurd & Houghton, 1873 (part I), 1876 (part II)). Glazebrook remembered that these two books were "almost all that were available" (*A History of the Cavendish Laboratory*, 44).

[87] C.V. Burton "Practical Physics (book review)," *Nature 31* (1885): 477-478.

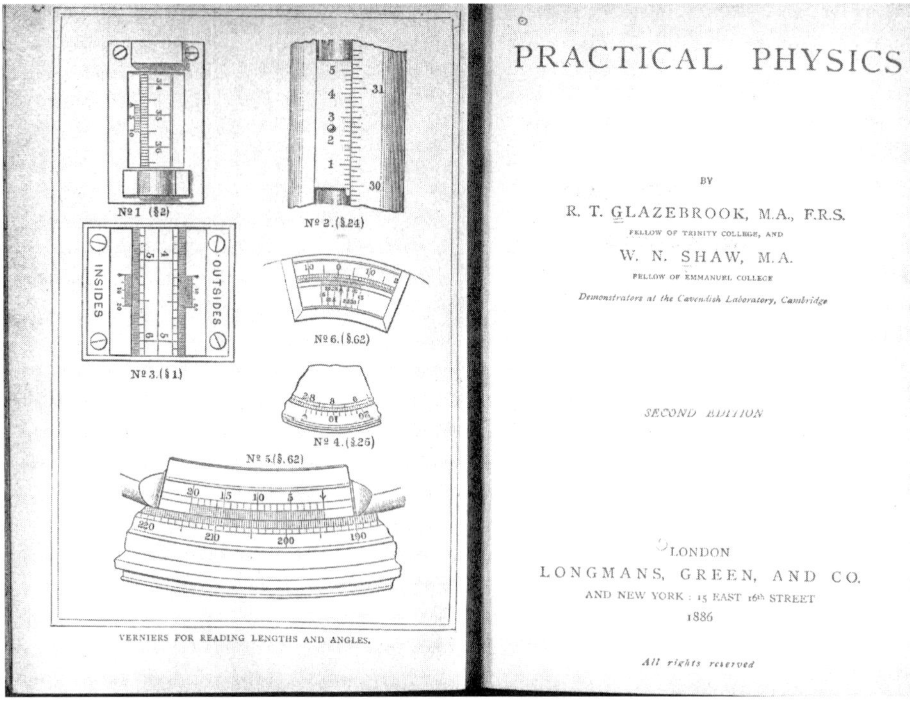

Figure 3.4. Frontispiece and title pages of Glazebrook and Shaw's Practical Physics (second edition).

For practical teaching purpose, therefore, it is an obvious advantage to divide the subject with direct reference to the apparatus required for performing the different experiments. We have endeavoured to carry out this idea by dividing the chapters into what, for want of a more suitable name, we have called 'sections,' which are numbered continuously throughout the book, and are indicated by black type of headings. Each section requires a certain group of apparatus, and the teacher knows that that apparatus is not further available when he has assigned the section to a particular student. The different experiments for which the same apparatus can be employed are grouped together in the same section, and indicated by italic headings . . .

The [proof] sheets, divided into the sections above mentioned, have been pasted into MS. books, the remaining pages being available for entering the results obtained by the students. The apparatus referred to in each book is grouped together on one of the several tables in one large room. The students are generally arranged in pairs, and before each day's work the demonstrator in charge assigns to each pair of students one experiment—that is, one section of the book. A list shewing the names of the students and the experiment assigned to each is hung up in the Laboratory, so that each member of the class can know the section at which he is to work. He is then set before the necessary apparatus with the MS. book to assist him; if he meets with any difficulty it is explained by the demonstrator in charge. The results are entered in the books in the form indicated for the several experiments. After the class is over the books are collected and the entries examined by the demonstrators. If the results and working are correct a new section is assigned to the student for the next time; if they are not so, a note of the fact is made in the class list, and the student's attention called to it, and, if necessary, he repeats the experiment. The list of sections assigned to the different students is now completed early in the day before that on which the class meets, and it is hoped that the publication of the description of the experiment will enable the student to make himself acquainted beforehand with the details of his day's work.

Adopting this plan, we have found that two demonstrators can efficiently manage two classes on the same day, one in the morning, the other in the afternoon, each containing from twenty-five to thirty students.[88]

Glazebrook and Shaw's distinctive textbook was made possible by their experiences at the Cavendish Laboratory: they had been testing every aspect of their procedures in their classes since the early 1880s. *Practical Physics* was widely used and became very popular; its fourth, revised edition was issued in 1893.[89]

In addition to *Practical Physics*, Shaw published *Practical Work at the Cavendish Laboratory: Heat* in 1886 and Glazebrook published *Mechanics: An Elementary Text Book, Theoretical and Practical* in 1895. These books further describe the kinds of experiments conducted in Cavendish courses. While *Practical Physics* described the organization and procedures of elementary courses, Shaw's *Heat* did the same in terms of advanced courses.[90] *Heat* delivered twelve "classical experiments" including Jolly's air thermometer, Matthiessen's method for determining the density of water at different temperatures, Joule's method for determining the point of maximum density of water, manufacture of an air thermometer for rapidly indicating comparatively high temperatures, an electric pyrometer, Laplace and Lavoisier's ice-calorimeter, and Bunsen's ice calorimeter.[91] *Heat* also provided valuable information about Cavendish apparatus. Glazebrook's *Mechanics*, a typical laboratory manual covering dynamics, statics, and hydrostatics, was based on his practical courses for medical students. Glazebrook carefully arranged experiments so that students could deduce theoretical consequences from experimental results. The student thus was being encouraged to "think about the physical meaning of the various steps he takes and not merely to employ certain rules and formulae in order to solve a problem."[92]

3.3.4. Finance

Lack of funds had been a chronic problem of the Cavendish Laboratory ever since its foundation. Although Rayleigh had established a fund for the purchase of

---

[88] Glazebrook & Shaw, *Practical Physics*, x-xi. Brackets added.

[89] Glazebrook & Shaw, *Practical Physics*, fourth revised edition (London: Longmans, Greens & Co., 1893). The alterations and additions to the fourth edition in 1893 included: geometrical representation of rates of variation, the enlargement of the chapter on magnetism, and new chapters on measurement of velocity and acceleration and electromagnetic induction. The authors thanked Newall and Searle for their contributions to the addition of new sections.

[90] W. N. Shaw, *Practical Work at the Cavendish Laboratory: Heat* (Cambridge: Cambridge University Press, 1886).

[91] The book contained twelve of the twenty-three experiments originally intended for the syllabus of advanced demonstrations in heat in 1884 at the Cavendish. For a complete list of the experiments, see Ibid., 5-6.

[92] R. T. Glazebrook, *Mechanics: An Elementary Text-Book, Theoretical and Practical* (Cambridge: Cambridge University Press, 1895), ix.

necessary apparatus, the Cavendish desperately lacked resources. In the first years of J.J.'s directorship, the University's annual grant to the Laboratory of about £250 did not increase, and it was used mainly to pay the wages of the Laboratory assistants (£214 of this amount, for example, went to salaries in 1892).[93] To pay for the apparatus needed for demonstration classes and research, J.J. relied on student fees.

Students ordinarily paid a fee of £1.1 to attend a lecture course and a fee of £3.3 to attend a demonstration course or to use space in the Laboratory. As the number of students taking Cavendish courses increased, so did the collected fees. In 1892, these fees totaled £1179; in 1893, the total rose a bit to £1240; and in 1894 rose again to £1409. Table 3.5 indicates that the Cavendish's expenditures for "Apparatus, Stores, Printing, & c." (£230 3s 6d in 1892) nearly equaled the University's entire grant to the Cavendish (£254 7s 6d in 1892).[94]

Student fees not only made possible the purchase of apparatus but also funded the building of a much-needed Laboratory extension. The need for more space was perceived as urgent in the late 1880s, and a special syndicate was appointed to seek the necessary means for a new building.[95] J.J., Glazebrook, and Shaw were members of the new syndicate. "We are making endeavours to get an addition to the Cavendish Laboratory," J.J. wrote to his old friend, R. Threlfall, who was then in Australia. "The University has given us the piece of land next the Laboratory running down Free School Lane."[96]

Emphasis was placed on the need for more space in which to conduct elementary demonstrations, but space for research and other purposes were equally wanted:

> Again, the space available for the advanced teaching and for original research is limited and the Instrument room is overcrowded with apparatus. A basement room or cellar in which operations requiring uniformity of temperature could be carried on, or apparatus needing exceptionally firm foundations could be set up, is highly desirable; while rooms arranged for various special experiments would be most useful.[97]

The syndicate came up with a clever solution. It proposed erecting not the entire proposed building at a cost of about £10,000, but part of that building, at a cost of about £4,000. This structure would "consist of ground floor which would contain the large laboratory, and a portion of the first floor which would form the Lecture rooms." J.J. suggested that £2,000 of the total cost be "supplied from the balance of the Laboratory fees which has accumulated during the past 10 years." Thus, the building's cost could be met with a grant of £2,000 from the Common University Fund. As J.J. pointed out in the discussion of the proposal, it was "the first occasion

---

[93] *CUR* (17 March 1893): 75.

[94] Beginning in 1893, professors were required to annually report their expenditures and the fees they received. These reports were published in the *CUR*.

[95] *CUR* (13 March 1888): 507; (24 February 1891): 558-559, 562; (28 November 1893): 235-236; (5 December 1893): 267; (23 January 1894): 377.

[96] CUL MSS ADD 7654 T30 (28 October 1893): J.J. to Threlfall.

[97] *CUR* (6 November 1894): 164-165 on 164.

Table 3.5. Receipts and Expenditure from Fees & Other Resources (1892)
[*CUR* (17 March 1893): 92-93]

Receipts			
	£	s.	d.
Fees for Demonstrations and Professor's Lectures	1038	9	0
Fees received for Courses of Lectures given by University and College Lecturers	111	6	0
Fees for Examinations in Practical Physics held in the Cavendish Laboratory	4	7	6
Interest on Lord Rayleigh's Fund	24	19	4
	1179	1	10

Expenditure			
Repaid to Emmanuel College and Trinity College for Lectures delivered in the Laboratory by Mr. Shaw and Mr. Glazebrook	74	4	0
Apparatus, Stores, Printing & c.	230	3	6
Wages (additional)	50	0	0
Demonstrators	480	0	0
Professor's share of Fees*	100	0	0
Balance in hand	244	14	4
	1179	1	10

* *This sum was used to provide an additional demonstrator, whose wages were not included in the University account.*

on which the department had been obliged to ask the University for pecuniary aid."[98] The proposal passed in November of 1894.

Another important event relating to finances during the first decade of J.J.'s tenure at the Cavendish was the establishment of the Clerk Maxwell Scholarship, the first research scholarship for graduate students in experimental physics. In early January of 1887, the Vice-Chancellor published a letter from the solicitor of the late Mrs. Maxwell, who had died on the previous December 12. In her will, Mrs.

[98] *CUR* (20 November 1894): 230.

Maxwell had designated some money for the foundation of a scholarship in honor of her late husband.

> By the tenth purpose of her will . . . that she was desirous of continuing the advancement of those sciences to which the researches of her late husband were chiefly directed left and bequeathed the sum of Five thousand pounds to be applied in founding a Scholarship in the Cavendish Laboratory at Cambridge to be called "The Clerk Maxwell Scholarship" the student to hold said scholarship to be chosen by the Professor of Experimental Physics and by the Lucasian Professor of the University of Cambridge and in case of any difference of opinion between them the final decision to rest with the Master of Trinity or any person appointed by him. The student to be so chosen must be, or must have been, a student in the Cavendish Laboratory foresaid and both in the mode of his election and in the conditions of Tenure of the Scholarship regard shall be had to the advancement by original research, of the Science of Electricity, Magnetism and Heat or to those Sciences especially which were advanced by the researches of her said husband. The Scholarship to be an University Scholarship and the Tenure of it shall be for three years.[99]

Mrs. Maxwell's wish was realized three years later, when the "Regulations for the Clerk Maxwell Scholarship" were issued.[100] To ensure that the elected student, who would receive £200 per year for three years, would commit himself fully to scientific research, J.J. added three important stipulations:

> 4) Any member of the University who has been a student for one Term or more in the Cavendish Laboratory shall be eligible for the Scholarship;
>
> 8) The Student so elected shall devote himself under the direction of the Cavendish Professor to original research in Experimental Physics within the University; he may however carry on his researches elsewhere if he has first obtained the written permission of the Cavendish Professor to do so;
>
> 11) The Student shall not during his tenure of the Scholarship systematically follow any business or profession or engage in any educational or other work which in the opinion of the Cavendish and Lucasian Professors would interfere with his duties as Student.

Because the amount of the scholarship was "very unusually large," J.J. believed that these restrictions were required to guide the beneficiary and also to persuade the members of the Senate. When, during discussion of the proposal, the qualification of the beneficiary was criticized as being "vague," J.J. replied that "the standing of the scholar was purposely left elastic, because in some exceptional cases the study of Physics might be better served by the election of some one who had not been studying in the Laboratory very long, rather than some one who had served his three years there."[101] J.J.'s comment meant that he did not want the scholarship to be used as an award for seniority at the Laboratory but instead should be bestowed upon the most talented young researcher. The regulations for the scholarship passed

[99] *CUR* (11 January 1887): 332-333.
[100] *CUR* (20 May 1890): 730-732.
[101] *CUR* (3 January 1890): 814.

the Senate in June, and the following year W. Cassie was selected to be the first Clerk Maxwell Scholar.[102]

Competition for the coveted scholarship was fierce. When Cassie left Cambridge to become professor of physics at Royal Halloway College in the fall of 1893, C. T. R. Wilson advanced himself as a candidate for the scholarship:

> I beg to offer myself as a candidate for the Clerk Maxwell Scholarship now vacant. The subject I would propose to investigate is the way in which a dissolved body distributes itself in its solvent when the top and bottom of the solution are kept at different constant temperature. Very few experiments appear to have been made on the subject, and it seems of considerable importance in connection with theories of solution and osmotic pressure.[103]

However, the future Nobel Laureate was defeated in his bid for the scholarship by W. C. D. Whetham and would wait two more years to be selected.

J.J. did his best to select the most talented of the many candidates, and in most cases he succeeded. The Clerk Maxwell scholars selected by J.J. became renowned scientists. They included C. T. R. Wilson (1895-1898), J. S. E. Townsend (1898-1901), H. A. Wilson (1901-1904), O. W. Richardson (1904-06), and F. Horton (1906-1909).

The establishment of the Scholarship was, overall, welcome news for Cavendish researchers, but it did create some problems. First, the amount of the scholarship, £200 per year, was certainly too high in comparison with the salaries of the Cavendish teaching staff: double the annual salary of a Cavendish demonstrator and three times the annual salary of an assistant demonstrator. What might have happened if the £200 had been divided into two, three, or four to benefit more researchers? A greater number of researchers would have been relieved of their teaching duties to concentrate on their research. The establishment of the Clerk Maxwell Scholarship would have been a truly great moment in the history of the Cavendish. However, the notion of supporting researchers was still a foreign one at Cambridge, where scholarships were awards for proven excellence. Because the Clerk Maxwell Scholarship was such an award, its purpose was seen as restricted to a very limited number of students, and what number could be more limited than one?

3.3.5. Instruments

Under J.J.'s leadership, the Laboratory's workshop capabilities and store of readily available apparatus were slowly but steadily developed. During the first decade of J.J.'s directorship, three men successively served as official Cavendish workmen: D. S. Sinclair, A. T. Bartlett, and W. G. Pye. J.J. added two lecture assistants to the staff to help the workmen (or the professor). In addition to these workman and assistants, who were paid by the University, J.J. employed a private assistant, E. Everett, who remained with him throughout his tenure and gave him

---

[102] *CUR* (17 June 1890): 977; (5 May 1891): 772.
[103] CUL MSS ADD 7654 W36 (8 November 1893): C. T. R. Wilson to J.J.

"most valuable and able assistance" in all his investigations.[104] Because J.J. was busy fulfilling his responsibilities as director and was "very awkward with his fingers," he needed the assistance of a skilled mechanic in almost all of his experiments.[105] For many years, Everett's principal job at the Cavendish was glassblowing, making the tubes which J.J. employed in his experiments on electric discharges through gases and in his later investigation of cathode rays. J.J. acknowledged Everett's expertise and pertinacity in this obituary notice:

> Everett took a very active and important part in the researches carried on in the Laboratory, by students as well as by the professor. The great majority of these involved difficult glass blowing, which was nearly all done by Everett, as it was beyond the powers of most of the students. In addition to this, he made all the apparatus used in my experiments for the more than forty years in which he acted as my assistant. I owe more than I can express to his skill and the zeal which he threw into his work. He was a very skillful glass blower, a quick worker, very pertinacious; if the first method failed he would try another and generally succeeded in finding one which would work. He was also an excellent lecture assistant, and was great help to me in my lectures at the Royal Institution. He took personal interest in the success of each experiment and spared neither time nor trouble to make it come off.
>
> In the early days of X-rays, before hospitals or medical men had any appliances for taking X-ray photographs, Everett and W. H. Hayles, another assistant at the Cavendish Laboratory and an expert photographer, organised a scheme for taking photographs at the Laboratory. Many medical men availed themselves of this, although the revelations made by the photographs as to the way in which bones had been set sometimes caused considerable embarrassment.[106]

The tubes used in J.J.'s experiments on cathode rays in the 1890s, and in particular those employed in the famous 1897 experiment, are preserved in the New Cavendish Laboratory and give ample proof of Everett's skill in creating workable devices out of unpromising materials. His tubes were relatively simple or even crude, and they signaled the opening of the renowned era of "sealing-wax and string." J.J. was invariably "on fire with an idea that he wished to explore" and in a great hurry.[107] For him, as George P. Thomson pointed out, "the apparatus existed for the experiment, never the experiment to justify the apparatus."[108]

Many instruments that J.J. and the other researchers used in their experiments were constructed in the workshop, but the Cavendish continued to purchase necessary teaching apparatus from outside sources. In the academic year of 1885-86, J.J. added to the Cavendish store a large bridge resistance box, a Wimshurst static electrical machine, a cathetometer, several standard resistance coils, a low resistance galvanometer, a small tangent galvanometer, six large Leyden jars,

---

[104] *A History of the Cavendish Laboratory*, 82.

[105] Strutt, *Life of J.J. Thomson*, 23.

[106] J.J. Thomson, "Mr. E. Everett," *Nature 132* (1933): 774.

[107] Strutt, *Life of J.J. Thomson*, 25.

[108] G. P. Thomson, "J.J. Thomson," *Nature 178* (1956): 1319. J.J. told his son that the trouble with experiments was "you had infinite labour in getting the apparatus to work, and when it did work the experiment was over too quickly."

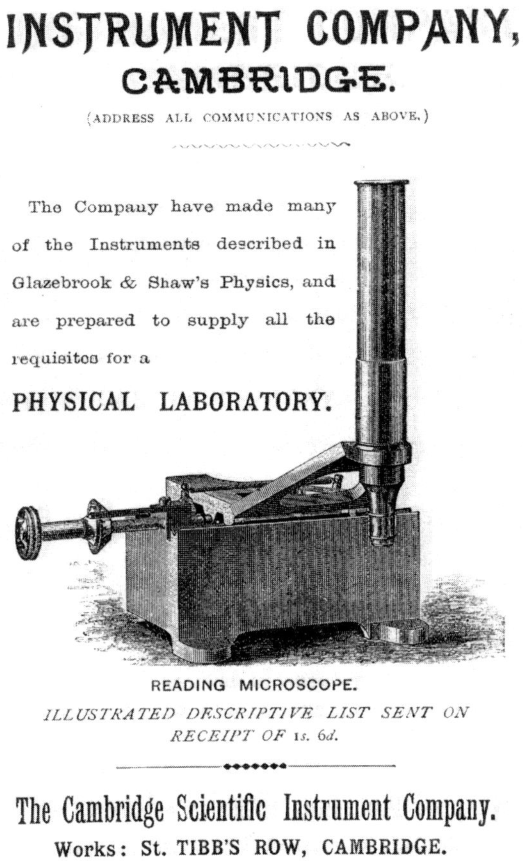

Figure 3.5. An advertisement of the Cambridge Scientific Instrument Company in Glazebrook and Shaw's Practical Physics, fourth edition (1893)

a commutator, and a Melloni radiant heat apparatus. In 1886-87, the Cavendish purchased a spectrophotometer, an Atwood machine, a mercury pump, and a whirling table. In 1887-88, the Cavendish purchased a two-horse power Otto Engine, wires, and fourteen storage cells. In 1888-89, the teaching apparatus was increased by purchases of an induction coil, a chronometer, two reading microscopes, a resistance box, a cylinder, an electrometer, a rotating disc of heavy glass for Villari's experiment, a set of Rowland's photographs of the solar spectrum, a large Wimshurst electrical machine, an optical bench, and a galvanometer. In 1889-90, the Cavendish was enriched by two reading microscopes, a high resistance galvanometer, a Deprez-D'Arsonval galvanometer, and a Brockie-Pell arc lamp. After 1890, J.J. stopped itemizing the new apparatus purchased and reported merely "considerable additions . . . to the stock of apparatus."[109]

During J.J.'s early tenure at the Cavendish, the Cambridge Scientific Instrument Company continued to supply the Cavendish with apparatus, particularly equipment for demonstration classes. An advertisement by the Company printed in the revised fourth edition (1893) of Glazebrook and Shaw's *Practical Physics* boasted of the Company's close relationship with the Cavendish Laboratory. The Cambridge Scientific Instrument Company also manufactured a few specially ordered instruments, such as the mirror and its mounting used in J.J.'s experiment on the propagation of the luminous discharge of electricity through a rarefied gas.[110] The Cavendish also ordered a few delicate instruments, for example, Jolly's air thermometer, from German instrument companies.[111]

3.4. Researchers and Researches

Under J.J., the Cavendish Laboratory continued to be the center of physics graduate studies at Cambridge, as it had been under Maxwell and Rayleigh. However, several noteworthy changes occurred during the first ten years of J.J.'s tenure. First, the number of research students stagnated and even decreased. Whereas the Cavendish under Maxwell and Rayleigh had consistently attracted an average of three or four new researchers each year, none arrived in 1892 and only one arrived in 1893. The number of new research students declined even as attendance at courses taught by J.J. and his assistants were rising, indicating that, at Cambridge, research and teaching were mostly separate enterprises (see Table 3.4).

A second important change was that the Cavendish ceased to be the center to which the high wranglers flocked. Although the Cavendish still attracted some high wranglers, notably W. H. Bragg, E. G. Gallop, J. J. Guest, W. Cassie, J. B. Peace, A. Brand, and R. H. D. Mayall, most of them came to the Cavendish only casually, spending a year or two and then leaving, often without producing any research papers.[112] Among the high wranglers at the Cavendish, only Bragg and Cassie were committed to science.

With high wranglers from the MT like J.J. or Glazebrook no longer representing the driving-force among Cavendish researchers, MT-NST and NST-only graduates were increasing in importance. Already the Cavendish had seen NST students like Shaw, Wilberforce, Threlfall, Fitzpatrick, Skinner, C. Chree, and Randell. These talented students were soon reinforced by such NST recruits as W. C. D. Whetham, G. F. C. Searle, R. S. Cole, and J. W. Capstick. From 1885 on, many NST graduates served as demonstrators or assistant demonstrators, which meant that they stayed at

[109] *CUR* (7 June 1886): 704; (26 May 1887): 749; (4 June 1888): 759; (29 May 1889): 785; (5 June 1890): 873; (8 June 1893): 971.

[110] J.J. Thomson, "On the Rate of Propagation of the Luminous Discharge of Electricity through a Rarefied Gas," *Proc. Roy. Soc. 49* (1891): 84-100 on 86.

[111] Jolly's air thermometer was supplied by Strollnreuther und Sohn in Munich. See Shaw, *Heat*, 7.

[112] Peace and Cassie, however, each wrote one research paper.

the Cavendish longer than before and influenced more undergraduates. This was a significant change not only because the Cavendish Laboratory was beginning to fill with students who were more oriented toward experimentation, but also because the Laboratory was failing to attract Cambridge's most talented students, the wranglers.

A third noticeable change at the Cavendish was the presence of a greater number of "outsiders." Visitors came from many different places. From within Britain came A. M. Worthington of Oxford University, O. Lodge of Liverpool, J. Monckman and C. V. Burton of the University of London. From Austria came K. Olearski; from the United States, H. F. Reid of Johns Hopkins University; from Poland, W. Natanson of the University of Cracow; from Australia, Miss F. Martin. Worthington and Olearski each wrote a research paper based on research performed at the Cavendish. No regulations however governed these non-Cambridge visitors, who stayed as guests for a few months or as long as a year.

The status of women students also had changed: beginning in 1882 women were admitted to take tripos examinations, including the MT or NST. The Cavendish Laboratory under Rayleigh was one of the first institutions to open its doors to women "on equal terms with men."[113] From the mid-1880s on, however, as Paula Gould has pointed out, women researchers became a somewhat more important group at the Cavendish.[114] For example, in 1890, MT examiners commended Miss P. G. Fawcett for attaining a rank "above the senior Wrangler."[115] In 1893, Miss H. G. Klaassen, who studied "cyclic magnetizing processes in iron" with Professor J. A. Ewing of Engineering Department, published her findings in *Philosophical Transactions* (1893).[116] The experiences of these two women at the Cavendish furthered their careers as lecturers for women students at Cambridge. Klaassen lectured on electricity and magnetism and Fawcett lectured on physics and astronomy. Miss I. Freund, who had worked in the Cavendish under Rayleigh, continued to lecture on chemistry.[117]

Despite many areas of progress, J.J.'s first ten years at the Cavendish were somewhat disappointing in terms of advancement of the Cavendish research tradition. The major impediment was the Cavendish's inability to recruit brilliant students. Under Maxwell, the Cavendish had attracted Schuster, Chrystal, and Glazebrook. Under Rayleigh, it had attracted J.J. Thomson. Although enlarging the teaching staff had provided a few researchers with financial security, this measure

---

[113] Crowther, *The Cavendish Laboratory*, 91.

[114] Paula Gould, "Women and the Culture of University Physics in Late Nineteenth-Century Cambridge," *BJHS 30* (1997): 127-149.

[115] *The Historical Register of the University of Cambridge to 1910*, 558. In those days, the ranks of women candidates for the MT and NST were published separately and compared to men's ranks, for example: "equal to 21 [of the men's rank]," or "between 27 and 28."

[116] J. A. Ewing & H. G. Klaassen, "Magnetic Qualities of Iron," *Phil. Trans. 184* (1894): 985-1039. Interestingly, in this paper the authors employed the word "we" instead of either "Professor Ewing" or "Miss Klaassen." Compare this use with that in Rayleigh and Sidgwick's paper of the early 1880s, when the authors used their individual names instead of the pronoun "we."

[117] Miss Klaassen lectured on elementary physics beginning in the Lent term of 1893 (*CUR* (17 January 1893): 384). For more information about Miss Fawcett's lectures, see *CUR* (16 January 1894): 367.

evidently was not enough to attract the most talented students. Nor did it convert less able students into stars. The Cavendish Laboratory was filled, but with students who, although more diverse, were also more ordinary.

From 1874 to 1884, twenty-five Cavendish researchers (excluding professors) produced papers; from 1885 to 1894, twenty-nine did. In addition to the senior group of researchers comprised by J.J., Glazebrook, and Shaw, a junior group of demonstrators and assistant demonstrators, in particular, Fitzpatrick, Wilberforce, Whetham, Skinner, and Searle, emerged as the main producers of papers. The papers by many Cavendish researchers published during this decade reflect the Cavendish's lack of outstanding scholars. Excluding J.J., Cavendish researchers during this decade produced eighty-six papers, only nine more than their predecessors of the previous decade who produced seventy-seven papers, and the number of researchers producing more than two papers rose only by three, from eight to eleven. No outstanding figure led the new group in authoring papers, as Schuster and Chrystal had done under Maxwell and Glazebrook and J.J. had done under Rayleigh. Instead, the number of jointly authored works increased, from eight in the decade prior to 1885 to twelve between 1885 and 1894. Of these twelve works, seven were co-authored by J.J.

In judging the quality of the works produced, publication in *Philosophical Transactions* serves as a useful index. Between 1874 and 1884, that periodical published sixteen papers by Cavendish researchers (including professors), nine of which were by research students—Glazebrook, J. E. H. Gordon, Schuster, and J.J. Thomson. Glazebrook contributed four of the papers and Gordon and J.J. each contributed two. During the following decade, the number of Cavendish papers published in *Philosophical Transactions* dropped to fourteen, seven of which were written by individual researchers (McConnel, Callendar, Cassie, Chree, Whetham, and Capstick). Only Whetham contributed more than one paper.

The distribution of the subject matter of these papers shows wide spectrum of interests, including electricity, magnetism, properties of matter, optics, heat, meteorology, electric discharge through gases, physical chemistry, and others.[118] Under Rayleigh, the main research topic at the Cavendish had been the determination of electrical standards but, under J.J., as can be seen in Table 3.6, no subject constituted a common interest. As J.J. transformed himself from

[118] Newall classified the works of this period (including J.J.'s works) as follows:

General Physics and Properties of Matter	13%
Heat	9%
Optics	9%
Electricity and Magnetism	20%
Conductivity of Gases	20%
Non-experimental	10%
Meteorology	2%
Reports and Summaries	7%
Miscellaneous	10%

(In *A History of the Cavendish Laboratory*, p. 118)

mathematician to physicist, the Cavendish Laboratory also leaned in the direction of experimentation. Under J.J., the Cavendish researchers produced fewer theoretical and mathematical studies and a greater number of experimental studies. As shown by their papers, Cavendish researchers were designing or improving apparatus, determining effects, or measuring constants. An exception who proved the rule was Wilberforce, who paid close attention to theory and mathematics and in a few experiments attempted to confirm Maxwell's electromagnetic theory of light.

Table 3.6. Number and Content of Research Papers at the Cavendish, 1885-1894

Name	*Individual Work*	*Co-Work*	*Main Topics*
J.J. Thomson	30	1/2 x 7	electromagnetism, electric discharge through gases
W. N. Shaw	12	1/2 x 1	electricity, meteorology
C. Chree	7		heat, elasticity, electricity, magnetostriction
R. T. Glazebrook	5	1/2 x 4	electrical standards, optics
L. R. Wilberforce	5		electricity, viscosity, vibration
W. C. D. Whetham	5		electrolysis, slippage
T. C. Fitzpatrick	4	1/2 x 2	electrolysis, specific resistance
S. Skinner	4	1/2 x 2	property of matter, Clark cell
J. Monckman	4	1/2 x 1	property of matter, electromagnetism
G. F. C. Searle	3	1/2 x 1	electromagnetism, magnetometer
H. F. Newall	2	1/2 x 3	property of iron
R. Threlfall	2	1/2 x 2	electricity, explosion
J. A. Fleming	2		electricity
J. C. McConnel	2		Nicole's prism
H. L. Callendar	2		temperature measurement, surface tension
R. H. Adie	2		physical chemistry
J. H. Poynting	1		mean density of the earth
A. M. Worthington	1		stretching of liquids
A. W. Ward	1		magnetic rotation of polarization
W. Coldridge	1		property of matter
H. G. Klaassen	1		thermo-electric properties
W. Cassie	1		thermo-electric properties
G. F. Emery	1		thermo-electric properties
J. B. Peace	1		electric discharge
J. W. Capstick	1		specific heat
P. G. Fawcett	1		electrical properties of gases
C. V. Burton	1		colour photography
K. Olearski	1		electrical properties of gases
A. H. Leahy	1		viscosity of gases
F. M. Turner		1/2 x 1	sound
Total: 30	104	12	

Another important change at the Cavendish under J.J. was the direct influence of the Professor on the research choices of students. Under Maxwell and Rayleigh, the Cavendish was a Laboratory to which research students often brought their own research topics for investigation. This tradition continued under J.J., with J.J. offering helpful advice to students completing studies of their own choosing. However, more and more, students selecting research topics began to rely on the Professor's suggestions. Some of these students, like Monckman, Olearski, Klaassen, and Fawcett, were outsiders, but others, like Chree and Adie, were Cambridge men. During this period, Cavendish researchers performed a significant number of electrolytic experiments not only because such studies were relatively easy to perform but also because they related to J.J.'s interests. This new trend was accelerated by the close contact J.J. maintained with his students. At the Cavendish, friendliness and enthusiasm had replace formality and distance,[119] and this new atmosphere reflected a change in the personalities not only of the Professor but also of the students: the air of hubris associated with the high wranglers was now missing from the Cavendish.

The major reason for the general stagnation at the Cavendish during the earlier years of J.J.'s tenure was that "Professor Thomson" was not yet the outstanding authority in the field of physics that he was soon to become. Glazebrook had been deeply influenced and stimulated by Rayleigh, and Fleming had abandoned his job to go to the Cavendish to work with Maxwell, but J.J. was not yet a charismatic leader who could attract outstanding outsiders and high wranglers. Researchers who specifically aimed to learn experimental skills usually sought out and relied on Glazebrook or Shaw, and most Cavendish students were of ordinary talent and came to the Laboratory to receive instruction or use facilities they needed. It was only toward the end of his first decade at the Cavendish that J.J. began to acquire fame as a foremost physical scientist and to attract students with outstanding potential and creative insight.

3.5. Was there a "Cavendish School" in 1894?

By the end of 1894, the Cavendish Laboratory was, from Cambridge University's point of view, a success. It was firmly rooted in Cambridge, and it was providing a teaching model for other University departments. Moreover, from the perspective of the University, its existence had been almost trouble-free: the Laboratory had asked little of the University and raised few problems in the Senate House.

However, the Cavendish Laboratory had not advanced the research tradition that Maxwell had envisioned in his Inaugural Lecture and that Rayleigh had worked to nurture. Several difficulties had become apparent. As its researchers became increasingly involved in experimental physics, their interests centered more and more on studies that measured "to another place of decimals," a pursuit that

---

[119] See collected obituaries for J.J. in *Nature 146* (1940): 351-357.

Maxwell had warned against. And the Laboratory failed to attract outstanding researchers within Cambridge. Even J.J.'s ambitious and effective reform of the physics curriculum had not stopped this trend.

There was another worrying indicator. Since 1891, the Royal Commission of the Exhibition of 1851 had awarded Science Research Scholarships to British science students under the age of thirty. It offered rare "opportunities for young graduates to continue their scientific training for 2 or 3 years under skilled direction in order to equip themselves for responsible positions in the scientific service of the country."[120] Although recipients of the scholarship could choose the school at which to continue science study, by the end of 1894, not one had gone to Cambridge to study physics.[121] For the most part, these scholarship holders went either to German universities or to Glasgow University, University College at London, the Royal College of Science, or Owens College. Many 1851 Research Scholars undoubtedly avoided the Cavendish Laboratory because Cambridge lacked any facility for providing a doctoral degree in physics. The Cavendish's failure to attract any 1851 Research Scholars was humiliating for Cambridge, and J.J. shared responsibility for this humiliation. If Rayleigh had remained Professor of Experimental Physics, the situation would have been different; J.J.'s qualities as professor were relatively untested.

In light of these problems, should J.J.'s first decade at the Cavendish, from 1885 to 1894, be considered a failure in terms of establishing a research school or maintaining a research tradition? Not necessarily. Under the disappointing surface pulsed a strong positive force, one that would soon prove pivotal in establishing the future "Cavendish School." That force was, of course, J.J. Thomson himself. By 1895 he had evolved into a leading physicist with considerable experience in both experimentation and the management of the Laboratory. Although he did not impose his influence on researchers of his own age, he gradually did influence the Cavendish's younger men and women, particularly by influencing their choices of research topics.[122] During his first decade at the Cavendish, J.J. introduced a pioneering topic of study, the electric discharge through gases, and in this field he

[120] *Record of the Science Research Scholars of the Royal Commission for the Exhibition of 1851, 1891-1929* (London, 1930), preface. The Scholarship was founded by a fund raised by the Royal Commission for the Exhibition of 1851. The scholarship ordinarily was tenable for two years (£150 per year) and could be renewed for a third year in special cases. The candidate was required to be a British subject who had been "a bona-fide student of Science for a term of three years in a University or College in which special attention [was] given to scientific study."

[121] J. R. Erskine-Murray (1894-96) and J. A. McClelland (1894-95) came to Cambridge (therefore to the Cavendish Laboratory) in 1895.

[122] See Isobel Falconer, "J.J. Thomson and 'Cavendish Physics,'" in Frank A. J. L. James (ed.), *The Development of the Laboratory* (London: MacMillan Press, 1989), 104-117 on 116. Falconer argued that among the "FRSs educated under Thomson," Chree, Threlfall, Callendar, Newall, and Searle were "those little influenced" whereas Schott and Whetham were "those strongly influenced." I do not entirely agree with this view. Nevertheless, it does support the conclusion that J.J. lacked authority among Cavendish researchers of his own age.

and a few other researchers at the Cavendish had accumulated valuable experience and insight.

J.J. also established, in 1893, the Cavendish Physical Society. This was an important contribution to the Cavendish research tradition. J.J. presided over meetings of the Society "fortnightly on Tuesday during the term time, in the lecture room of the Cavendish Laboratory with the Professor in the chair."[123] The society's main activity was presentation of colloquia or seminars at which "recently published papers in Physics" were discussed or at which Cavendish researchers gave accounts of the results of studies before publishing those results. At these meetings, J.J. often presented his own results to the younger researchers.

Another contribution by J.J. to the Cavendish research tradition was his continuance of the daily tea hour instituted by Rayleigh. Tea became "the best time in the Laboratory day" and an important, meaningful ritual in the life of the researchers.[124]

Thus, intentionally or fortuitously, it was J.J. Thomson who laid the foundations for the successes of the Cavendish over the decade to come. Both J.J. and the Cavendish were transformed between the years of 1885 and 1894, and neither would change much during the remainder of J.J.'s tenure. By the end of 1894, the Cavendish Laboratory had acquired certain essential characteristics: the leadership of a mature, widely respected scientist; organizationally stronger teaching and research; and experimentally oriented research students. The two factors responsible for its subsequent success, a prestigious professor with outstanding students, would be secured during the next five years with the arrival of "Advanced Students" and the discovery of the electron. The emergence of the "Cavendish School" was not far in the future.

---

[123] Strutt, *Life of J.J. Thomson*, 40. See also *A History of the Cavendish Laboratory*, 90.

[124] Strutt, *Life of J.J. Thomson*, 53-54. During tea, no limitation was placed on subjects for conversation, which included politics, fiction, sports, and matters of general scientific interest.

CHAPTER 4

THE EMERGENCE OF THE CAVENDISH SCHOOL, 1895-1900

> It was fortunate that the new University regulations, which caused a large influx of students to the Laboratory bent upon original research, came just at the time when the discovery of the Röntgen rays gave us a very powerful method of investigating the phenomena attending the discharge of electricity through gases.
>
> J.J. Thomson[1]

> The short period under review [1895 to 1898] is one of the most interesting and important in the history of research in the Cavendish Laboratory . . . Amongst other discoveries it witnessed within its walls the final proof of the nature of the cathode rays, the advent of the negative corpuscle, or electron, as a definite entity, the experimental proof of the character of the conduction of electricity through gases, and the initial analysis of the radiations from radioactive matter.
>
> Ernest Rutherford[2]

4.1. The 1895 Regulation

During the closing decades of the nineteenth century, it became evident that some of the exclusive traditions of the Universities of Oxford and Cambridge required radical reform to meet the challenges of changing times. One obvious defect was these Universities' lack of graduate programs, which were flourishing in the German-speaking countries and which had already proven their effectiveness and value, particularly in the sciences. Because teaching positions at other universities increasingly were being filled by doctors of science (or Ph.D. holders), graduate students from English-speaking countries were flocking to German universities or British universities where doctoral degrees were available. The problem had political ramifications: was it right for an Australian graduate student to use a British government scholarship to fund his study of electromagnetism at a German university? As the significance of the situation became recognized, a crucial reform in Cambridge's regulations began to be formulated. This reform soon would prove to be one of the most significant events in the history of the Cavendish Laboratory.

---

1 *A History of the Cavendish Laboratory*, 93.
2 Ibid., 159.

In February of 1894, Cambridge University's "Council on Post-Graduate Study" released its first report on "the advisability of encouraging the residence in the University of Graduates of other Universities who may desire to pursue in Cambridge a course of advanced study or research."[3] In this report, the Council recommended that the University establish two new degrees, a Bachelor of Letters (Litt. B.) and a Bachelor of Science (Sc. B.). A graduate of one of certain recognized universities or a qualified Cambridge graduate could apply for one of these new degrees after presenting an original dissertation, for examination and approval, to the degree committee of one of the special boards of studies. The Council further recommended that a distinctive student who had made an "original contribution to the advancement of science" could be granted the degree of "Doctor Designate in Science," and a student who had made similar contributions to learning could be awarded the degree of "Doctor Designate in Letters." This new scheme, the Council thought, would make Cambridge "attractive to Graduates of other Universities, who naturally desire that a course of study pursued by them in Cambridge should receive some mark of University recognition."

The ensuing discussion of the report, which took place on February 15, was heated.[4] In the University Senate, supporters of the Council's new scheme faced strong opposition. H. Lawrence argued that the reform would have "a great political advantage, as it would help to bring English-speaking peoples together and would promote good feeling both between England and America and England and her colonies." G. M. Humphry argued that introducing a greater number of advanced and older students to Cambridge would be advantageous for both professors and younger students. H. Jackson supported the reform as necessary to prevent able students from being lost to German universities, pointing out that many American students just passed through England, "either having spent or about to spend a year in studying at a German University." J.J. reported that many foreign students expressed "the great desire" for higher degrees at Cambridge and that ratification of the proposal would be much welcomed at the Cavendish Laboratory. The opposition, however, suggested that graduates of other universities might achieve the proposed degrees more easily than Cambridge graduates, causing "unpleasant feelings to Cambridge students comparing their own position with the special privilege of outside students." One member of the University Senate thought the reform would cause Cambridge to become overcrowded with "degree hunters." Some argued that the new degrees might not solve the problems they were designed to address: the distinction, competence, and reputation of professors, and the development of advanced instruction for Cambridge students. W. Bateson, basing his comments on his experience at the morphological laboratory under Professor F. M. Balfour, suggested that "what attracted [graduate students] to Cambridge was that they could see methods applied and investigations carried on which they could hardly see anywhere else in the world." A. Marshall agreed that Cambridge had need of "some

---

[3] *CUR* (6 February 1894): 425-427.

[4] *CUR* (20 February 1894): 492-498. Brackets added.

system analogous to the German one [seminar and Privatdocent]" to serve Cambridge students, but he was unsympathetic to the notion of attracting graduate students to Cambridge from elsewhere.

The divide in the Senate indicated that the Council's proposal would not be passed without revision. In October, therefore, the Council issued a second report, this time recommending the appointment of a syndicate "to consider the proposals for . . . special degrees for advanced study and research."[5] The Council also suggested relaxing the requirements for Cambridge graduates wishing to enter the new degree programs. To support its agenda, the Council included in its report a communication from the Secretary of State for the Colonies, which contained the following two letters from Canada:

> 8th Oct. 1894
>
> I enclosed herewith a copy of a letter I have received from the Minister of Education of Ontario, who informs me, as you will observe, that the arrangement at present in operation at the Universities of Oxford and Cambridge have the effect of sending Canadian graduates who desire to take a post-graduate course to the Continent in preference to the English Universities.
>
> I think this is very much to be regretted on every account and I should be extremely obliged if you could induce the authorities of the English Universities to dispense with the preliminary examination, which under the circumstances does not seem to be necessary, and thus afford Canadian graduates the same facilities that are offered them at Berlin, Leipzig and elsewhere.
>
> Charles Tupper, the High Commissioner for Canada

> 13th Sept. 1894
>
> It is quite the usual thing for graduates of the Universities of Ontario to take a post-graduate course on the Continent either at the English or at the German Universities. They are, however, placed at some disadvantage in taking such a course at Oxford and Cambridge as compared with the facilities afforded at Berlin, Leipzig, and elsewhere. The German Universities admit graduates of our Canadian Universities to their courses without any preliminary examination. At Oxford and Cambridge however a preliminary examination is required, the effect of which is to force our students to prefer the German post-graduate course to the English one. I think if this circumstance were pointed out to the authorities at Oxford and Cambridge they would relax a regulation which is vexatious to the Canadian students and which places them in foreign associations which are certainly not as desirable in many respects as the association of the British University. If such relaxation were obtained I am quite sure many Canadians would prefer remaining in England. Do you think anything could be done in the direction I indicate?
>
> Geo. W. Ross, the Minister of Education of Ontario

---

[5] *CUR* (23 October 1894): 80-86. The regulations allowed Cambridge graduates to enter the program "if they [had] passed a Tripos Examination (Part I or Part II in the case of a divided Tripos)." The report included four appendices: Appendix 1 contained suggested provisions for the degrees of Litt.B. & Sc.B.; Appendix 2 contained the Oxford resolution for the new degrees; Appendix 3 contained the Scottish universities' regulations for the research students; and Appendix 4 contained the two letters from the Secretary of State for the Colonies. The council also reported that "a scheme having the same object has recently come into operation at the University of Harvard" (p. 80).

The message from Canada was clear: Britain was losing able students from its own empire to Germany, her chief rival on the Continent. To emphasize the need for reform, the Council reissued these two letters several times. Because Britain was acutely aware of the expansion and growth of the German empire, discussion of the reform of Britain's two most venerable universities, Oxford and Cambridge, was not confined to academic circles. The proposed reform was a political matter, and few dared to object to a scheme clearly in the best interests of Britain. Even the most conservative of reform opponents had to give up their resistance. Just two weeks after the Council's release of its second report, a proposal to appoint a syndicate for the reform was passed.[6] J.J. was elected as one of the twelve members of the new Advanced Study and Research Syndicate.

In February of 1895, one year after the publication of the Council's first report, the new syndicate published its first report.[7] Despite the push for the radical reform, it recommended a compromise between the reform and conservative positions. As usual, Cambridge was resisting change. The syndicate had sacrificed the radical proposal of instituting new degrees in favor of awarding non-Cambridge degree candidates with the B.A. degree and/or a "Certificate of Research." The syndicate also had made a number of other important revisions to the 1894 proposal. The new proposal called for a distinction between Cambridge and non-Cambridge students: "Advanced Students who enter for a Tripos Examination and those who pursue a more independent course of study or research." Second, it included more stringent prerequisites for non-Cambridge applicants to the graduate program. Third, it lengthened the period of Cambridge residence required of degree candidates from three to "at least six" terms. In other words, the Syndicate aimed to promote the supervision and assimilation of outsiders. As the Vice-Chancellor noted in the discussion of this report, although few objected in principle to the prospect of bringing advanced students to Cambridge from elsewhere, "great diversity of opinion" still existed concerning the methods to be used in doing so and the number of such students to be admitted.[8]

The elimination of the proposed new baccalaureates in letters and in science, however, was largely offset by the establishment of the new certificate of research. Most advanced students, especially those in the sciences, came to Cambridge to perform research and not to sit for a tripos. According to the new proposal, a certificate of research was available to a Cambridge graduate or an advanced student from outside who pursued "under supervision a course of research in the University" and "submitted a dissertation which shall have been adjudged to be of distinction as an original contribution to learning or as a record of original

---

[6] *CUR* (13 November 1894): 208. The Syndicate would "consider (1) the best means of further help and encouragement to persons who desire to pursue courses of advanced study or research within the University, (2) what classes of students should be admitted to such courses, (3) what academic recognition, whether by degrees of otherwise, should be given to such students, and upon what conditions."

[7] *CUR* (5 March 1895): 594-598.

[8] *CUR* (19 March 1895): 666.

research," provided that the student had resided in Cambridge for two terms or longer.[9] If the student had resided in Cambridge for at least six terms, he or she would also receive a B.A. degree. In other words, an advanced student at Cambridge could receive a B.A. degree plus a certificate in two years without taking any examinations. The student could then proceed "under the usual conditions to the Degree of M.A. and to other Degrees in the University." It would be an attractive plan because the student would be able to concentrate on research for two years, with few interruptions, receive degrees for that research from one of the most prestigious universities in the world, and then go on to earn a master's degree from that same university. This second proposal was reviewed promptly and the final third report was passed by the Senate, without significant modification, in June of 1895.[10] The corresponding statutes for advanced students were altered and then approved by the Queen in Council.[11]

The new regulations directly impacted the Cavendish Laboratory by making its status as a graduate school official. The first beneficiaries of this change were the Cambridge graduates at the Cavendish, who previously had not received any official University recognition for their work. As the number of foreign students studying at the Cavendish increased, the character of the institution also changed. Although a German Ph.D. still had more value than a second bachelor's degree and Cambridge certificate of research, the Cambridge affiliation was powerful enough to attract a greater number of students from English-speaking countries. Under the new regulations, non-Cambridge men and women, once in the minority at the Laboratory, no longer were regarded as guests. Nor did they follow John A. Fleming's example of pursuing the NST. They now were regular members of the University.

However, the new regulations could not ensure a constant flow of able students to the Cavendish. The most effective magnet for attracting students to the Laboratory was still the Professor himself: it was "any special distinction on the part of the Professor or any original and important method of work" that induced "the best students from all sides" to come to Cambridge.[12] J.J. Thomson's fame as a physicist and research projects were now capable of attracting talented students from Cambridge and beyond. Among these talented newcomers was J.J.'s future successor as Cavendish professor, Ernest Rutherford.

4.2. J.J. Thomson and the Newcomers

4.2.1. J.J. and the First Wave of Advanced Students

When the first of advanced students, E. Rutherford and J. S. Townsend, arrived at the Cavendish Laboratory in the Michaelmas Term of 1895, they were warmly

---

[9] *CUR* (5 March 1895): 595. This residence requirement was added to the Syndicate's amended third report. See *CUR* (11 June 1895): 942.

[10] *CUR* (11 June 1895): 940-943.

[11] *CUR* (2 October 1895): 2-4; (26 May 1896): 799-800.

[12] *CUR* (20 February 1894): 494.

welcomed by J.J. He was then in his forties and had earned both the respect of his colleagues and the admiration of his students. Honors gradually had come to him, and his reputation had spread outside Cambridge.[13] As he wrote Threlfall in April of 1896, "I have got my hands full just at present. I have to give the Rede Lecture this term, preside over Section A [of the British Association at Liverpool] in September & lecture in America in October."[14] J.J.'s selection as Cambridge University delegate to Princeton University's celebration of its 150th anniversary was a clear sign of J.J.'s growing stature within Cambridge. At Princeton, his increasing international fame was marked by his receipt of an honorary doctorate of laws and his delivery of lectures on the discharge of electricity through gases, which were later published in a book.[15]

The first group of advanced students, who arrived at the Cavendish between 1895 and 1900, came mostly from English-speaking countries. From various parts of the British Isles came Townsend, J. A. McClelland, G. B. Bryan, W. Craig Henderson, S. W. Richardson, G. A. Shakespear, J. H. Vincent, H. A. Wilson, J. B. B. Burke, R. S. Willows, R. L. Wills, C. G. Barkla, V. J. Blyth, and J. A. Cunningham; from New Zealand came Rutherford; from Australia came J. J. E. Durack; from the United States came J. Zeleny; and from Canada came J. C. McLennan, W. C. Baker and J. Patterson. Together with the guest researchers, V. Novák from Bohemia, P. Langevin from France, I. Nabl from Austria, Miss E. Neumann from Germany, and C. D. Child, R. B. Owens, J. E. Almy and E. Rhoads from the United States, the advanced students created a new cosmopolitan atmosphere in which Cambridge graduates at the Laboratory met "men of widely different training and experience, of different points of view in political, social, and scientific questions, of very different temperaments."[16]

The newcomers were "not popular in all quarters."[17] Cambridge graduates viewed them as "competitors for the limited facilities in the way of apparatus and the services of the workshop, and also (some may possibly have feared) for the attention and sympathy of the Professor."[18] J.J. endeavored to remove the distinction between Cambridge and non-Cambridge men at the Laboratory, but the newcomers nevertheless detected unease among their Cambridge counterparts. Some Cambridge men viewed the newcomers' origins with contempt, and Rutherford later recalled

---

[13] During the 1890s, J.J. received the following honors: Honorary Doctor, Dublin University (1892); Royal Medal, Royal Society (1894); President of the Cambridge Philosophical Society (1894); Honorary Member of the Manchester Literary and Philosophical Society (1895); Rede Lecture (1896); President of Section A of the British Association at Liverpool (1896); Honorary Doctor of Laws, Princeton University (1896); Foreign Correspondent of Royal Academy of Sciences of Turin (1896). For a complete list of his awards, see Strutt, *Life of J.J. Thomson*, 288-291.

[14] CUL MSS ADD 7634 T36 (17 April 1896): J.J. to R. Threlfall. This quotation was printed in Strutt's *Life of J.J. Thomson*, 54. Brackets added.

[15] J.J. Thomson, *The Discharge of Electricity through Gases* (Westminster: Archibald Constable & Co., 1898).

[16] J.J. Thomson, *Recollections*, 138.

[17] Strutt, *Life of J.J. Thomson*, 60.

[18] Ibid., 60-61.

that "one or two demonstrators with the ancient prejudice that no good things can come from the Colonies" always passed his door "with a snigger."[19] Furthermore, while Cambridge graduates identified each other by their ranks in the triposes, the newcomers lacked these identifying labels because they had not, and would not, take these examinations. As a result, the newcomers' relationships with the demonstrators were very different from those of the Cambridge graduates who depended heavily on their instruction to earn high marks. Worse, the newcomers often were better researchers than the Cambridge demonstrators and other senior researchers, who at times suffered from envy.

To survive under these inhospitable conditions, the newcomers grouped together.[20] The adversity they faced was somewhat alleviated by two factors: J.J.'s

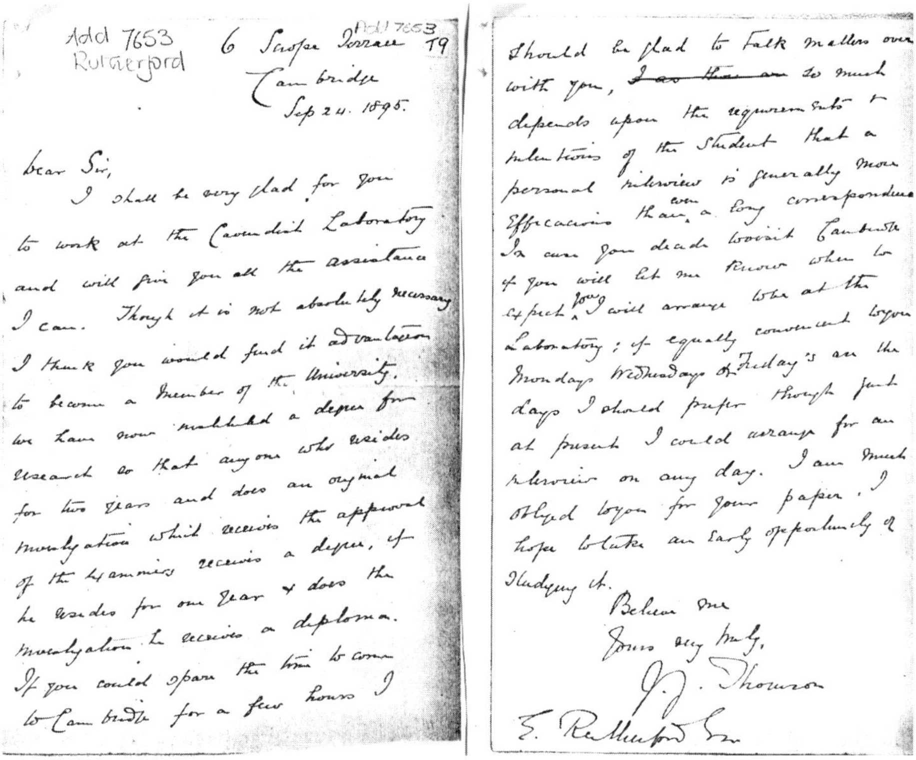

6 Scroope Terrace
Cambridge
Sep 24. 1895.

Dear Sir,

I shall be very glad for you to work at the Cavendish Laboratory and will give you all the assistance I can. Though it is not absolutely necessary I think you would find it advantageous to become a member of the University. We have now established a degree for research so that anyone who resides for two years and does an original investigation which receives the approval of the Examiners receives a degree, if he resides for one year & does the investigation he receives a diploma. If you could spare the time to come to Cambridge for a few hours I should be glad to talk matters over with you, ~~I see these are~~ so much depends upon the requirements & intentions of the Student that a personal interview is generally more efficacious than even a long correspondence. In case you decide to visit Cambridge if you will let me know when to expect you I will arrange to be at the Laboratory; if equally convenient to you Mondays Wednesdays or Friday's are the days I should prefer though just at present I could arrange for an interview on any day. I am much obliged to you for your paper. I hope to take an early opportunity of studying it.

Believe me
Yours very truly,
J. J. Thomson

E. Rutherford Esq.

Figure 4.1. J.J. Thomson's letter welcoming Rutherford to the Cavendish (CUL MSS ADD 7653, T9) [Courtesy of the Cambridge University Library].

[19] A. S. Eve, *Rutherford: Being the Life and Letters of the Rt Hon. Lord Rutherford, O. M.* (New York: Macmillan, 1939), 14.

[20] W. Craig Henderson told Strutt that Rutherford, Townsend, McClelland and Henry were his "constant companions while I remained at the Laboratory." The group later included Langevin, Zeleny, and H. Wilson, all of whom were non-Cambridge research students. See Strutt, *Life of J.J. Thomson*, 62-63.

concern for their welfare and their own remarkable talent. The professor attempted to ensure that the newcomers would not be viewed as "intruders" but as full and equal members of the Cavendish Laboratory.[21] He urged the newcomers to join one of the colleges rather than elect to be non-collegiate members of the University community. He gave them more attention in the Laboratory, where they were generally neglected by the demonstrators and other senior researchers. At first J.J. even tried to exempt the newcomers from paying the Laboratory fee, but he retreated from this position because "outside opinion was so strong against this."[22] Nevertheless, J.J. successfully used his influence to secure financial aid for some of the newcomers: Rutherford received the Coutts Trotter Fellowship in 1898, and Townsend was elected a Trinity College Fellow in 1899.

The regular afternoon tea that J.J. held in his room helped set a tone of community among the Cambridge graduates and the newcomers. H. A. Wilson's recollections reflect the friendly atmosphere of this daily social event:

> We also all went to his office for afternoon tea. At tea all sorts of subjects were discussed; for example, rugby footfall, in which J.J. was keenly interested. He usually knew several of the players on the University team personally. Another popular topic was golf. J.J. frequently played golf on Saturday afternoons and sometimes told stories about his games. He once said that he drove a ball against a strong wind and the ball went about 50 yards against the wind and then rose up very high and was blown back over his head so that it finally landed about 20 yards behind him! Someone, I think it was Professor Bumstead of Yale, said, "Is that a fact?," whereupon everyone, including J.J., roared with laughter.[23]

The second factor contributing to the breaking down of the barrier between insiders and outsiders at the Cavendish was the exceptional talent characterizing the first wave of newcomers. The following excerpts from Rutherford's letters to his fiancé in New Zealand illustrate how J.J.'s support and the talent of the students combined to eradicate obstacles:[24]

> Cambridge, 3 Oct. 1895
>
> . . . I went to the Lab. and saw Thomson and had a good long talk with him. He is very pleasant in conversation and is not fossilized at all . . . We discussed matters in general and research work, and he seemed pleased with what I was going to do . . . My success here will probably depend entirely on the research work I do. If I manage to do some good things, Thomson would probably be able to do something for me. I am very glad I came to Cambridge. I admire Thomson quite as much as I thought I would, which is saying a good deal. They have both been very kind to me, as you may judge from what I have written.

> Trinity College, 28 Nov. 1895
>
> You have heard me talk of the Physical Society, J.J.'s pet society. Well, he has asked me to give an account of some of my work before it, and I am to occupy the whole

---

[21] Strutt, *Life of J.J. Thomson*, 61.

[22] Ibid.

[23] AIP MSS MB 578 (Biographical Notes of H. A. Wilson), 9. H. A. Wilson's recollections probably refer to the year 1904 or 1905, when Bumstead worked in the Cavendish. Brackets added.

[24] Eve, *Rutherford*, 15-29.

meeting. I am to show some experiments for the interest of the vulgar. Usually it is only well-known people, Profs. and such like, who shine before the society, so I appreciate the honour of being asked. It is my chance of getting a little lift up on the scientific ladder, and I intend to make as good use of it as possible. . . . A great thing about it is that it is a sign J.J. thinks a fair amount of my work, and if one gets a man like J.J. to back one up, one is pretty safe to get any position for which he will put himself out to aid you.

8 Dec. 1895

. . . In my last letter I told you that J.J. had asked me to give an account of my work at the Physical Society. I am the first member of the Cavendish who has given an original paper before it, so I may consider the honour is greater than I can bear . . . Instead of taking only part of the time, J.J. stuck the following announcement on the notice board "A method of measuring waves along wires and determination of their period" with experiments by E. Rutherford, and left me to fill up the whole time. The term 'with experiments' rather knocked me sideways, but I went to work and rigged up a good few interesting experiments, which all came off very well. I had quite a distinguished audience including J.J. and Mrs. J.J. and several other ladies—Sir G. Stokes--a good few lecturers and demonstrators besides the usual vulgar herd. I think the paper was quite a success and J.J. was pretty pleased, I think. No one but myself made any remarks, as it was rather beyond most of them. I really had to give a lecture and did not read anything at all. My friends all reckoned I did very well indeed so I suppose I may consider it a success.

. . . I am at present investigating a new detector I have for electrical waves of high period and find it works very well . . . J.J. is very interested and comes round very often and gives what help he can. By the way I have not told you I will be publishing some of my works before long. I spoke to J.J. about it and he said I had better send it to the Royal Society. As only the best papers, or at any rate the papers of eminent men, are chiefly found there, I have nothing to complain of. As a matter of fact very few papers are recommended by Thomson for the Royal Society.

Cambridge, 15 Jan. 1896

. . . I remarked before that J.J. is very interested in my work and makes all sorts of suggestions to help me in getting a clear space for signalling . . . Townsend, my special friend, has not been successful so far in research, but J.J. has appointed him part-time demonstrator which is very good indeed. I could be offered a similar position, only my schol. will not allow me to do so. If I keep on as I am going, there will be no difficulty in getting enough to keep me here even if my schol. is not continued. Townsend and I have placed the Research Student on a pinnacle of honour. I by my paper and work on waves, and T. by being appointed a demonstrator after three months here. The three demonstrators are all extremely friendly now they see we have made a strong position for ourselves and I grimly rejoice for they did not take any notice of us the first two months although they knew we were strangers and had no friends in Cambridge. My paper before the Physical Society was a heavy blow to their assumed superiority, and now they all offer to help us in any way they can and tell me confidentially about their own little researches—so wags the world.

Trinity College, 21 Feb. 1896

. . . He [Sir R. Ball] reckons the work I have done shows the wisdom of the University in providing for research students and the fame of Townsend and me has travelled to all the colleges, so that they are all opening their doors in their anxiety to welcome the research student of which we are the first examples. J.J. I believe openly declares that the new Research Students are a great success, and as I am at present the most prominent in that respect, I take a little of the praise unto myself . . . J.J.'s experiments [in the Physical Society] did not go off very well in the lecture, so he got a

> bit mad, and made those sit up who asked silly questions. I was too wise to do so, but some fellows think they show their smartness by asking questions, whether he had tried this, and that, and so on, and J.J.'s dander rose and he turned the laugh against them most skilfully [sic.]—whereat I rejoiced, for many of them are my enemies. I am of the opinion that the demonstrators regard research students here with very little favour and try and put little obstacles in our path, but verily we rise superior to their machinations and they gnash their teeth with envy.

Rutherford's vivid descriptions of his early days at Cambridge offer proof that his talent and hard work, combined with J.J.'s solicitous care, enabled him to withstand his initially inhospitable treatment at the Cavendish and, eventually, to achieve the immense success that he did.

Following Rutherford's example, other newcomers persevered to gain the acceptance of Cambridge scholars. In 1898, Townsend became the first non-Cambridge man to receive the Clerk Maxwell Scholarship. In 1900, H. A. Wilson was awarded the Allen Scholarship (£200 for one year), and in 1901 he succeeded Townsend as Clerk Maxwell Scholar when Townsend became Professor of Physics at Oxford. As discussed in Section 4.4, many of the newcomers were prolific writers who published in prestigious journals like *Philosophical Transactions*, *Philosophical Magazine*, and *Proceedings of the Royal Society*. Only one Cambridge man of this period, C. T. R. Wilson, could compete with them.

It was natural for the newcomers to be far more dependent on "Professor Thomson" than the Cambridge graduates. Bypassing the elementary courses taught by demonstrators largely because they would not sit for triposes, the newcomers often enrolled in J.J.'s lectures on advanced topics even when they had no special obligation to do so. Rutherford's lecture notes indicate that he attended J.J.'s lectures on the discharge of electricity through gases (1897) and probably Stoke's or Glazebrook's courses on "wave theory, with applications to acoustics and optics, also the viscosity of fluids (1896-97)."[25] The newcomers also were receptive to J.J.'s advice about research topics, and a small number of them began to assist J.J. in his researches. It was with the help of such a group that J.J. carried out his research on the nature of X-rays and made his discovery of the electron. Among these assistants, J.J. found "confidants" who enjoyed the privilege of "constant intercourse" with him.[26] One of those confidants, H. A. Wilson, shared a small space with Rutherford and Langevin "in a corner of the room on the ground floor next to the room in which [J.J.] had his apparatus."[27] John Zeleny, who moved from Berlin to Cambridge as soon as he heard the news of J.J.'s discovery of the electron, also received J.J.'s good-natured encouragement. "I hardly think at present," J.J. wrote, "we have any in the new lot as good as Zeleny and [H. A.] Wilson. The amount of glass they break at present is appalling."[28]

---

[25] CUL MSS, Add. 7653, NB 4 & 5.
[26] Strutt, *Life of J.J. Thomson*, 50.
[27] AIP MSS MB 578 (H. A. Wilson), 7.
[28] CUL MSS ADD 7653, T11 (22 November 1898), J.J. to Rutherford.

4.2.2. J.J., Advanced Students, and the Discovery of the Electron

When the news of Röntgen's discovery of X-rays arrived at Cambridge in the last days of 1895, J.J. immediately made use of the mysterious rays in his investigations of electric discharges through gases. The newly discovered rays were particularly useful in this research because gases exposed to them were ionized immediately and only small voltages were required.[29] J.J. quickly reported the value of the rays in producing discharges "when they fall upon electrified bodies."[30]

In many ways, the discovery of the electron was a result of J.J.'s new style of research. The discovery was, first of all, the product of an orchestrated collaborative

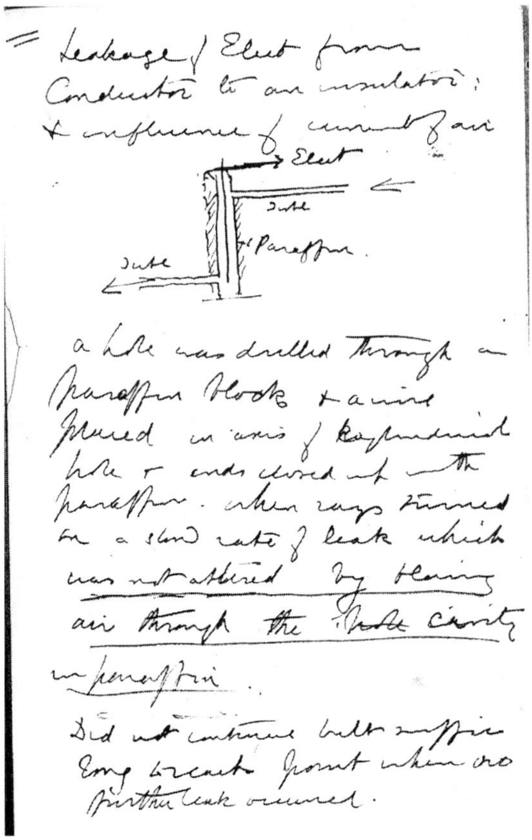

Figure 4.2. Note from a J.J.-Rutherford experiment on X-rays (CUL MSS ADD 7654 NB 39) [Courtesy of Cambridge University Library].

---

[29] See J.J. Thomson, *Recollections*, 325.

[30] J.J. Thomson, "On the Discharge of Electricity produced by the Röntgen Rays, and the Effects produced by these Rays on Dielectrics through which they pass," *Proc. Roy. Soc. 59* (1896): 274-276.

effort which became the point of departure for more extensive collaborative researches at the Cavendish. When J.J. began his research on X-rays, he frequently worked with his research students, mostly the group from outside Cambridge University. In his first study of X-rays, J.J. was assisted by McClelland and Everett, whom he thanked, in his paper of February 1896, "for the assistance they have given me in carrying out these experiments."[31] In March, J.J. and McClelland published a joint paper on the investigation of the passage of electricity through dielectrics exposed to X-rays.[32] "At the beginning of the Easter Term of 1896," J.J. and Rutherford joined in an intense collaborative study of X-rays and their ionizing effects, with impressive results. Through "a two-way flow of concepts and practice," they quickly discovered several characteristics of ionization by X-rays.[33] In these professor-student collaborations, it is likely that Rutherford and McClelland carried out most of the routine procedures. For example, in the Thomson-Rutherford collaboration, Rutherford mostly kept the notebook on the experiments and J.J. interpreted the experiments.[34] The results of these joint efforts and of other experiments suggested by J.J. to Cavendish researchers reinforced J.J.'s belief that atoms might be divisible into two oppositely charged parts and that ionization was the dissociation of these oppositely charged parts. J.J. first read his celebrated paper, "Cathode Rays," on Friday, April 30, 1897, at one of the weekly meetings of the Royal Institution. In it, he proposed that cathode rays are negatively charged particles called "corpuscles," that the atoms of the elements are made up of such corpuscles, and that the mass of the corpuscles is 1000 times less than that of a hydrogen atom.[35]

The story of J.J.'s discovery of the electron need not be repeated here.[36] One

---

[31] Ibid., 276.

[32] J.J. Thomson & J. A. McClelland, "On the Leakage of Electricity through Dielectrics traversed by Röntgen Rays," *Proc. Camb. Phil. Soc. 9* (1896): 126-140.

[33] J.J. Thomson & E. Rutherford, "On the Passage of Electricity through Gases exposed to Röntgen Rays," *Phil. Mag. 42* (1896): 392-407. For this collaborative work, see Isobel Falconer, "Transfer of Concepts & Practices between J.J. Thomson and His Research Students," Paper for the meeting on the transfer of concepts and practice in the physical and mathematical sciences, (Maison Francaise, Oxford, 23 May 1996). Falconer pointed out:

> J.J. learnt that a less fundamental approach may be more productive than a fundamental one; and he acquired the basis of a mathematical formulation of ionisation theory which he built on extensively . . . Rutherford gained practical expertise in designing and performing discharge experiments; he learnt how to work in the comparison of orders of magnitude; he learnt ionisation theory; however, he resisted Thomson's fundamental concepts of the way ionisation worked because they did not have sufficiently clear experimental consequences.

[34] CUL MSS ADD 7654, NB 39.

[35] J.J. Thomson, "Cathode Rays," *Not. Proc. Roy. Inst. 15* (1896-98): 419-432. A slightly modified version was published later in *Phil. Mag. 44* (1897): 293-316.

[36] There are many detailed accounts of the discovery of the electron. See J.J. Thomson, *Recollections*, 325-341; *A History of the Cavendish Laboratory*, 161-172 (written by Rutherford); Strutt, *Life of J.J. Thomson*, 76-96; D. Anderson, *The Discovery of the Electron* (Princeton: 1964); Isobel Falconer, "Corpuscles, Electrons and Cathode Rays . . .'," 241-276; S. B. Sinclair, "J.J. Thomson and the Chemical Atom: from Ether Vortex to Atomic Decay," *Ambix 34 (1987): 89-116*; Stuart M. Feffer, "Arthur

point, however, ought to be stressed. Although J.J. secured extensive help from his research students in the series of experiments that led to the great discovery, only he was able to synthesize all the experimental results into a theory. The Cavendish Laboratory was fortunate to have such a talented theorist as its director. As Falconer has rightly pointed out, J.J.'s main concern until 1896 was not the nature of cathode rays, but rather the broader issues of "a quest for unity" between electricity and matter and "an ultimate theory" for conduction and ionization.[37] His 1895 paper, "The Relation between the Atom and the Charge of Electricity Carried by It," demonstrated J.J.'s broad perspective even before his investigation of cathode rays. In that article, J.J. emphasized the importance of "the connection between ordinary matter and the electrical charge on the atom . . . which must be closely related to a

Figure 4.3. J.J. Thomson giving a lecture demonstration with tubes (ca. 1909) [Courtesy of the Cavendish Laboratory]

Schuster, J.J. Thomson, and the Discovery of the Electron," *HSPS 20* (1989): 33-61; N. Robotti, "J.J. Thomson at the Cavendish Laboratory: the History of an Electric Charge Measurement," *Annals of Science 52* (1995): 265-284; E. A. Davis & I. J. Falconer, *J.J. Thomson and the Discovery of the Electron* (London: Taylor & Francis, 1997); Per F. Dahl, *Flash of the Cathode Rays: A History of J.J. Thomson's Electron* (Bristol: Institute of Physics, 1997); Jed Z. Buchwald & A. Warwick (eds.) *Histories of the Electron* (Cambridge, Mass.: MIT Press, forthcoming).

[37] Falconer, "Corpuscles, Electrons and Cathode Rays. . . ," 252.

good many of the most important chemical as well as electrical phenomena."[38] He predicted that "a complete explanation of this connexion would probably go a long way towards establishing a theory of the constitution of matter as well as of the mechanism of the electric field." The importance of J.J.'s role as a theorist at that time cannot be too strongly emphasized. Although J.J. had learned a great deal about experimentation during the previous ten years, he still was principally a "Cambridge Mathematician" who never pursued "another place of decimals" in his experiments. Rather, he usually was satisfied with obtaining the correct order of magnitude, leaving the acquisition of more precise results to his research students.

Cavendish research students in general, but particularly those who worked directly with J.J., benefited from the Professor's broad perspective. The first group of advanced students worked on problems intimately related to J.J.'s interests, receiving from the Professor considerable advice, criticism, and encouragement, which they frequently acknowledged in their papers. Newcomers such as Townsend, Rutherford, H. A. Wilson, and Zeleny, sometimes joined by Cambridge graduates, for example C. T. R. Wilson, worked to determine the velocity of ions and the charge of the electron. C. T. R. Wilson's early model of the cloud chamber added a very effective tool to researches which demanded ion counts.

The relationship between "Professor Thomson" and his advanced students was always a two-way street. J.J. provided his students with ideas, and they reciprocated in kind. For example, J.J.'s 1898 article on the charge of ions produced by X-rays directly acknowledged his debts to the work of Rutherford and C. T. R. Wilson:

> The theory of the method used is as follows: —By measuring the current passing through a gas exposed to Röntgen rays and acted upon by a known electromotive force, we determine the value of the product *nev*, where *n* is the number of ions in unit volume of the gas, *e* the charge on an ion, and *v* the mean velocity of the positive and negative ions under the electromotive force to which they are exposed.
>
> Mr. Rutherford (*Phil. Mag.* vol. xliv. p. 422, 1897) has determined the value of *v* for a considerable number of gases; using these values, the measurement of the current through a gas gives us the product *ne*; hence if we can determine *n*, we can deduce the value *e*.
>
> The method I have employed to determine n is founded on the discovery made by C. T. R. Wilson (*Phil. Trans.*, A, 1897, p. 265) that when Röntgen rays pass through dust-free air a cloud is produced by an expansion which is incapable of producing cloudy condensation when the gas is not exposed to these rays.[39]

In an 1899 article about the mass of ions, J.J. strengthened his argument by citing a finding made by Townsend:

> Thus I have shown (*Phil. Mag.* Dec. 1898) that when a gas is ionized by Röntgen rays, the charges on the ions are the same whatever the nature of the gas: thus we get the same charges on the ions whether we ionize the hydrogen or oxygen. This result has

[38] J.J. Thomson, "The Relation between the Atom and the Charge of Electricity carried by It," *Phil. Mag.* *40* (1895): 511-544 on 512.

[39] J.J. Thomson, "On the Charge of Electricity carried by the Ions produced by Röntgen Rays," *Phil. Mag.* *46* (1898): 528-545 on 528.

been confirmed by J. S. Townsend ("On the Diffusion of Ions," *Phil. Trans.* 1899), who used an entirely different method.[40]

For the Cavendish Laboratory, the arrival of the advanced students was an exceptionally timely occurrence.

4.3. Organization

Although the influx of newcomers dramatically altered the demography of the Laboratory, their arrival had little effect on management. Characteristically, J.J. did not attempt to create any new posts for the newcomers, which might have incurred the wrath of the University's conservative members. Instead, he endeavored to find the ablest research students suitable positions within the existing framework. The internal structure of the Cavendish was now firmly established.

Between 1895 and 1900, the teaching staff changed little (see Table 3.2). Except for one assistant demonstratorship, the key posts were occupied by Wilberforce, Searle, Skinner, and Fitzpatrick, all veterans who had spent in excess of ten years at the Laboratory as research students or instructors. These old hands usually took charge of the regular courses leading to the traditional examinations, such as the NST or the first examination for the medical degree. In addition, Shaw and Whetham regularly taught the elementary courses for candidates for these degrees. Advanced and specialized courses frequently were delivered in the Laboratory. J.J. taught courses on the discharge of electricity through gases, the properties of matter, and on electromagnetism. Glazebrook continued to deliver advanced courses on optics and electrical measurements, and Shaw delivered advanced courses on heat and electrical measurement. Junior staff members lectured on their own specialties: Capstick on sound, Wilberforce on polarized light and mathematics, Skinner on measurements in physical chemistry, and Whetham on physical chemistry, particularly solutions and electrolysis. The increase of these specialized courses proved that the Cavendish had advanced to the status of a mature research institution: in 1893-94, teaching staff members provided four advanced lecture courses in addition to the five courses taught by the professor; by 1898-99, the staff was providing eight advanced or specialized courses in addition to the professor's.

The Laboratory's financial situation also changed very little during the period. The University contributed nearly the same small sum for wages and maintenance that it had been providing for the previous twenty years, with some additional funding for exceptional circumstances. For example, the Senate in 1899 passed a recommendation to pay, from the Common University Fund, "the further sum of £445. 6s. 2d. required to meet the expenditure on the new building of the Cavendish Laboratory" and also agreed to pay £22. 9s 8d for the cost of altering drains in the laboratory.[41] The Museum and Lecture Rooms Syndicate agreed to pay for the

---

[40] J.J. Thomson, "On the Masses of the Ions in Gases at Low Pressures," *Phil. Mag. 48* (1899): 547-567 on 564.

[41] *CUR* (16 February 1897): 521; (9 November 1897): 183.

purchase of "a new dynamo to replace the old one which had been running for 18 years and was worn out."[42] Even during the final years of the nineteenth century, Cambridge University still did not offer financial support to research in the sciences. To balance the Cavendish budget, J.J. often sought donations from outside the University. In 1899, he expressed "much pleasure in announcing that a benefactor who wishes to remain anonymous has presented £100 to the Laboratory for the purchase of instruments."[43]

The main source of funding for the Cavendish was student fees, and it was upon these that the purchase of necessary apparatus and equipment almost totally depended. The growth in enrollment of undergraduates at the Cavendish thus meant an increase in total receipts. During the four years between 1896 and 1899, the Cavendish's receipts annually totaled about £1600-1700, of which "Fees for Demonstrations and Lectures" amounted to approximately £1500. Of this total, J.J. managed to save about £200 per year to meet future needs.

The Clerk Maxwell Scholarship continued to assist the most able Cavendish research students. As beneficiaries of the Clerk Maxwell Scholarship, J.J. wisely chose younger, less established researchers who, not surprisingly, later became Cavendish demonstrators or assistant demonstrators. C. T. R. Wilson, who was awarded the scholarship from 1895 to 1898, was succeeded by Townsend, the first newcomer to receive this coveted award. Because the amount of the scholarship was an exceptionally generous £200, a Clerk Maxwell Scholar was well able to concentrate on his work. It was while being supported by this scholarship that Wilson began to build his eponymous cloud chamber and that Townsend carried out his research on ionization.

The design, construction, maintenance, and renovation of instruments became more crucial during "the days of rays."[44] Because the Cavendish had very limited resources with which to purchase instruments, most experimental apparatus was made by researchers or constructed by the workshop attendant or by Everett. The Cavendish was twice blessed in having attracted not only talented experimentalists with a bent toward gadgeteering, but also the able workshop attendant, William G. Pye and J.J.'s assistant, Everett, both of whom built the apparatus required by J.J. for his experimentation. Pye "had received his training in Mr. Horace Darwin's workshop at the Cambridge Scientific Instrument Company." "His tenure of office as attendant," H. F. Newall said, "was marked by the excellent organization of the workshop department of the Laboratory at a time when much apparatus was needed for the increasing classes under Fitzpatrick."[45] Pye also constructed much of the apparatus required by researchers, including, for example, the instruments used by G. A. Shakespear to investigate the relation between temperature and elasticity.[46]

---

[42] *CUR* (5 June 1899): 986.
[43] Ibid.
[44] A. Keller, *The Infancy of Atomic Physics: Hercules in His Cradle* (Oxford: Clarendon Press, 1983), chapter 6.
[45] *A History of the Cavendish Laboratory*, 116.
[46] G. A. Shakespear, "The Application of an Interference Method to the Investigation of Young's Modulus for Wires, and its Relation to Changes of Temperature and Magnetization; and a further

The story of the development of C. T. R. Wilson's cloud chamber, the instrument used most frequently in Cavendish investigations on the properties of atoms, sub-atoms, ions and electrons, illustrates the financial constraints existing at the Laboratory. In this special apparatus, "in which all danger of the entrance of dust from the outside is avoided," as C. T. R. reported, "a very sudden and perfectly definite increase in volume is produced."[47] With this impressive piece of apparatus, C. T. R. succeeded in realizing his dream of reproducing clouds in the laboratory room, and acquired the nickname, "Cloud Wilson," which distinguished him from H. A. Wilson.[48]

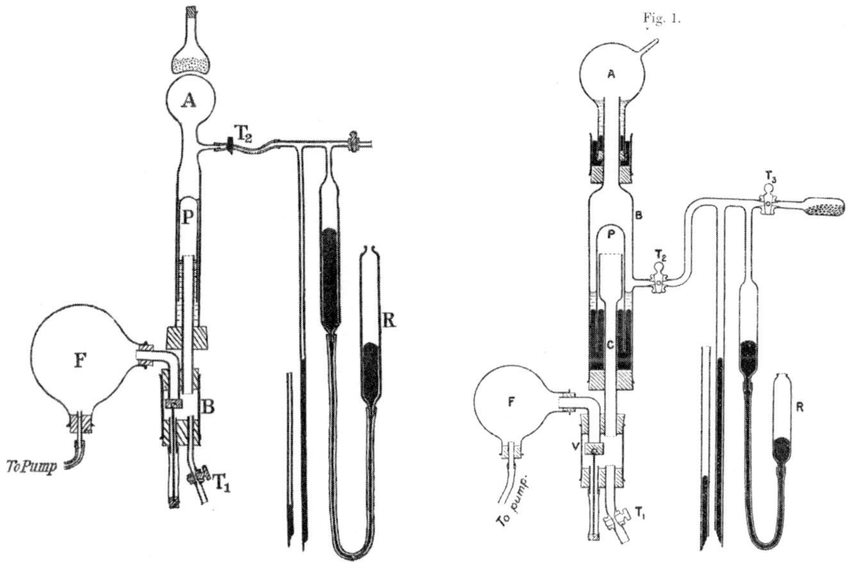

Figure 4.4. Two early models of C. T. R. Wilson's expansion apparatus [from "On the Action of Uranium Rays on the Condensation of Water Vapour," Proc. Camb. Phil. Soc. 9 (1897): 334; "On the Condensation Nuclei produced in Gases by the Action of Röntgen Rays, Uranium Rays, Ultra-Violet Light, and other Agents," Phil. Trans. 192 (1899): 405].

Application of the same Method to the Study of the Change in Dimensions of Iron and Steel Wires by Magnetization," *Phil. Mag. 47* (1899): 539-556 on 556.

[47] C. T. R. Wilson, "On the Formation of Cloud in the Absence of Dust," *Proc. Camb. Phil. Soc. 8* (1895): 306.

[48] In his Nobel Lecture of 1927, Wilson vividly recalled how he first got an idea of recreating clouds in the laboratory while he spent a vacation "on the summit of Ben Nevis, the highest of the Scottish hills" in September of 1894. See C. T. R. Wilson, "On the Cloud Method of Making Visible Ions and the Tracks of Ionizing Particles," in *Nobel Lectures in Physics, 1922-1941* (River Edge, NJ: World Scientific, 1965), 194-215. See also Peter Galison & Alexi Assmus, "Artificial Clouds, Real Particles," in D. Gooding, T. Pinch & S. Schaffer (eds.), *The Use of Experiment* (Cambridge: Cambridge University Press, 1989), 224-274; and Strutt, *Life of J.J. Thomson*, 98.

Despite its near-magical powers, the apparatus required only inexpensive materials, primarily several long, very thin glass tubes and some cylindrical glass vessels. Its design and assembly, however, demanded considerable time and dexterity. J.J. vividly described Wilson's labor:

> This work of C. T. R. Wilson, proceeding without haste and without rest since 1895, has rarely been equalled as an example of ingenuity, insight, skill in manipulation, unfailing patience and dogged determination. Those who were not working at the Cavendish Laboratory during its progress can hardly realise the amount of work it entailed. For many years he did all the glass-blowing himself, and only those who have tried it know how exasperating glass-blowing can be, and how often when the apparatus is all but finished it breaks and the work has to be begun again. This never seemed to disconcert Wilson; he would take up a fresh piece of glass, perhaps say "Dear, dear," but never anything stronger, and begin again. Old research students when revisiting the Laboratory would say that many things had altered since they went away, but the thing that most vividly brought back old reminiscences was to see C. T. R. glass-blowing.[49]

Wilson's indefatigable efforts became legendary and deeply influenced the attitude of the other researchers in the Laboratory. It was neither surprising nor a matter of luck that the Cavendish Laboratory became a major center for technical developments in physics research during the next few decades.[50]

4.4. Researchers and Researches

The period between 1895 and 1900 was one of the most productive and celebrated in the entire history of the Cavendish Laboratory. Although the number of researchers and the number of papers they produced increased dramatically, the most influential changes were qualitative: formal admission of advanced students, increasing influence of J.J.'s work on research undertaken by those students, and formation of a core research group around the Professor.

The changing demography of the researchers was the first sign of a break with the past, and the main cause for the change, as has been discussed, was the influx of advanced students from other universities. The ratio of outsiders (advanced students and guests) to Cambridge graduates newly entering the Cavendish was surprisingly high: 3:1 in 1895, 6:3 in 1896, 5:2 in 1897, 6:6 in 1898, 1:0 in 1899, and 9:4 in 1900. The influx of newcomers contributed to the steady increase of researchers at the Cavendish. Because most Cambridge graduates now earned the NST qualification, the number of MT graduates at Cavendish was on the decline. The

---

[49] J.J. Thomson, *Recollections*, 419. See also Wilson's own recollection in "Reminiscences of My Early Years," *Notes and Records of the Royal Society of London 14* (1960): 163-173. Wilson remembered, "I spent every day of the Long Vacation there in 1895, being generally quite alone in the whole Laboratory (p. 167)."

[50] In the "new" Cavendish Laboratory, some apparatus are exhibited in the corridor. In 1980 Isobel Falconer classified these apparatus into 15 sections (Isobel Falconer, *The Cavendish Laboratory: An Outline Guide to the Museum*, a pamphlet (Cambridge, 1980)). For the early apparatus and instruments, see also *Empires of Physics: A Guide to the Exhibition* (Cambridge: Whipple Museum of the History of Science, 1993).

absence of wranglers at the Cavendish enhanced, in effect, the reputations of advanced students. The Cambridge graduates of considerable fame, C. T. R. Wilson and R. J. Strutt, had taken the NST and not the MT. As Andrew Warwick observed, "the very different styles of physics practiced in the MT and NST" would become evident in the new century, and this separation would "not constitute a useful division of labour between experimental and mathematical physics."[51]

The number of successful studies completed at the Laboratory, of which the first notable sign was an increase in the number of research papers produced, also enhanced the reputation of advanced students and Laboratory alike (Table 4.1).

Table 4.1. Number and Content of Research Papers, 1895-1900

Name	*Individual Work*	*Co-Work*	*Main Topic*
J.J. Thomson	22	1/2 x 3[A]	electric discharge through gases, cathode rays, ionization
C. T. R. Wilson	8		condensation nuclei, properties of rays
J. S. Townsend	8		electrical properties of gases, ionization
J. H. Vincent	7		photography
S. Skinner	6	1/2 x 1	electrical batteries
E. Rutherford	5	1/2 x 1	properties of rays
J. A. McClelland	5	1/2 x 1	properties of rays, photography
W. C. D. Whetham	5		electrolysis, ionization
G. F. C. Searle	5		electricity, hysteresis
H.A. Wilson	5		charged clouds, solidification
R. J. Strutt	4		cathode rays
W. N. Shaw	3		meteorology, potentiometer
J. W. Capstick	3		specific heat, cathode rays
J. Henry	2	1/2 x 1	properties of rays, magnetic deflection
J. R. Erskin-Murray	2		electricity
E. B. H. Wade	2		solution
J. Zeleny	2		electric discharge through gases
W. C. Henderson	2	1/2 x 1	electromagnetism, evaporation
R. S. Willows	2		thermo-electricity
H. S. Allen, W. C. Baker, G. B. Bryan, F. Martin, J. C. McLennan, J. Monckman, V. Novák., R. B. Owens, S. W. Richardson, G. A. Shakespear, S. W. J. Smith, R. L. Wills ----> each published one paper			
Total: 31	110	4	

A: One with McClelland, one with Rutherford, one with Skinner.

[51] A. Warwick, "Cambridge Mathematics and Cavendish Physics: Cunningham, Campbell and Einstein's Relativity, 1905-1911. Part II: Comparing Traditions in Cambridge Physics," *Studies in History and Philosophy of Science 24* (1993): 1-25 on 2 and 23.

During the five years between 1895 and 1900, Cavendish researchers (excluding J.J.) produced ninety-two papers, exceeding by six the number of papers produced during the previous decade (1885-1894). The number of Cavendish researchers publishing multiple papers also increased: fourteen produced more than two papers during that five-year period as compared to eleven in the previous decade: six published more than five papers during this short period as compared to five in the previous decade.

The quality of these papers also was more impressive than before, as indicated by the prestigious journal, *Philosophical Transactions*, which published thirteen of them between 1895 and 1900, with C. T. R. Wilson contributing three, Townsend two, and Whetham two. Some of the papers were as important as the one detailing the celebrated discovery of the electron, notably Rutherford's work on radioactivity, Townsend's work on ionization, and C. T. R. Wilson's work on condensation nuclei. The Cavendish workers had become more productive researchers and they often cultivated their own areas of research.

A survey of this research also suggests that changes had taken place at the Cavendish Laboratory. More and more, students acknowledged in their papers the "suggestions," advice, criticism, and encouragement of "Professor J.J. Thomson." Such expressions of gratitude had been rare under Maxwell and Rayleigh, and even under J.J. in his younger days, and they indicated the dependence of the students on J.J.' s advice when selecting their research topics. C. T. R. Wilson remembered that J.J.'s mind was "full of new ideas" during that period, and that J.J. had "a remarkable instinct for knowing which were the problems most worth working on, and the general nature of the methods of investigation most likely to succeed."[52] H. A. Wilson remembered that J.J.

> . . . always had all sorts of ingenious suggestions for getting over experimental difficulties and making new experiments. I think one reason why he was so successful at research was that he could form a clear picture of an apparatus in his mind before getting it made and see what changes in it were required to make it work properly, before having it made.[53]

When Glazebrook and Shaw eventually departed, J.J.'s authority at the Cavendish became definite and complete. The students' increased dependence on the Professor's advice naturally meant that their studies were influenced by his interests and concerns. The areas in which he was interested in fact were exciting and important areas to be investigated, and thus it was with good conscience that J.J. could recommend that his students pursue investigations of properties of the electron, X-rays, uranium radiation, and the relation of electron theory to ionization theory. As a result, C. T. R. Wilson, Rutherford, Townsend, McClelland, J. W.

---

[52] C. T. R. Wilson, "Reminiscences of My Early Years," 168. He said that J.J. "never tried to influence me at all in the way I design my experiments but I could see from the degree of his cordiality when he thought I was getting results quickly and when he thought I was spending too much time perfecting my method."

[53] AIP MSS MB 578 (H. A. Wilson), 8.

Capstick, J. Henry, Zeleny, R. B. Owens, and R. J. Strutt investigated the nature of various rays, including X-rays, cathode rays, uranium rays, and ultra-violet light. Townsend, McClelland, Whetham, H. A. Wilson, Zeleny, F. Martin, J. Monckman and others worked on ionization and related topics. These researchers shared common interests, knew each other's works quite well, and relied on each other's findings. For example, when investigating condensation nuclei, C. T. R. Wilson used Rutherford's value of the velocity of the ions as well as the results of Townsend's and H. A. Wilson's experiments.[54] When investigating the ion, J.J., Townsend, and H. A. Wilson liberally employed C. T. R. Wilson's condensation method. Rutherford employed Zeleny's experiments showing that the velocity of a negative ion was greater than that of a positive ion.[55] The net result of this shared understanding was a rapid growth in the research students' productivity.

Maxwell's laissez-faire policy had not succumbed to the new circumstances, however. J.J.'s core group of researchers was relatively small, and research students remained free to select research topics and methods. Immediately after finishing his experiments on ultra-violet light discharge in gases, Rutherford turned his attention to newly discovered uranium radiation.[56] J. H. Vincent published a series of papers on photographic methods for recording ripples.[57] Skinner worked on electrical batteries such as the Clark cell, tin-chromic cell, and carbon-consuming cell.[58] Whethan continued his research on electrolysis.[59] Searle investigated the additional electrical mass of a moving charge, the theory of motion of a charged ellipsoid, and hysteresis loss.[60] Capstick continued his research on the ratio of the specific heat of some compound gases.[61] W. Craig Henderson and J. Henry carried out a delicate experiment to detect the motion of the æther.[62] E. B. H. Wade devised a sensitive method for determining the boiling points of liquids.[63] V. Novák systematically

[54] C. T. R. Wilson, "On the Condensation Nuclei produced in Gases by the Action of Röntgen Rays, Uranium Rays, Ultra-violet Light, and other Agents," *Phil. Trans. 192* (1899): 403-453 on 452.

[55] E. Rutherford, "Uranium Radiation and the Electrical Conduction produced by it," *Phil. Mag. 47* (1899): 108-163 on 148.

[56] *A History of the Cavendish Laboratory*, 183. See "Uranium Radiation and the Electrical Conduction produced by it," *Phil. Mag. 47* (1899): 109-163;

[57] J. H. Vincent, "On the Photography of Ripples—Second Paper," *Phil. Mag. 45* (1898):191-197; "Third Paper," *Phil. Mag. 46* (1898): 290-296; "Fourth Paper," *Phil. Mag. 47* (1899): 338-344.

[58] S. Skinner, "The Clerk Cell when producing a Current," *Phil. Mag. 39* (1895): 375-376; "The Tin-Chromic Chloride Cell," *Phil. Mag. 39* (1895): 444-447; "The Carbon-Consuming Cell of Jacques," *B.A. Report* (1898): 804.

[59] W. C. D. Whetham, "The Coagulative Power of Electrolytes," *Phil. Mag. 48* (1899): 474-477.

[60] G. F. C. Searle, "Problems in Electric Convection," *Phil. Trans. 187* (1896): 675-713; "On the Steady Motion of an Electrified Ellipsoid," *Phil. Mag. 44* (1897): 329-341; "A Method of Measuring the Loss of Energy in Hysteresis," *Proc. Camb. Phil. Soc. 9* (1895-98): 2-6.

[61] J. W. Capstick, "On the Ratio of the Specific Heats of Some Compound Gases," *Phil. Trans. 186* (1895): 567-592.

[62] W. C. Henderson and J. Henry, "Experiments on the Motion of the Æther in an Electromagnetic Field," *Phil. Mag. 44* (1897): 20-26.

[63] E. B. H. Wade, "On a New Method of determining the Vapour Pressures of Solutions," *Proc. Roy. Soc. 61* (1897): 285-287; "On a New Method of determining the Vapour Pressures of Solutions," *Proc. Roy. Soc. 62* (1897-98): 376-385.

examined depression of the freezing-point and of the conductivity of solutions.[64] Shakespear worked on the interference method for investigating Young's modulus for wires.[65] Nevertheless, fewer researchers now ventured beyond the boundary of the Cavendish's central subjects: ionization, radiation, and the structure of matter and of the atom. As the years went by, the number of independent projects undertaken at the Laboratory decreased, and the diversity of the Cavendish research projects declined even as the diversity of researchers performing them increased.

4.5. The Emergence of the Cavendish School

In his biography of J.J. Thomson, R. J. Strutt listed three essential ingredients necessary to the formation of a scientific school:

> The first is a stimulating leader, one who is not only abounding in energy and ideas, but also one who can without too great an effort throw himself into the difficulties of others. This requires a peculiar kind of versatility not always easily combined with great powers of concentration on any one line of thought. Then again it is necessary to have a productive line of investigation opening up: lastly it is necessary to have the right kind of pupils. These must be men of the not very common kind of ability which makes a scientific investigator: they must not be too young: and they must be provided with the means of subsistence while the work goes on.[66]

Although Strutt was neither historian nor sociologist, his comments are insightful. J.J. certainly was a stimulating and versatile leader who pursued "a productive line of investigation" and acquired an entourage of talented pupils. However, Strutt may have underestimated some features vital to the formation of the Cavendish School. These were the contributions of Cambridge University and the historical evolution of the Cavendish Laboratory. The Cavendish Laboratory was part of Cambridge's educational enterprise, and thus the University deeply influenced the everyday lives of the Laboratory researchers. As Strutt correctly emphasized elsewhere, prior to the admission of advanced students from outside the University, Cambridge's system had been ineffective. The influence of the newcomers when they finally were admitted greatly affected the future of the Laboratory.[67] Nonetheless, the success of the Cavendish Laboratory did occur within the walls of Cambridge. Advanced students undoubtedly contributed to the Laboratory's rising fame as a center for experimental physics, but they also benefited from the contributions of the University. One can also point to the contributions of C. T. R. Wilson, the NST graduate who produced more research

---

[64] V. Novák, "Specific Electric Conductivities and Freezing-points of Solutions of Water in Formic Acid," *Phil. Mag. 44* (1897): 9-20.
[65] G. A. Shakespear, "The Application of an Interference Method to the Investigation of Young's Modulus for Wires, and its Relation to Changes of Temperature and Magnetization; and a further Application of the same Method to Study of the Change in Dimensions of Iron and Steel Wires by Magnetization," *Phil. Mag. 47* (1899): 539-556.
[66] Strutt, *Life of J.J. Thomson*, 58.
[67] Ibid., 58-59.

papers than most of the newcomers. And, of course, J.J. Thomson himself was a bona fide product of Cambridge and an example of the best that the University offered.

It is, however, useful to contemplate the relationship of the Cavendish Laboratory's early years to the emergence of the Cavendish School in the years between 1895 and 1900. The early years hardly can be viewed as an "incubation" period for the "sudden birth" of the illustrious Cavendish School. In fact, I want to assert the reverse: the emergence of the Cavendish School was partly the result of fortuitous historical coincidence. How is it that Rutherford, the second choice for the 1851 Exhibition Scholarship, arrived at the Cavendish a few months before the discovery of X-rays?[68] Why did C. T. R. Wilson start "in 1894-95" a life-long investigation that would produce an instrument extremely useful to research in the Laboratory? Certainly J.J. had no intention of creating his own school with his talented students. In fact, there never was such an entity as "Thomson's School." However, such coincidences as led to the emergence of the Cavendish School could converge to create a consequential historical event because the locus where they conjoined was well prepared. The Cavendish Laboratory, with its particular history and with J.J. Thomson as its head, was such a place. Thus, the factors that gave birth to the Cavendish School were, it seems, both coincidental *and* contingent.

For the Cavendish researchers, the feeling that they shared something in common, that they were members of a school, was exciting and valuable. When they began to identify themselves as members of the "Cavendish School," they also decided to record the establishment of that school. The first symbolic gesture of this identity was the Cavendish annual photograph, which was instituted in 1897 and which show J.J. surrounded by his research students (Figure 4.5). In more than one physicist's career, appearance in the Cavendish annual photograph provided evidence of the claim that he or she had indeed worked in the Cavendish Laboratory.

The Cavendish research students celebrated the genesis of their school at the annual "Cavendish Laboratory Dinner." W. Craig Henderson remembered the first of these dinner parties as follows:

> It was at one of these meetings [of the advanced students] that the idea was first raised of having a Cavendish Laboratory Dinner. It was at once acted on and the first dinner was held before the Christmas Vacation in 1897 and was so successful that this function became thereafter an annual event. There is an error in J.J.'s account of this first dinner. He says it was held in December 1898 at Bruvet's restaurant in Sidney Street. I have the menu card before me which shows that the date was December 9th, 1897 and the place of meeting was the Prince Wales' hotel. The only Toasts (apart from loyal Toast) were 'Our Guests' proposed by Townsend, the reply being given by J.J., and 'Our old Universities' which, I see, I proposed. J.J. presided, and was as happy as a

---

[68] Norman Feather, *Lord Rutherford* (Glasgow: Blackie & Sons, 1940), 32-33. The scholarship was usually assigned "once every two or three years" to the University of New Zealand, and first choice in 1895 was a chemist named J. S. Maclaurin. Since Maclaurin could not receive the scholarship because of a "family problem," Rutherford, the second choice, was awarded the scholarship and was sent to Cambridge in 1895 as he had hoped.

> sand-boy, and at the Laboratory on the following day remarked that he had no idea that the Laboratory held such a nest of singing birds.[69]

This first annual dinner, like the first annual photograph, was initiated by the Cavendish advanced students in the year that J.J. discovered his "corpuscles." It is understandable that these newcomers should wish to identify themselves with the laboratory headed by the celebrated discoverer. The Cambridge old guard, however, seemed less interested in the event. After the second annual dinner party, held in 1898, Townsend wrote to Rutherford:

> We had a Research Students Dinner. This time it went off very well. Some of the chaps wanted us to let the demonstrators join, but the old birds who received such "kindness" from the men in the Cavendish absolutely refused. So we only had J.J.[70]

The old pride of Cambridge men as well as animosity toward the outsiders were still alive but would soon disappear with the rapid growth of the Cavendish School.

S.W.Richardson. C.T.R.Wilson. J.Henry. J.McClelland. L.Blaikie.

E.B.H.Wade. G.B.Bryan. W.Craig-Henderson. Prof. J.J.Thomson. J.S.Townsend. E.Rutherford.

Figure 4.5. The first Cavendish annual photograph (1897) [Courtesy of the Cavendish Laboratory].

---

[69] Strutt, *Life of J.J. Thomson*, 64. Brackets added. See also *A History of the Cavendish Laboratory*, 96-97.
[70] CUL MSS ADD 7653 T71 (27 December 1898): Townsend to Rutherford.

From 1897 on, the existence of the Cavendish School was easily recognized. The years from 1895 to 1900 marked not only a high point at the Cavendish but also a turning point in the development of the Cavendish Laboratory. At the turn of the century, the Cavendish was emerging as a premier center for the study of experimental physics, particularly atomic physics. Few other research institutes could boast as many researchers and as charismatic a leader as J.J. Thomson. As a research institute and as a graduate school, the Cavendish was firmly established and able to attract young, able, ambitious researchers from all over the world.

Excerpts from some letters written to Rutherford at McGill University by his former student, R. K. McClung at the Cavendish, illustrate the changed atmosphere at the Cavendish.[71]

> Oct. 10th, 1901
>
> Dear Prof. Rutherford,
>
> I have been in Cambridge over a week now .... There are two or three other new men starting to work in the Lab. besides myself. I am just getting a small piece of apparatus made to start on as soon as it is ready. Prof. Thomson asked me if I had anything to start on and I mentioned to him the work which you suggested to me before I left. He however seemed anxious that I should try an experiment on thorium that had suggested itself to him from a paper of Crookes which appeared in the Proc. of Roy. Soc. in 1887.

> Dec. 11th. 1901
>
> Dear Prof. Rutherford,
>
> ... I have become pretty well acquainted with most the men in the Lab. now and have found them all very nice indeed. They have been very kind in making me feel at home among them and have been very obliging indeed in giving me any assistance that I required. There are eleven 1851 men in the Lab. this term, this being the first term here for five of us. I am the only "foreigner" out of the five, the other from being all graduates of English universities.
>
> ... I have been taking lessons in glass-blowing from Everett during the term and have learned considerable in that line. I done quite a lot of glass-blowing in connection with my experiment which I would not have thought of attempting when I came first. I have found the lessons very useful indeed.

> Oct. 11th, 1902
>
> ... There is quite an addition this term to the number of research students. There are about six or eight new ones coming in. I have not met them yet. Miss Brooks arrived this week. I just met her today. We have one new man from Australia this term from Melbourne University, and also a couple of men from the United States. I think there will be something like twenty four research people in the Lab. this term.

> Mar. 4th 1903
>
> ... I am anxious to obtain sufficient results to send in for the degree this next term and also to have a sufficiently satisfactory report to send in to the Commissioners next June in order to apply for a renewal for third year. If I should succeed in getting a renewal I should like to put in another year here unless something better and more permanent should turn up in the meantime ... The rumored report that J.J. had accepted

[71] CUL MSS ADD 7653 M3, M4, M7, M8: R. K. McClung to Rutherford.

> the offer of Columbia University to go to New York created quite a sensation in the Cavendish as well as elsewhere but it was quite a relief to know that it was only a report and that he had not accepted.

Compared to Rutherford's experience at the Cavendish, as recorded in letters written only few years earlier, McClung's entry to Cambridge was easy indeed. Advanced students from outside Cambridge no longer were greeted with scorn by Cambridge men, nor did they feel the need to justify their stay at the Cavendish. Clearly, Cavendish research students now regarded themselves not merely as a collection of individuals but as a community. They were members of the Cavendish School, an illustrious group headed by one of the most prominent physicists in the world. The newcomers' easy assimilation into the Cavendish community, their strong desire to remain as long as possible at the Laboratory, and their professor's active involvement in their research, all marked the Laboratory's entrance into a new era. "The Cavendish" no longer referred to a building but to a research center.

CHAPTER 5

J.J. THOMSON'S LEADERSHIP AND THE DEVELOPMENT OF THE CAVENDISH SCHOOL, 1901-1914

> The theory is not an ultimate one; its object is physical rather than metaphysical. From the point of view of the physicist, a theory of matter is a policy rather than a creed; its object is to connect or co-ordinate apparently diverse phenomena, and above all to suggest, stimulate and direct experiment. It ought to furnish a compass which, if followed, will lead the observer further and further into previously unexplored regions.
>
> J.J. Thomson[1]

> I feel I owe a great deal to the Laboratory and to you [J.J.] personally for whatever success has attended my work. It was in the Cavendish that I was first incubated with the spirit of research & it was there I made my first investigations under your guidance. The work for which the [Nobel] Prize is awarded was begun in the Cavendish Laboratory and I recall that the α particle was born and cherished in one of the lower rooms.
>
> E. Rutherford[2]

> It was his boundless enthusiasm, his endless fertility in suggestion, and his unequaled knowledge of the literature that made him such an inspiring teacher . . . We might perhaps laugh at his little peculiarities; but we knew he was a great man, and we all loved him.
>
> R. J. Strutt[3]

5.1. J. J. Thomson's Research in the New Century

During the first two decades of the new century, J.J. Thomson researched a number of subjects, including corpuscles (electrons), electric discharge and conduction, ionization, radioactivity, radiation, atomic structure, and positive rays and their applications in chemistry. He published more than 80 papers and seven

---

[1] J.J. Thomson, *The Corpuscular Theory of Matter* (London: Archibald Constable & Co., 1907), 1.
[2] CUL MSS ADD 7654 R68 (18 January 1909): Rutherford's reply to J.J.'s telegram congratulating him on his receipt of the 1908 Nobel Prize in Chemistry.
[3] Strutt, *Life of J.J. Thomson*, 150-151.

books, including three textbooks that he co-authored with his old friend, J. H. Poynting. In Table 5.1 these works are classified into five general categories: (a) corpuscles and ionization, (b) radioactivity and radiation, (c) atomic structure, (d) positive rays and their applications in chemistry, and (e) other fields. All these works served J.J.'s search for the ultimate relationship between electricity and matter.

Although the peak of J.J.'s interest in corpuscles, electric discharge and conduction, and ionization had passed with the turn of the century, he continued to show interest in these phenomena until the 1920s. J.J. redetermined the charge of corpuscles, measured the velocity of secondary cathode rays, determined the number of corpuscles in an atom, worked on the nature of electric discharge and conduction, and refined ionization theory. After J.J. and H. A. Wilson independently measured the charge of corpuscles in 1903,[4] J.J. shifted his focus to the applications of corpuscles, particularly in thermionics. Many Cavendish researchers were soon attracted to this new field when, in a 1901 lecture at

Table 5.1. J.J. Thomson's Research Papers during 1901-1918

Subject	*1901-02*	*03-04*	*05-06*	*07-08*	*09-10*	*11-12*	*13-14*	*15-18*
cathode rays/ corpuscles/ ionization/ electric discharge & conduction	7	3	4	3	3	1	1	4
radioactivity radiation/light	2	5	3	5	2	2	1	3
structure of the atom	0	2	2	0	0	1	5	2
positive rays/ its application in chemistry	0	0	1	6	4	7	6	0
electricity, magnetism, etc.	1	2	0	2	5	0	0	4
	10	12	10	16	14	11	13	13

[4] J.J. Thomson, "On the Charge of Electricity carried by a Gaseous Ion," *Phil. Mag.* 5 (1903): 346-355; H. A. Wilson, "A Determination of the Charge on the Ions produced in Air by Röntgen Rays," *Phil. Mag.* 5 (1903): 429-441.

the Royal Institution, J.J. discussed the "existence of free corpuscles or negative electricity in metals" and pointed out that "the corpuscles disseminated through the metal will do more than carry the electric current, they will also carry heat from one part to another of an unequally heated piece of metal."[5] In his 1903 book, *Conduction of Electricity through Gases*, and in its subsequent two editions in 1906 and 1927/1933, J.J. well summarized the research conducted in this area by himself and the Cavendish researchers in the early twentieth century.[6] This book became a standard text for the various researchers at the Cavendish, who often used it as a source of research topics. Figure 5.1 shows the second edition's table of contents. For this edition, J.J. rewrote "a considerable part" of the first edition, adding new material and deleting "some matter fully treated by Rutherford."[7]

During this period, radioactivity and radiation were other major research topics for J.J., who performed experiments to examine induced radioactivity, tested the presence of radioactivity in ordinary substances, and reviewed the nature of radium. J.J.'s correspondence with Rutherford during the early 1900s indicates that J.J. closely followed his distinguished pupil's work on radiation. J.J. was particularly concerned about the source of radioactive energy and, in a 1903 article about radium published in *Nature*, he suggested the following interesting idea:

> The changes we are considering are changes in the configuration of the atom, and it is possible that changes of this kind may be accompanied by the liberation of very large quantities of energy. Thus, taking the atomic weight of radium as 225, if the mass of the atom of radium were due to the presence in it a large number of corpuscles, each carrying the charge of 3.4 by 10^{-10} electrostatic units of negative electricity, and if this charge of negative electricity were associated with an equal charge of positive, so as to make the atom electrically neutral, then if these positive and negative charges were separated by a distance of 10^{-8} cm., the intrinsic energy possessed by the atom would be so great that a diminution of it by 1 per cent would be able to maintain the radiation from radium as measured by Curie for 30,000 years.[8]

Although many former and present Cavendish researchers, particularly Rutherford, enthusiastically cultivated the study of radioactivity. J.J. remained a minor player in this pursuit, as demonstrated by the number and depth of the papers he published on this topic.

His investigations into the nature of radiation, especially X-rays and γ-rays, were more serious. J.J. showed great interest in these mysterious rays, produced several stimulating papers on the topic, and encouraged his students to join the debate on the nature of these rays. His primary goal was to build plausible theories

---

[5] J.J. Thomson, "The Existence of Bodies Smaller than Atoms," *Not. Proc. Roy. Inst. 16* (1901): 574-586 on 582.
[6] The third edition was written by J.J. and his son, G. P. Thomson. The first volume of the third edition appeared in 1927 and the second in 1933.
[7] J.J. Thomson, *Conduction of Electricity through Gases,* second edition (Cambridge: Cambridge University Press, 1906), vii.
[8] J.J. Thomson, "Radium," *Nature 67* (1903): 602.

TABLE OF CONTENTS.

Figure 5.1. Table of Contents of Conduction of Electricity through Gases, second edition (1906).

explaining radiation and light, which made some of his work highly theoretical and even speculative.[9] Although J.J. was one of the first to advance the impulse theory

[9] See, J.J. Thomson, "On a Theory of the Structure of the Electric Field and its Application to Röntgen Radiation and to Light," *Phil. Mag. 19* (1910): 301-313; "On the Theory of Radiation," *Phil. Mag. 20* (1910): 238-247; "Radiant Energy and Matter," *Engineering 91* (1911): 319-321, 353-354, 386-388, 421-422, 454-455; "The Unit Theory of Light," *Proc. Camb. Phil. Soc. 16* (1912): 643-652.

of X-rays, he nonetheless considered other possible explanations and endeavored to build a grand theory uniting all the different opinions that had been put forward. The result was a swing between Newtonian corpuscular theory and Maxwellian electromagnetic theory, or between Planck's quanta and J.J.'s old etherean vortex ring model. In a 1912 paper, J.J. pointed out the difficulties as follows:

> On the unit theory of light, radiant energy is supposed to have a molecular structure and to be made up of a finite number of units . . . In every case in which this conception of the unitary character of light is helpful we are concerned, I think, with transformation of energy, while when we are dealing with purely optical phenomena, such as interference and the like, the same conception leads us into great difficulties, and in the present state of the subject at any rate is a hindrance and not a help.[10]

Thirteen years later J.J. caricatured the battle between the two competing interpretations as "something like one between *a tiger and a shark*, each is supreme in its own element but helpless in that of the other."[11] Sixty years later, historian of science, Bruce R. Wheaton, adapted this image for the title of a book, *The Tiger and the Shark: Empirical Roots of Wave-Particle Dualism*.[12]

During the early decades of the twentieth century, J.J. centered his major research on positive rays and their applications. Beginning in 1905, he spent most of his time and energy studying positive rays and devoted more than a quarter of his papers to this topic. In a 1909 paper, he announced his goals in pursuing this line of research: "The most important questions to be settled as to the nature of positive electricity are: —(1) Does a definite unit of positive electricity exist? (2) If so, what is the size of the unit?"[13] To measure the deflection of positive rays under the influence of electric and magnetic fields, J.J. used a new apparatus in which at first a fluorescent Willemite screen at the end of the tube, and from 1910 on a photographic plate within the tube, was positioned to record the trace of positive rays (Figure 5.2). From these distinctive traces, which had parabolic shapes, J.J. could calculate the values of e/m for the corresponding positive rays.[14]

J.J.'s search for a "definite unit of positive electricity" did not progress as smoothly as his prior research on corpuscles. It was not simply because J.J. was "in his middle fifties, and increasingly involved in the public affairs of science, besides

---

[10] J.J. Thomson, "The Unit Theory of Light," 643.

[11] J.J. Thomson, *The Structure of Light: The Fison Memorial Lecture 1925* (Cambridge: Cambridge University Press, 1925), 15. Emphasis added.

[12] Bruce R. Wheaton, *The Tiger and the Shark: Empirical Roots of Wave-Particle Dualism* (Cambridge: Cambridge University Press, 1983).

[13] J.J. Thomson, "Positive Electricity," *Phil. Mag. 18* (1909): 821-845 on 821.

[14] For earlier research on positive rays before 1910, see J.J. Thomson, "Some experiments on Canalstrahlen," *Proc. Camb. Phil. Soc. 13* (1905): 212-214; "Rays of Positive Electricity," *Not. Proc. Roy. Inst. 18* (1907): 577-592; "On Rays of Positive Electricity," *Phil. Mag. 13* (1907): 561-575; "Rays of Positive Electricity," *Phil. Mag. 14* (1907): 295-359; "The Carriers of Positive Electricity," *Not. Proc. Roy. Inst. 19* (1908): 171-201; "Positive Rays," *Phil. Mag. 16* (1908): 657-691; "On the Carriers of the Positive Charges of Electricity emitted by Hot Wires," *Proc. Camb. Phil. Soc. 15* (1908): 64; "Positive Electricity," *Phil. Mag. 18* (1909): 821-845.

directing the ever-growing Laboratory."[15] The difficulties presented by his research on positive rays, as Falconer has pointed out, were much more complicated.[16] In his early positive ray experiments, J.J. could not secure low pressure and low potential at the same time, and the Willemite screen was not sensitive enough to identify several parabolas. These difficulties were resolved only after 1910, when F. W. Aston became J.J.'s research assistant and greatly improved the apparatus (Figures 5.2b & 5.2c).

J.J.'s preoccupation with theory provided another hindrance to his success. When he discovered that the maximum value of e/m was always that of the H^+ ion (despite different conditions), J.J. jumped to the conclusion that H^+ might be a fundamental constituent of all atoms.[17] When he found that, within a certain limit, the velocity of positive rays was almost the same as that of negative rays (against his conjecture that positive rays would be much slower than negative rays) and that the velocity of the secondary cathode rays was independent of that of the primary rays, J.J. rather hurriedly explained these anomalies by proposing a neutral doublet model "as a working hypothesis."[18] "Neutral doublets," J.J. suggested, consisting of a positive unit and corpuscles, would be "an intermediate stage between the ion and the neutral molecule," and therefore could explain various ionic phenomena.[19] This theory, however, survived only about two years.

With the 1910 improvements in the apparatus and also in research techniques, J.J. could see more parabolas with improved clarity.[20] Nevertheless, he quietly abandoned his original plan to discover the unit of positive electricity, perhaps because a reasonable interpretation of the experimental results seemed too complex. Instead he focused on a new direction: experimental technique and its applications in chemistry. He and Aston discovered in the parabola pictures the first non-radioactive isotope to be found—the isotope of neon—, and they also worked hard to prove the existence of H_3.[21] In his book, *Rays of Positive Electricity and the Application to Chemical Analyses*, published in 1913, J.J. compiled "some account of the experiments on Positive Rays which have been made at the Cavendish Laboratory during the last seven years" and attempted to "induce others, and especially chemists," to apply the use of "Positive Rays to chemical analysis."[22] "Many problems in Chemistry," he argued, "could be solved with far greater ease

---

[15] Crowther, *The Cavendish Laboratory*, 162-163.

[16] Isobel Falconer, "J.J. Thomson's Work on Positive Rays," *HSPS 18:2* (1988): 265-310.

[17] J.J. Thomson, "On Rays of Positive Electricity," *Phil. Mag. 13* (1907): 561-575.

[18] J.J. Thomson, "Positive Rays," *Phil. Mag. 16* (1908): 685.

[19] J.J. Thomson, "Positive Electricity," *Phil. Mag. 18* (1909): 828.

[20] J.J. Thomson, "Rays of Positive Electricity," *Phil. Mag. 19* (1910): 424-435; "Rays of Positive Electricity," *Phil. Mag. 20* (1910): 752-767; "A new Method of investigating the Positive Rays," *Proc. Camb. Phil. Soc. 16* (1910): 120 ; "Rays of Positive Electricity," *Phil. Mag. 21* (1911): 225-249.

[21] See J.J. Thomson, "A New Method of Chemical Analysis," *Not. Proc. Roy. Inst. 20* (1911): 140-148; "Further Experiments on Positive Rays," *Phil. Mag. 24* (1912): 209-253; "Some Further Applications of the Method of Positive Rays," *Not. Proc. Roy. Inst. 20* (1913): 591-600.

[22] J.J. Thomson, *Rays of Positive Electricity and Their Application to Chemical Analyses* (London: Longmans, Green and Co., 1913), v.

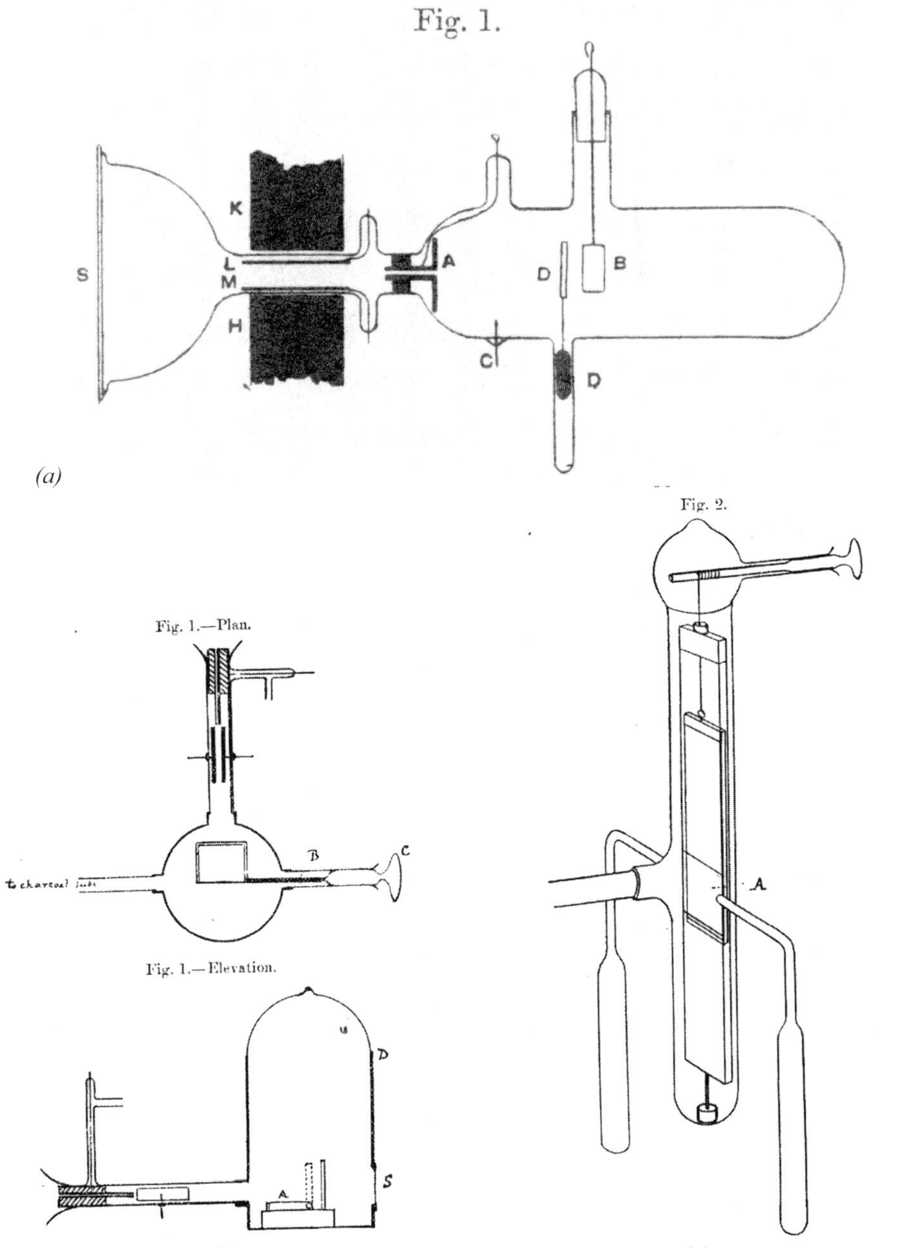

Figure 5.2. J.J.'s apparatus for positive rays research. (a): "Positive Rays," Phil. Mag. 16 (1908): 658; (b) & (c): "Rays of Positive Electricity," Phil. Mag. 21 (1911): 226 & 228.

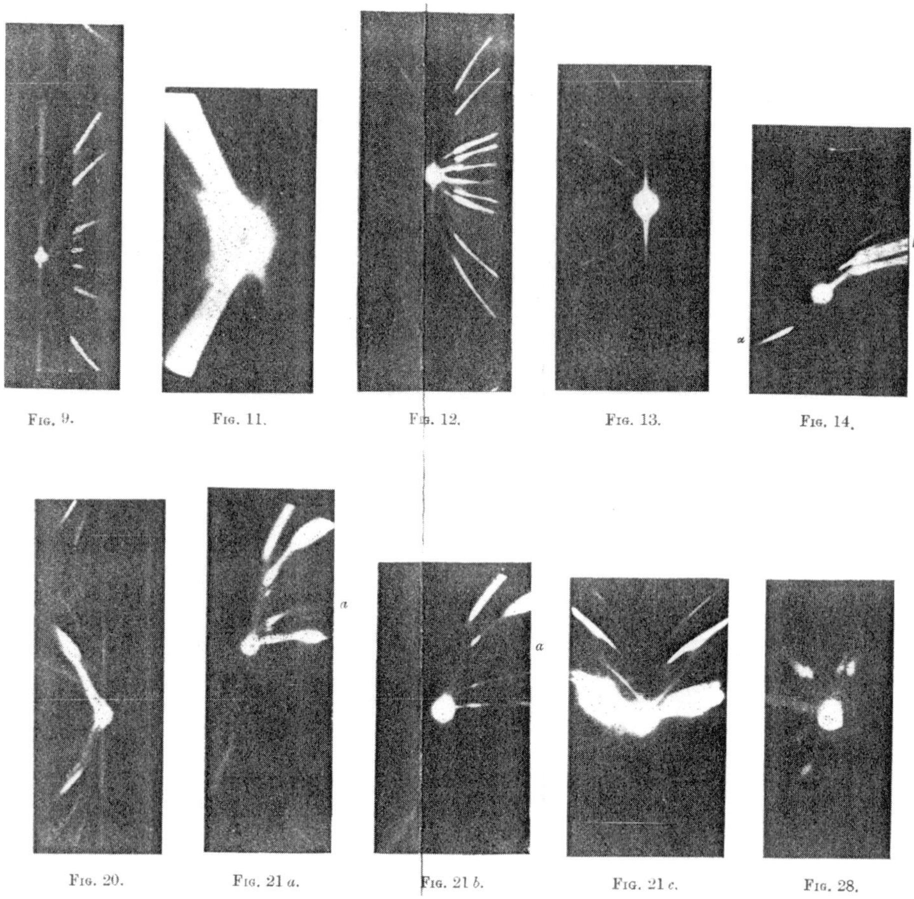

Figure 5.3. Some early parabola pictures taken using the new apparatus. From "Further Experiments on Positive Rays," Phil. Mag. 24 (1912): 209-253.

by this than by any other method." In this book, J.J. concentrated on his experiment and its results and devoted little attention to his atomic model.

It was natural for others to expect the discoverer of corpuscles and the pioneer investigator of positive rays to attempt to explain the atom's structure and, as Heilbron correctly pointed out, J.J. became the true pioneer of this effort.[23] Two of J.J.'s books and more than dozen of his papers contained lengthy discussions of atomic structure; the titles of four of his papers directly addressed "the structure of

[23] J. L. Heilbron, "Lectures on the History of Atomic Physics, 1900-1922," in C. Weiner (ed.), *History of Twentieth Century Physics: Proceedings of the International School of Physics <<Enrico Fermi>>, Course 57* (New York: Academic Press, 1977), 40-108.

the atom" or "atomic structure."[24] In *Electricity and Matter* (1904), J.J. devoted an entire chapter to the "constitution" of the atom, in which he employed A. M. Mayer's concept of floating magnets to illustrate the arrangement of negative corpuscles within a positive sphere.[25] In *The Corpuscular Theory of Matter* (1907), J.J.'s discussion of the arrangement and number of corpuscles in the atom accounted for two fifths of the book. J.J.'s research on positive rays had begun with his attempt to derive a more complete picture of atomic structure by discovering the counterpart of negative corpuscles. In 1904 he enthusiastically reported to Rutherford:

> I have been working hard for some time at the structure of the atom, regarding the atom as built up of a number of corpuscles in equilibrium or steady motion under their mutual repulsion and a central attraction . . . I really have hopes of being able to work out a reasonable theory of chemical combination and many other chemical phenomena.[26]

J.J. initially regarded the atom as mechanically stable. Thinking that the atom might be comprised by a large number of negatively electrified corpuscles enclosed in a sphere of uniform positive electrification, he focused his efforts on discovering how "geometrically" negative corpuscles might be distributed in such a uniformly positive sphere. Mayer's concept of floating magnets provided an analogy useful in explaining an atom in which the corpuscles form several "rings." This so-called "plum-pudding" model, J.J. believed, could explain such chemical phenomena as valency, periodic properties, radioactivity, and "widening of lines in spectra."[27]

Because J.J. originally regarded the mass of the atom as mainly the sum of the mass of its corpuscles, he concluded that the number of corpuscles in an atom must be huge. By 1906, however, experimental work on the dispersion of light by gases, the scattering of X-rays by gases, and the absorption of β rays clearly indicated that the atom contained far fewer corpuscles than J.J. had first predicted. Experimental results showed that the number of corpuscles in the atom were nearly equal to its atomic weight. If its corpuscles were so few in number, what accounted for the atom's weight? Because the mass of a corpuscle was known to be about 1,800 times less than the mass of a hydrogen atom, J.J. proposed, logically, that "the mass of the

---

[24] See J.J. Thomson, "On the Structure of the Atom: an Investigation of the Stability and Periods of Oscillation of a Number of Corpuscles arranged at equal Intervals around the Circumference of a Circle; with Application of the Results to the Theory of Atomic Structure," *Phil. Mag. 7* (1904): 237-265; "The Structure of the Atom," *Not. Proc. Roy. Inst. 18* (1905): 49-63; "On the Structure of the Atom," *Phil. Mag. 26* (1913): 792-799 & 1044. For other papers and books that directly address atomic structure, see J.J. Thomson, "On the Vibrations of Atoms containing 4, 5, 6, 7 and 8 Corpuscles and on the Effect of a Magnetic Field on such Vibrations," *Proc. Camb. Phil. Soc. 13* (1904): 39-48; "On the Number of Corpuscles in an Atom," *Phil. Mag. 11* (1906): 769-781; *Electricity and Matter* (New York: Charles Scribner's Sons, 1904), especially chapter 5; *The Corpuscular Theory of Matter* (London: Archibald Constable & Co., 1907), especially chapters 6 and 7; *The Atomic Theory* (The Romanes Lecture, 1914) (Oxford: Clarendon Press, 1914). Many of J.J.'s discussions about positive rays closely relate to concerns of atomic structure.

[25] J.J. Thomson, *Electricity and Matter*, 115-122.

[26] CUL MSS ADD 7653 T23 (18 February 1904): J.J. to Rutherford.

[27] J.J. Thomson, "A Theory of widening of Lines of Spectra," *Proc. Camb. Phil. Soc. 13* (1906): 318-321.

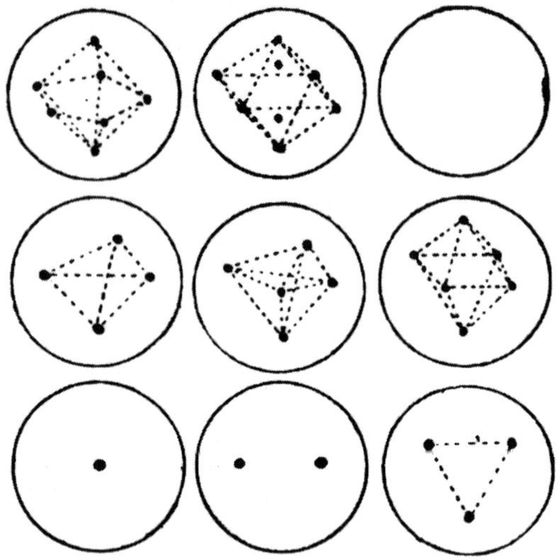

FIG. 1.

Figure 5.4. J.J.'s suggested arrangement of corpuscles in an atom. From "The Structure of the Atom," Not. Proc. Roy. Inst. 18 *(1905): 50.*

carrier of unit positive charge is large compared with that of the carrier of unit negative charge."[28] His plum-pudding model clearly needed revision, and he devoted great effort to arriving at an understanding of a positive unit with a definite mass. J.J.'s next five or six years of intensive research on positive electricity, beginning in 1905, produced disappointing results at least in terms of his search for the positive "unit" corresponding to negative corpuscles, and he finally abandoned his efforts to apply the "positive" part to his atomic model. He still endeavored to develop his atomic model, however, using negative corpuscles as the only "real" entities in the atom. In his Romanes Lecture of June 1914, J.J. argued:

> The most pressing need at this stage of the Atomic Theory is the exploration by experiment of the *distribution of electrons in the atom*; when we know this distribution we may be able to see how we must modify the accepted laws of electrical action to make them applicable to these small charges.[29]

The nucleus, which Rutherford and his students discovered in their α-scattering experiments of 1911, never entered J.J.'s atomic model.

In the new century, J.J. also wrote on various other subjects, including "momentum in the electric field," "the rotation of the plane of polarization by solutions," "the electric theory of gravitation," "the relation between matter and

---

[28] J.J. Thomson, "On the Number of Corpuscles in an Atom," *Phil. Mag. 11* (1906): 774.

[29] J.J. Thomson, *The Atomic Theory*, 24. Emphasis added.

ether," and "the dynamics of a golf ball."[30] In collaboration with J. H. Poynting of Victoria University, Manchester, J.J. co-authored textbooks on the properties of matter (1902), heat (1904), and electricity and magnetism (1914). Along with a previous textbook on sound (1899), these extremely popular books, which ran into several editions, were intended "for the use of students who lay most stress on the study of the experimental part of Physics, and who have not yet reached the stage at which the reading of advanced treatises on a special subject is desirable."[31]

5.2. J.J. Thomson's Leadership and the Cavendish School

5.2.1. J.J.'s Intellectual Leadership

J.J.'s prolific and diverse work greatly influenced the researches of Cavendish researchers in the new century. Cavendish researchers considered him a preacher of new ideas and a trusted advisor, and many of them frequently consulted his books and papers as major references. Many Cavendish researchers acknowledged that their efforts had been initiated at the "suggestion of Professor J.J. Thomson," and nearly all their papers concluded with an expression of gratitude for J.J.'s benevolent influence and "kindly interest and advice." Aston, who closely collaborated with J.J. on positive-ray research, summarized J.J.'s intellectual leadership as follows:

> Working under him never lacked thrills. When results were coming out well, his boundless, indeed childlike, enthusiasm was contagious and occasionally embarrassing. Negatives just developed had actually to be hidden away for fear he would handle them while they were still wet. Yet, when hitches occurred, and the exasperating vagaries of an apparatus had reduced the man who had designed, built and worked with it to baffled despair, along would shuffle this remarkable being, who, after cogitating in a characteristic attitude over his funny old desk in the corner, and jotting down a few figures and formulae in his tiny, tidy handwriting, on the back of somebody's fellowship thesis, or an old envelope, or even the laboratory cheque book, would produce a luminous suggestion, like a rabbit out of a hat, not only revealing the cause of the trouble, but also the means of a cure. This intuitive ability to comprehend the inner working of intricate apparatus without the trouble of handling it appeared to me then, and still appears to me now, as something verging on the miraculous, the hall-mark of a great genius.[32]

---

[30] J.J. Thomson, "On Momentum in the Electric Field," *Phil. Mag. 8* (1904): 331-356; "On the Theory of the Rotation of the Plane of Polarization by Solutions," *Proc. Camb. Phil. Soc. 14* (1907): 313-317; *On the Light thrown by recent Investigations on Electricity on the Relation between Matter and Ether* (The Adamson Lecture) (Manchester: University of Manchester Press, 1908); "On the Electric Theory of Gravitation," *Proc. Camb. Phil. Soc. 15* (1908): 65-69; "The Dynamics of a Golf Ball," *Not. Proc. Roy. Inst. 19* (1909): 795-810.

[31] J. H. Poynting and J. J. Thomson, *A Text-Book of Physics: Sound* (London: Charles Griffin & Co., 1899); *Properties of Matter* (London: Charles Griffin & Co., 1902); *Heat* (London: Charles Griffin & Co., 1904); *Electricity and Magnetism* (London: Charles Griffin & Co., 1914).

[32] F. W. Aston, "J.J. Thomson," *The Times* (4 September 1940). See also Strutt, *Life of J.J. Thomson*, 174-175.

J.J.'s influence on the work of Cavendish researchers was both direct and indirect. Sometimes he would straightforwardly suggest research topics. J. Patterson, for example, began investigating changes in the electric resistance of metals when "it was suggested by Prof. Thomson that I should make a series of experiments on some of the non-magnetic metals."[33] To W. A. D. Rudge, J.J. suggested that "it would be of interest to observe the effect produced by placing a piece of wire-gauze in the path of the cathode discharge."[34]

More commonly, however, J.J.'s influence was felt indirectly, through careful study of his published papers and books, including *Conduction of Electricity through Gases* (1903), *The Corpuscular Theory of Matter* (1907), and even *Notes on Recent Researches in Electricity and Magnetism* (1893) where Cavendish researchers found ideas, possible research topics, theories, and data. E. M. Wellisch noted in one of his papers, for example, "It is of interest to record that Prof. J. J. Thomson has recently advanced the theory that . . ."[35] Similarly, R. K. McClung stated in one of his papers that "it has been suggested by J. J. Thomson, in speaking of this point in his recent book on the *Conduction of Electricity through Gases* . . ."[36]

J.J.'s influence naturally affected Cavendish researchers investigating areas in which J.J. had worked. Starting in 1897, a large number of Cavendish researchers attempted to confirm the existence of electrons or to apply research on electrons to other phenomena, such as ionization.[37] Townsend's application of electron theory to ionization in gases at the turn of the century was a line of research that quickly won the approval of Cavendish researchers and that continued to flourish at the Laboratory well into the first decade of the twentieth century. In independent research, H. A. Wilson redetermined the charge of the electron in 1903. Thermionics especially attracted a large number of Cavendish researchers, prominently headed by O. W. Richardson. The researchers, C. T. R. Wilson, Patterson, and A. Wood worked on another popular research subject of the early 1900s, spontaneous ionization. K. Przibram, J. L. Glasson, R. Whiddington, and others also worked on the electron-related subjects until 1914.

J.J.'s two major research topics in the twentieth century, positive rays and atomic structure, also attracted some Cavendish researchers. J. Kunz, C. T. Knipp, H. Smith, and Aston performed research on the nature and characteristics of positive rays.[38] P. D. Innes, following a suggestion by J.J., examined the theory of

---

[33] J. Patterson, "On the Change of the Electric Resistance of Metals when placed in a Magnetic Field," *Phil. Mag. 3* (1902): 643-656 on 645.

[34] W. A. D. Rudge, "On the Difference of Potential between the Terminals of a Vacuum Tube," *Proc. Camb. Phil. Soc. 12* (1903): 155-162 on 155.

[35] E. M. Wellisch, "The Passage of Electricity through Gaseous Mixtures," *Proc. Roy. Soc. 82* (1909): 500-517 on 517.

[36] R. K. McClung, "The Relative Amount of Ionisation produced in Gases in Röntgen Rays of Different Types," *Phil. Mag. 8* (1904): 357-373 on 357.

[37] A brief summary of this line of research could be found in *A History of the Cavendish Laboratory*, 195-220, 227-249.

[38] J. Kunz, "An Abrupt Limit of Distance in the Power of the Positive Rays to produce Phosphorescence," *Phil. Mag. 14* (1907): 614-617; C. T. Knipp, "Rays of Positive Electricity from the Wehnelt Cathode,"

atomic disintegration by carrying out experiments on the velocity of the cathode particles emitted by various metals.[39] J. A. Crowther experimentally examined J.J.'s theoretical hypothesis concerning the scattering of rapidly moving electrified particles and "calculated the number of electrons contained in atoms of different elements."[40] J.J.'s influences on these investigations were more direct than his influences on investigations of electrons, but they were fewer and less organized.

J.J. also directly influenced research on the nature of X-rays and γ-rays. After Röntgen's discovery, J.J. became a leading British supporter of the impulse theory, which became a popular research topic at the Cavendish in the twentieth century, and one in which J.J. became deeply involved.[41] The first recruit of the new topic was C. G. Barkla, who evolved from research on secondary X-rays to become "an early advocate of the pulse hypothesis."[42] In 1906, with the assistance of G. W. C. Kaye, J.J. found "a very intimate relation" between ionization by secondary X-ray radiation and atomic weight, an achievement that stimulated R. D. Kleeman, a former student of W. H. Bragg (another former Cavendish researcher) at Adelaid, to investigate the close resemblance between X-rays and γ rays.[43] During the first and second decade of the new century, many Cavendish researchers and alumni, including N. R. Campbell, Crowther, and Whiddington, entered the field. Eventually Barkla and W. H. Bragg became the leaders of two competing views on the nature of X-rays (or γ-rays), with Barkla supporting the pulse hypothesis and Bragg proposing that X-rays consist of neutral material particles.[44]

J.J.'s influence on Cavendish researchers extended far beyond even the numerous examples given above. In communicating with the researchers, he never relied on intermediaries but always made direct contact with each individual, keeping aware of each researcher's studies and stopping at each researcher's bench every day to observe, make suggestions, and enjoy conversations, often lengthy ones. Like Maxwell, J.J. preferred that the Cavendish researchers choose their own problems and carry out their own studies, but he always was ready to propose

---

Phil. Mag. 22 (1911): 926-933; H. Smith, "A Comparison of the Positive Rays with the Spectrum of the Positive Column in a Mixture of Helium and Hydrogen," *Phil. Mag. 30* (1915): 805-811; F. W. Aston, "Sir J. J. Thomson's New Method of Chemical Analysis," *Scientific Progress* (July, 1912): 48.

[39] P. D. Innes, "On the Velocity of the Cathode Particles emitted by Various Metals under the Influence of Röntgen Rays, and its bearing on the Theory of Atomic Disintegration," *Proc. Roy. Soc. 79* (1907): 442-462.

[40] J. A. Crowther, "On the Scattering of Homogeneous β-Rays and the Number of Electrons in the Atom," *Proc. Roy. Soc. 84* (1910): 226-247.

[41] Wheaton, *Tiger and Shark*, 20-29.

[42] Ibid., 44. See also C. G. Barkla, "Secondary Radiation from Gases subject to X-rays," *Phil. Mag. 5* (1903): 685-698.

[43] J.J. Thomson, "On Secondary Röntgen Radiation," *Proc. Camb. Phil. Soc. 14* (1906): 109-113. See R. D. Kleeman, "On the Secondary Cathode Rays emitted by Substances when exposed to the γ Rays," *Phil. Mag. 14* (1907): 618-644. Kleeman published many papers on the subject. His 1909 paper specifically supported J.J.'s "modified aether pulse theory." (Kleeman, "On the Velocity of the Cathode Rays ejected by Substances exposed to the γ Rays of Radium," *Proc. Roy. Soc. 82* (1909): 128-145.)

[44] For more information about this controversy, see Wheaton, *Tiger and Shark*, part II.

interesting topics when asked. He once told his biographer, Robert Strutt, that he had no trouble finding good research topics, but only "reasonably easy problems for beginners."[45]

5.2.2. The Emergence of Research Subgroups and a New Cavendish Style

The Cavendish researchers' high productivity under J.J.'s leadership brought the Laboratory to a new stage of development in the early years of twentieth century. Several semi-independent subgroups emerged, each concentrating on a different topic, such as radioactivity, thermionics, condensation nuclei, or spontaneous ionization. More often than not, a subgroup was formed in response to an idea put forward by J.J. Because he neither organized nor controlled these subgroups, J.J.'s contribution to their emergence was indirect, but he warmly encouraged their progress.

The development of research subgroups at the Cavendish was important and unusual. In light of the Laboratory's laissez-faire tradition, it also was unexpected. No Cavendish directors ever had imposed specific research programs on Laboratory researchers and, under J.J.'s directorship, there never had been what Peter Galison later called "the Cavendish research program."[46] Any Cambridge man who wanted to research physics could come to the Laboratory to use space and apparatus and, as an additional premium, gain advice from the Professor and the Cavendish demonstrators. Cooperative efforts by Cavendish researchers had been temporary and co-works relatively few. Because the Cavendish had no established system for supplying researchers with financial support to carry out their experiments, most Cavendish researchers independently pursued the resources they needed, and this responsibility contributed greatly to their independence from one another and even from the Professor. Suddenly, however, in the new century, cooperation at the Cavendish increased and cooperative efforts enjoyed greater longevity, even though no significant systematic changes had occurred at the Laboratory. Notably, the research subgroups formed at the Cavendish during this time were temporary groups organized on a voluntary basis. They did not bear official names or use separate rooms, and they were not yet regarded as official organizations.

Very active among these research subgroups was the one devoted to studying radioactivity. Before the new century, research on radioactivity had not been very popular at the Laboratory, partly because J.J. and many of the researchers were busy investigating corpuscles. One exception had been Rutherford, who in 1898 had distinguished α-rays and β-rays.[47] However, as the new century began, the nature of radioactivity attracted the research efforts of a number of Cavendish researchers, including J. C. McLennan, W. Makower, Miss J. M. W. Slater, H. L. Cooke,

---

[45] Strutt, *Life of J.J. Thomson,* 150.

[46] Peter Galison, *Image and Logic: A Material Culture of Microphysics* (Chicago: University of Chicago Press, 1997), 97.

[47] E. Rutherford, "Uranium Radiation and the Electrical Conduction produced by it," *Phil. Mag. 47* (1899): 109-163.

Campbell, Wood, Crowther, J. Satterly, and several others. In addition, some researchers, particularly C. T. R. Wilson, Townsend, R. J. Strutt, J. J. E. Durack, T. Noda, and Kleeman were interested in the application of radioactive rays to their studies of ionization. Although the radioactivity subgroup was large and its members produced a correspondingly large number of papers, this subgroup's achievements were nonetheless limited in significance. The subgroup lacked a Rutherford or Curie, and the Laboratory failed to secure powerful radioactive material useful for experimentation.

Even more active than the radioactivity group, and much more successful, was the thermionics subgroup. In the late nineteenth century, Elster and Geitel had systematically investigated the phenomena in which incandescent metals ionize

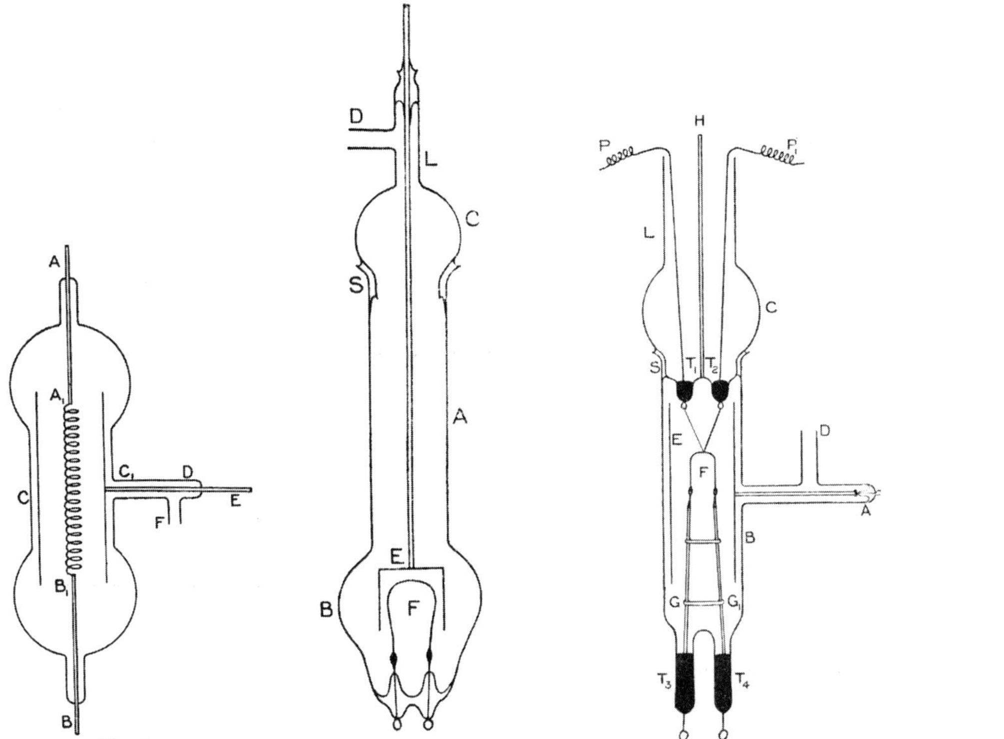

Fig. 1. Fig. 6. Fig. 7.

(a) (b) (c)

Figure 5.5. Some apparatus that O. W. Richardson employed in his thermionic experiments. Platinum wire (a) and carbon wire (b, c) were inserted in the tubes. From "The Electrical Conductivity imparted to a Vacuum by Hot Conductors," *Phil. Trans. 201 (1903): 507, 518, 519].*

surrounding air.[48] J.J.'s 1900-01 observation that electric currents in metals were being carried by negatively charged ions moving at definite velocity under a given electric field immediately sparked the interest of several Cavendish researchers.[49] Owen W. Richardson began his lifelong study of the subject that he reported "was directly suggested by the theory of the mechanism of the conduction of electricity in metals recently forward by Professor J. J. Thomson."[50] Richardson's series of experiments with hot platinum wire and other metal substances produced results supporting J.J.'s theoretical hypothesis that metals contain large numbers of corpuscles which move freely, or nearly so, inside them.[51] Richardson published thirteen papers (including one co-authored paper) on this subject before leaving the Cavendish in 1906 to teach physics at Princeton University. H. A. Wilson, G. Owen, J. A. Cunningham, F. Horton, and G. W. Todd also worked on thermionics. Research on thermionics was so popular at the Laboratory that, in the eyes of researchers of the day, it "seemed likely to develop into a branch of physics with almost the same importance as radio-activity."[52]

Another active, but much smaller, subgroup was concerned with condensation nuclei. C. T. R. Wilson had initiated a systematic investigation of condensation nuclei, which he documented in a series of papers published in the late 1890s and early 1900s. J.J.'s influence on C. T. R.'s research was clear from the beginning. In an 1897 paper, C. T. R. employed J.J.'s idea that free ions, "in virtue of the charge they carry," encourage the process of condensation.[53] After J.J. pointed out that, "if negative ions were to differ in their power of condensing water around them from the positive, then we might get a cloud formed round one set of ions and not round the other," C. T. R. compared the condensation efficiency of positive and negative charges.[54] His cloud condensation method—the "primordial" cloud chamber—proved a very useful tool for the study of ions and electrons (for example, in determining the charge of electrons), which soon was employed by many Cavendish researchers and by J.J. himself. In 1899, J.J. wrote to Rutherford, "I see you tried some experiments & did not get a cloud by expansion—did you use Wilson's apparatus for getting very sudden expansion? The ordinary ways of producing

---

[48] J.J. Thomson & G. P. Thomson, *Conduction of Electricity through Gases*, third edition, vol. 1, 339-341.
[49] J.J. Thomson, "Indications relatives à la constitution de la matière," *Rapports Congrès de physique 3* (1900): 138; "The Existence of Bodies Smaller than Atoms," *Not. Proc. Roy. Inst. 16* (1901).
[50] O. W. Richardson, "On an Attempt to detect Radiation from the Surface of Wires carrying Alternating Currents of High Frequency," *Proc. Camb. Phil. Soc. 11* (1901): 168-178 on 168.
[51] See O. W. Richardson, "The Electrical Conductivity imparted to a Vacuum by Hot Conductors," *Phil. Trans. 201* (1903): 497-549.
[52] *A History of the Cavendish Laboratory*, 232.
[53] C. T. R. Wilson, "On the Action of Uranium Rays on the Condensation of Water Vapour," *Proc. Camb. Phil. Soc. 9* (1897): 333-338 on 337. For J.J.'s work, see J.J. Thomson, "On the Effect of Electrification and Chemical Action on a Steam-jet, and of Water-Vapour on the Discharge of Electricity through Gases," *Phil. Mag. 36* (1893): 313-327.
[54] C. T. R. Wilson, "On the Comparative Efficiency as Condensation Nuclei of Positively and Negatively," *Phil. Trans. 193* (1899): 289-308 on 289. See also J.J. Thomson, "On the Charge of Electricity carried by the Ions produced by Röntgen Rays," *Phil. Mag. 46* (1898): 528-545 on 533-534.

expansion are useless in many cases where Wilson's method gives a fog."[55] In the early 1900s, C. T. R. shifted his attention to atmospheric electricity and spontaneous ionization, but in 1911 returned to his old subject. Although he was the originator of research on condensation nuclei, as well as the major player in this field, he was not the only Cavendish researcher who investigated condensation nuclei. G. Owen, J. H. Vincent, K. Przibram, T. H. Laby, and a few others also immersed themselves in elaborating on the condensation method.[56]

J.J. more often than not provided the first impetus for the formation of a research subgroup at the Laboratory, but the developments made by these subgroups were made largely independent of him. Each subgroup had its own leader or leaders, each of whom was more knowledgeable and experienced than J.J. in its particular field of interest. The radioactivity subgroup turned to Kleeman, Crowther, and Campbell, and Richardson's influence on the thermionics subgroup continued even after he left the Cavendish. The condensation nuclei group looked to C. T. R. Wilson, frequently cited his prior works on the subject, and expressed gratitude for his "most helpful discussion."[57] Although J.J. remained a patron of the Cavendish subgroups, and sometimes introduced newcomers to them, the emergence and development of these semi-independent subgroups signaled that the Cavendish school had entered a new phase of maturation.

The increasing influence of the professor on student works and the emergence of the Cavendish research subgroups indicated another direction of growth: more and more, the Cavendish researchers studied closely related subjects in areas of special interest to the Professor: ionization, electrons, radioactivity, radiation, thermionics, and so on. This is not to say that the Cavendish Laboratory had discarded its traditional laissez-faire policy of research freedom: this tradition was alive and thriving. Instead, the researchers who chose to enter fields in which J.J. demonstrated special interest selected these fields because they held promise. Even beginning researchers and visiting researchers, who were understandably more dependent on J.J than other Cavendish researchers, were able to go in independent directions when they did not agree with the Professor's advice. For example, R. K. McClung, who at the Cavendish was continuing research on the rate of recombination of ions in air that he had started at McGill University under Rutherford, rejected J.J.'s advice of "try[ing] an experiment on thorium that had suggested itself to him from a paper of Crookes which appeared in the *Proceedings of the Royal Society* in 1887."[58]

---

[55] CUL MSS ADD 7653 T12 (23 July 1899): J.J. to Rutherford.

[56] G. Owen, "On the Condensation Nuclei produced in Air and Hydrogen by heating a Platinum Wire," *Phil. Mag. 6* (1903): 306-315; J. H. Vincent, "The Action of Ultra-Violet Light on Moist Air," *Proc. Camb. Phil. Soc. 12* (1903): 305-311; T. H. Laby, "The Supersaturation and Nuclear Condensation of Certain Organic Vapour," *Phil. Trans. 208* (1908): 445-474.

[57] T. H. Laby, "The Supersaturation and Nuclear Condensation of Certain Organic Vapours," 474.

[58] CUL MSS ADD 7653 M3 (10 October 1901): R. K. McClung to Rutherford. Brackets added. See also R. K. McClung, "The Relation between the Rate of Recombination of Ions in Air and the Temperature of the Air," *Phil. Mag. 6* (1903): 656-666.

The more orchestrated research performed at the Cavendish in the early years of the twentieth century led the Laboratory to a further stage of development: the accumulation of valuable knowledge in a limited number of specific fields. At the end of the nineteenth century, the traditionally weak cooperation of Cavendish researchers and the wide variety of research topics pursued by them, along with researchers' frequent departures from the Laboratory, had made it almost impossible for specific knowledge and skills to accumulate at the Laboratory. One exception had been the line of research that resulted in measurement of the ohm, a pursuit that had been launched by Lord Rayleigh and was continued by Glazebrook until he left for the National Physical Laboratory. However, the discovery of the electron and the successful application of electron theory to ionization studies at the turn of the century precipitated more frequent exchanges of knowledge and instruments among the Cavendish researchers. This new trend was greatly enhanced in the new century by the emergence of the semi-independent research subgroups. In previous years the Professor often had collaborated with his students on specific researches: thus, McClelland and Rutherford *assisted J.J.* in his research on gaseous discharges under the influence of X-rays, which eventually led to the discovery of corpuscles. Now, researchers *collaborated with each other* to explore new fields.

It is therefore noteworthy that in their research reports of this period, Cavendish researchers increasingly cited other researchers' works and increasingly acknowledged J.J. and each other for sharing data and loaning instruments. Barkla thanked J.J. for suggesting "the use of the Rutherford detector in a measurement of the velocities along wires of various diameters and materials."[59] G. F. C. Searle, one of the Cavendish's two senior demonstrators, noted that R. L. Wills had employed some of Searle and T. G. Bedford's apparatus "in a series of experiments made by the ballistic method, at the Cavendish Laboratory, upon the effect of temperature upon hysteresis."[60] H. A. Wilson thanked C. T. R. Wilson for the "kindly" loan of the expansion apparatus C. T. R. had used to determine the charge of corpuscles.[61] Wood expressed gratitude to Campbell for having "kindly placed" at his disposal a summary of Campbell's results.[62] W. Wilson reported using, in his research on the absorption and reflection of homogeneous β-particles, "the recent remarkable photographs of the paths of β-particles through gases, made by C. T. R. Wilson," probably before these photographs were published in the same journal as Wilson's article.[63] J. Patterson expressed his thanks to C. F. Mott for "kindly placing his

---

[59] C. G. Barkla, "The Velocity of Electric Waves along Wires," *Phil. Mag. 1* (1901): 652-666 on 653.
[60] G. F. C. Searle, "The Ballistic Measurement of Hysteresis," *Electrician 49* (1902): 100-103 & 219-222 on 221. See also R. L. Wills, "Effect of Temperature on the Hysteresis Loss in Iron," *Phil. Mag. 5* (1903): 117-133 on 117.
[61] H. A. Wilson, "A Determination of the Charge on the Ions produced in Air by Röntgen Rays," *Phil. Mag. 5* (1903): 429-441 on 432.
[62] A. Wood, "Spontaneous Ionisation of Air in closed Vessels and its Causes," *Phil. Mag. 9* (1905): 550-576 on 565.
[63] W. Wilson, "On the Absorption and Reflection of Homogeneous β-particles," *Proc. Roy. Soc. 87* (1912): 310-325 on 316.

apparatus for making the films at the author's disposal."[64] Campbell and Wood used the specimen of tin foil previously "employed by Crowther in his researches on the β rays of uranium."[65]

The scale of sharing of data and apparatus indicated by these acknowledgments was quite new at the Cavendish. However, despite increasing cooperation among the Cavendish researchers, the number of research papers written jointly by these researchers did not rise during the transition years between the two centuries, but instead slightly decreased, from 16 (1885-1900) to 14 (1901-1914), illustrating the strange prevailing mixture at the Cavendish of new cooperative effort and traditional Cambridge individuality.

5.2.3. J.J.'s Charisma

J.J.'s unique personality contributed to the consolidation of his leadership during the early years of the twentieth century. His manner never had been authoritarian. On the contrary, he had been very close to his students, who freely called him "J.J." or "J.J.T." instead of "Professor Thomson." The Cavendish annual photograph and annual dinner party commemorated not only the 1897 birth of the Cavendish school but also the intimate relationship between the students and their professor.

Letters to J.J. from his former students revealed deep respect and love for him, appreciation for his unique personality, and gratitude for his kindness. George C. Jaffé wrote that "the time which I spent in Cambridge working under your guidance was certainly the happiest of my life."[66] L. L. Campbell wrote that he would "always be deeply grateful to you for the courtesies and kindness extended me" and recalled that the years he and his wife spent at the Cavendish were "a bright and cherished period in our lives."[67] When in 1915 the War Office (of Britain) adopted an invention of G. A. Brodsky of Kiev, he wrote:

> I have this opportunity for thanking you most heartily for your very kind assistance which enabled me to give a material shape to my idea . . . Small as my achievement is on this occasion I owe it entirely to you, the old Lab., and the few old Cambridge friends who encouraged my attempt. It is the first real success of my restless life.[68]

The following account by H. A. Bumstead, a visiting professor from Yale, nicely summarizes the unique bond between the Professor and his students:

> Another thing which struck me as paradoxically fine was the relation between the Professor and his students. The great admiration and reliance with which 'J.J.' was regarded by everybody was unquestionable: yet in matters of detail there was no subserviency. I saw a good many men, while I was there, following their own courses,

---

[64] J. Patterson, "On the Effect of Magnetic Field on the Resistance of Thin Metallic Films," *Proc. Camb. Phil. Soc. 11* (1901): 118-119 on 119.
[65] N. Campbell & A. Wood, "The Radioactivity of the Alkali Metals," *Proc. Camb. Phil. Soc. 14* (1906): 15-21 on 18.
[66] CUL MSS ADD 7654 J2 (9 November 1909): G. C. Jaffé to J.J.
[67] CUL MSS ADD 7654 C5 (4 July 1915): L. L. Campbell to J.J.
[68] CUL MSS ADD 7654 B67 (9 July 1915): G. A. Brodsky to J.J.

> and I remember well the frank tone prevailing at the meetings of the Cavendish Society, where the Professor's theories or experiments were not spared in the general criticism.[69]

J.J. also enjoyed the loyalty of his teaching staff, as shown by a letter to him from senior demonstrator Searle, who began teaching physics at the Cavendish in 1891. In 1910, Searle had been seriously considering applying for a professorship at Bristol, possibly his last chance to leave the Cavendish.

> My dear Thomson,
> I think you will wish to know my views as to Bristol. I had put off the consideration of the question till I felt stronger.
> There are very few posts in which the opportunities are greater than in my present post. The title of professor and the increase of salary are no real temptation for me. If I applied for the post it would be less because I wanted what might be considered a more important post than because it seemed best to endeavour to sever my connexion with the Cavendish.
> But things have been looking a good deal brighter lately. In allowing me to undertake the Experimental Lectures for the Mathematical Tripos you showed that I had your confidence and in appointing Bedford to the other class you showed that you had regard both to the interests of the students and to the worth of the unfailing devotion and the ability with which Bedford has served for so many years . . .
> I believe that I may take your recent attitude as an indication that you desire for the future that the work of the demonstrators shall be as much a matter of interest and concern to you as the research work and that you are anxious that as far as possible any obstacles in the way of the demonstration work should be removed. I think you have come to see that the absence of "complaints" from those who shrink from complaining is no evidence that things are entirely satisfactory.
> With this belief in my mind, I look forward to the future with a new hope and I believe that I shall be doing the right thing in not making an application for Bristol. If my hope has any foundation, the strain of the work would be very greatly diminished.[70]

Admiration, respect, loyalty, and love often led J.J.'s students and colleagues to seek his advice on confidential matters. On such an occasion in 1901, J.J. advised Rutherford:

> As I cabled to you I think you had better stand for the Edinburgh Chair if you wish to return to England. I do not think the chances of your getting the post very promising as Knott has been doing the work for some time and has a good deal of great influence . . . ; at the same time I think the candidature will do you good as it will let people know that you are willing to leave Montreal and your name will naturally come up if other vacancies occurred.[71]

Four years later, J.J. wrote Rutherford, "With regard to [the] King's College post I will tell you all I know about it & try to be as unbiased as possible although my keen desire to have you back in England makes this difficult."[72] In 1919, Barkla

---

[69] *A History of the Cavendish Laboratory*, 226.

[70] CUL MSS ADD 7654 S33 (11 August 1910): Searle to J.J. Searle had previously tried to get a suitable post outside Cambridge. In 1899, for example, he had applied for the University of Glasgow position vacated by Lord Kelvin. Other Cavendish men, including Townsend, C. T. R. Wilson, McClelland, and Walker also applied for this post but failed.

[71] CUL MSS ADD 7653 T16 (12 April 1901): J.J. to Rutherford.

[72] CUL MSS ADD 7653 T27 (1 October 1905): J.J. to Rutherford.

asked J.J. whether he might apply for the Cavendish Professorship to succeed his former mentor: "It is perhaps the only time I shall have to make the choice between staying here for life & going elsewhere to a better climate & a more leisured life.... I don't wish anyone to know I have written to you about the matter."[73] J.J. seems to have responded with a frank letter which persuaded his Nobel Prize-winning former pupil not to apply for the position. Barkla immediately wrote back, "When I wrote to you I had no idea that the appointment to the Cavendish Chair was to be made so hurriedly. Your second letter gives so little time for consideration that I cannot feel prepared to risk the future on a decision made in a day . . . Your letter places the position so clearly . . . "[74]

The most extraordinary illustration of the high regard in which J.J. was held by his students can be found in the total of 14 affectionate songs they wrote about him and performed at annual Cavendish Laboratory dinners of different years. One of the most popular songs was as follows:

The Dons of the Day[75]
Air: Father O'Flynn
Lyrics: A. A. Robb

1. Of Dons we can offer a charming variety
All the big pots of the Royal Society,
Still there is no one of more notoriety
Than our professor, the pride of us all.
Here's a health to Professor J.J.!
May he hunt ions for many a day,
And take observations,
And work out equations,
And find the relations
Which forces obey.

2. Our worthy professor is always devising
Some scheme that is startlingly new and surprising,
In order to settle some question arising
On ions and why they behave as they do.
Thus, when he wants to conclusively show
Some travel quickly and some travel slow
He brings into action
Magnetic attraction
And gets a deflection
Above and below.

3. All preconceived notions he sets at defiance
By means of some neat and ingenious appliance,
By which he discovers a new law of science
Which no one had ever suspected before.
All the chemists went off into fits,
Some of them thought they were losing their wits,
When quite without warning
(Their theories scorning)
The atom one morning
He broke into bits.

4. When the professor has solved a new riddle,
Or found a fresh fact, he's fit as a fiddle,
He goes to the tea-room and sits in the middle
And jokes about everything under the sun.
Then if you try to look grave at his jest,
You'll burst off the buttons which fasten your vest,
For when he starts chaffing,
Though tea will be quaffing,
You cannot help laughing
Along with the rest.

---

[73] CUL MSS ADD 7654 B18 (11 March 1919): Barkla to J.J.
[74] CUL MSS ADD 7654 B19 (26 March 1919): Barkla to J.J.
[75] See John Satterly, "The Postprandial Proceedings of the Cavendish Society," *The American Physics Teachers* 7 (1939): 179-185, 244-248.

Chorus:

Here's health to Professor J.J.
May he hunt ions for many a day,
And take observations,
And work out equations,
And find the relations
Which forces obey.

This cordiality of relationship between professor and student was not to be found in any other laboratory, institute, or university of the time and perhaps cannot be found in any such institution today. As shown in the fourth verse of "The Dons of the Day," J.J.'s students even celebrated the professor's ability to lighten their day with a stream of good-humored jokes.

J.J. was not a perfect figure, however. He was noted for peculiarities; not the least of which was his "pure absence of mind," which, although often the source of laughter, sometimes resulted in consequences that were far from humorous, as can be seen in an anecdote provided by J.J.'s biographer, Robert Strutt.[76] In about 1899, according to Strutt, Townsend observed that if the electromotive force were sufficiently large, the current (at low pressure) could exceed the saturation value. Townsend, noting that A. Stoletow's photoelectric experiment of 1890 had produced a similar result, proposed the idea that a fast-moving negative ion collides with a neutral molecule and ionizes that molecule. The original ion thus breeds new ions, causing a rapid increase of the current. When Townsend shared this idea with other Cavendish researchers and with J.J., the Professor disputed it. Eventually, however, Townsend's pursuit of this line of research led him to the theory of ionization by collision, which J.J. then adopted as a useful hypothesis and employed in two papers, "Ionisation of Gases in the Electric Field" (February 1900) and "The Genesis of the Ions in the Discharge of Electricity through Gases" (September 1900).[77] Townsend openly complained that J.J. had not properly acknowledged his contribution and, in a letter to his old friend Rutherford, expressed some hard feelings:

> J.J. has been like a weathercock on all this subject during the last 9 months. You will notice that [when] I communicated my paper myself I don't know how much of it he agrees with, as he stuck vehemently to his theory of layers as an explanation of my results and Stoletow's. The layer theory is I believe absolutely rot . . .
>
> After I published a preliminary account of my theory in *Nature*, J.J. published a paper in the *Phil. Mag.* on the genesis of ions, without referring to my paper and mentioning some of the things I had told him. I was very much put out about it particularly as he would not accept one word of my theory when I was doing my experiments . . . When I proposed the collision theory & showed that it gave a mathematical coincidence he put in his paper that the collisions would produce the effect observed by Stoletow but never mentioned that I had proved it, and was the first to suggest it.

---

[76] Strutt, *Life of J.J. Thomson*, 128-130.

[77] J.J. Thomson, "Ionisation of Gases in the Electric Field," *Proc. Camb. Phil. Soc. 10* (1900): 380; "The Genesis of the Ions in the Discharge of Electricity through Gases," *Phil. Mag. 50* (1900): 278-283.

However it does not much matter, and I dare say he stuck it into his paper without thinking, but it caused some amusement in the Cavendish.[78]

In his biography of J.J., Strutt devoted considerable space to this incident to save his hero from the taint of plagiarism, arguing that the idea of ionization by collision was not first developed by Townsend, although it was he who proved it both mathematically and experimentally [79] Strutt relied heavily on J.J.'s absentmindedness as a defense against the charge: "That Stoletow's work carried this implication was pointed out by Townsend to J.J. and myself in a triangular conversation, and J.J. had clearly forgotten this when he drafted the paper of September 1900."[80] To add more weight to his explanation, Strutt quoted a passage in a letter from Searle:

J.J.' s mind worked in strange ways. He could not always remember how an idea had got into his mind. I did not meet this phenomenon much myself, but it caused difficulties to a good many people. He would be told by someone or would read somewhere some new idea. Later on he would find the idea floating in his mind and he would suppose the idea was original to himself and would treat it as if it were. This gave the appearance that he was claiming as his own ideas which others had already published. I am sure he was unconscious that he was using the work of others.[81]

A letter from J.J. to Rutherford, dated December 21, 1899, supports the position of Searle and Strutt: in it J.J. treated Townsend's idea as his own and failed to mention Townsend's name.[82] J.J.'s charisma, however, was so powerful that even such potentially serious incidents did not become a source of discontent among the Cavendish researchers but remained a source of "amusement."

In the new century, J.J.'s charisma and his strong intellectual leadership became the keystones of his directorship. J.J.'s remarkable leadership qualities gave the Cavendish something unique to offer in its competition with other universities for outstanding researchers. The Cavendish always had been less attractive than its German counterparts for a number of reasons. German laboratories offered superior science, a Ph.D. system, a more amiable attitude toward foreigners, a tradition that permitted researchers to circulate freely among universities and, perhaps most important, much better financial support.[83] Without any system for attracting and financially supporting able researchers, the Cavendish had been at a clear disadvantage when compared to German teaching and research laboratories. Even in the new century, American and Japanese scientists overwhelmingly preferred German universities to British ones.[84] The Professor's good humor, which

[78] CUL MSS ADD 7653 T73 (14 January 1900): Townsend to Rutherford. The year must be 1901 rather than 1900.

[79] Strutt, *Life of J.J. Thomson,* 115-119.

[80] Ibid., 118.

[81] Ibid., 118-119.

[82] CUL MSS ADD 7653 T13 (21 December 1899), J.J. to Rutherford.

[83] See Jungnickel & McCormmach, *Intellectual Mastery of Nature,* vol. 2.

[84] According to James R. Bartholomew, 69% of Japanese physicists went to Germany for advanced studies between 1869-1914, while Britain attracted much fewer. See James R. Bartholomew, *The Formation of Science in Japan* (New Haven: Yale University Press, 1989), 71.

contributed so much toward creating a friendly atmosphere at the Cavendish, along with the Professor's thoughtful attention to the Cavendish researchers' interests, were valuable premiums that attracted researchers in fields in which the Cavendish was accumulating knowledge. Once admitted to the Cavendish, students were no longer individual researchers but members of a research school that shared, along with ideas and instruments, a unique ethos.

5.2.4. The Growth of the Cavendish School

In the new century, the status of the Cavendish School began a rapid rise. The first indicator was a continuous influx of researchers into the Cavendish Laboratory. At the turn of the century, about a dozen researchers studied at the Cavendish at any given time; by 1910, this number had risen to 25, and by 1912, had risen again to over 30. The quality of the new researchers was high, as indicated by the large number of 1851 Exhibition Scholars who chose to study at the Cavendish between 1891 and 1914. During this period, 54 of the 90 recipients of the 1851 Exhibition scholarship who studied physics in the U.K. elected to study at the Cavendish.[85]

Cavendish researchers and alumni soon began to dominate major British scientific journals. Between 1901 and 1910, the prestigious journal, *Philosophical Transactions*, delivered 26 works that had been produced at the Cavendish, a significant increase over such publication in the past, which included only 5 Cavendish papers in the 1870s, 16 in the 1880s, and 15 in the 1890s. Also during the first decade of the century, roughly one third of the physics papers published by *Philosophical Magazine* were authored by Cavendish researchers or former researchers. *Proceedings of the Royal Society* published fewer papers by Cavendish researchers, but this number increased steadily in the new century, especially after 1907. While such statistics demonstrated the rising status of the Cavendish in the scientific communities of Britain and the world, the dominant position of the Cavendish Laboratory within the Cambridge physics community was unmistakably illustrated by its representation in the *Proceedings of the Cambridge Philosophical Society*, in almost every volume of which works by Cavendish researchers were in the majority.

As the Cavendish advanced into the new century, the number of its researchers and alumni who reaped major honors also increased. Before 1900, only three of J.J.'s pupils, Callendar, Chree, and Threlfall, had become Fellows of the Royal Society. By 1919, this number had risen to sixteen and included C. T. R. Wilson, Rutherford, Townsend, Strutt, Searle, H. A. Wilson, Barkla, Richardson, and G. I. Taylor. Some of J.J.'s pupils also received Hughes, Royal, Copley, or Rumford awards for their achievements.[86]

An astounding number of Nobel Prizes ultimately would be awarded to Cavendish researchers. In 1904, Lord Rayleigh became the first British winner in

---

[85] Roy M. Macleod & E. Kay Andrews, "Scientific Careers of 1851 Exhibition Scholars," *Nature 218* (1968): 1011-1016 on 1015.

[86] J.J. Thomson, *Recollections*, 435-437.

physics and, two years later, J.J. received the coveted award. In time, six of J.J.'s former pupils would receive the Nobel Prize: Rutherford (Chemistry, 1908); W. L. Bragg (with his father, W. H. Bragg, Physics, 1915), Barkla (Physics, 1917), C. T. R. Wilson (Physics, 1927), Richardson (Physics, 1928), and Aston (Chemistry, 1922). Of the ten British scientists who received Nobel Prizes for physics before 1945, only one, P. A. M. Dirac (1933), never had been a Cavendish researcher. Before 1945, eleven Cavendish researchers (nine in physics and two in chemistry) received Nobel Prizes. Perhaps no other research institution had been so productive during the early years of the twentieth century.

In Britain and its dominions, Cavendish alumni began to dominate academic positions in physics. In 1898, Rutherford was appointed Macdonald Professor of Physics at McGill University in Montreal, where he began the research that eventually brought him the Nobel Prize. In 1900, J. A. McClelland left the Cavendish for University College of Dublin. Also in that year, Townsend left the Cavendish to occupy the newly established Wykeham Chair of Experimental Physics at Oxford. In 1903, Barkla moved to the University of Liverpool, then in 1909 succeeded to H. A. Wilson's chair at King's College, and in 1912 became Professor of Natural Philosophy at the University of Edinburgh. In 1905, H. A. Wilson became Professor of Physics at King's College, London, then moved to McGill University, then Rice Institute of Houston, and finally to the University of Glasgow. In 1906, Richardson became Professor of Physics at Princeton University. In 1907, Rutherford returned to England to succeed Schuster at Victoria University of Manchester. In 1910, E. F. Burton was appointed to a position at his alma mater, the University of Toronto. Other Cavendish researchers found positions at a variety of colleges, universities, technical schools, and military academies. According to J.J.'s own list of "universities and learned societies in which my pupils have held professorship," Cavendish researchers occupied more than seventy academic positions in twelve countries.[87] J.J.'s leadership had been handsomely rewarded.

5.3. Organization

5.3.1. Physics Teaching in the New Century

The make-up of the Cavendish teaching staff remained quite stable throughout the first two decades of the new century. One noticeable change, however, was the establishment of a new University Lectureship in Experimental Physics. In 1900, Shaw, Assistant Director of the Cavendish and the last of the Cavendish old guard, left the Laboratory to take charge of the Meteorological Council. Once again, as he had done when Glazebrook resigned the demonstratorship, J.J. wisely used the departure of Shaw as an opportunity to increase the Laboratory's number of teaching posts.[88]

---

[87] Ibid. England, Scotland and Wales were counted as one country, Ireland another. Newfoundland was included as part of Canada.
[88] *CUR* (1 May 1900): 762.

The Assistant Directorship of the Cavendish was allowed to lapse and the stipend of £50 attached to this position was "devoted to the second University Lectureship." Thus, two new lecturers, Searle and L. R. Wilberforce, were elected.[89] The Cavendish entered the twentieth century with two university lecturerships, three demonstratorships, and two assistant demonstratorships for teaching experimental physics.

These teachers delivered lectures on heat, light, sound, mechanics, electricity, and magnetism to candidates for the MT and the NST. Courses specifically intended for medical students were regularly offered by Fitzpatrick (for more than three decades) and by Horton. For advanced students, J.J. offered courses on such subjects as structure of the atom, discharge of electricity through gases, advanced electromagnetism, and even "some applications of recent researches in Physics to Chemistry" (Michaelmas Term, 1911). H. E. Watson's careful notes of J.J.'s advanced course on "electromagnetic momentum and themodynamics of radiation with particular attention to Einstein's views, cathode rays, number of corpuscles in the atom, etc." indicates that J.J.'s lecture was quite comprehensive.[90] Searle continued Glazebrook's line of teaching on "Electrical and Magnetic Measurements," C. T. R. Wilson taught light, and Whetham regularly lectured on "Thermodynamics of Physics and Chemistry" and "Theory of Electrolytic Dissociation." It is interesting to note that no regular courses were offered on radioactivity, a field of high interest to many Cavendish researchers. Such courses were limited to a 1903 "course of eight lectures on The Chemical Aspects of Radio-activity" by F. Soddy and a 1905 and 1906 Easter Term course on "Electrolysis, Radio-activity, etc." by Whetham.[91] In addition, Larmor, who became the new Lucasian Professor in 1903, offered instruction on the mathematics and theory of electromagnetism, thermodynamics, and optics.

As the new century progressed, the number of Cambridge students attending demonstration classes at the Cavendish during the academic year steadily increased, growing from 641 in 1901-02 to 752 in 1907-08 and 842 in 1913-14. This rise demonstrated the success of physics teaching at the Cavendish, but it also exacerbated the Laboratory's chronic need for more space. As early as 1900 Searle had written Rutherford, "When I got back from Canada I found that my class had grown so much that a division had been necessary, the new section working under Skinner in the M. B. room."[92] Meanwhile the number of researchers working at the Cavendish during the summer vacation delivered a slightly different message. In the first ten years this number grew slowly until peaking at 53 in 1909 and thereafter declined, descending to 25 in 1913, an unmistakable sign that the research conditions at the Laboratory had begun to deteriorate, despite the Cavendish's flourishing physics teaching program.

---

[89] *CUR* (29 May 1900): 935; (5 June 1900): 994; (12 June 1900): 1022.
[90] Archive for History of Quantum Physics, microfilm no. 78, section 4.
[91] *CUR* (20 October 1903): 67.
[92] CUL MSS ADD 7653 S49 (25 March 1900), Searle to Rutherford.

The success and stability of the physics teaching program in the early decades of the twentieth century owed greatly to two teachers, Searle and C. T. R. Wilson. Both taught at the Cavendish for decades and were responsible for giving instruction in basic knowledge and technique to several generations of students who were more than often "absolutely fresh to the [experimental] work."[93] Just as Glazebrook and Shaw had symbolized Cavendish teaching in the nineteenth century, Searle and C. T. R. Wilson symbolized the continuation of Cavendish teaching into the new century.

Searle was the Laboratory's most influential teacher for several generations.[94] He had been appointed Demonstrator in 1890 and University Lecturer in 1900. Although "Searle's Class" was the practical class for the NST, Part I, his teaching methods were anything but ordinary. As George P. Thomson recalled:

Figure 5.6. George Frederick Charles Searle and Charles Thomson Rees Wilson. They trained several generations of Cavendish men and women [Courtesy of the Cavendish Laboratory].

[93] CUL MSS ADD 7653 S50 (14 September 1905): Searle to Rutherford. Searle wrote that he had been busy with the Long Vacation class because it had 25 pupils with no previous experience.

[94] Searle was the twenty-eighth wrangler in 1887, and he took the Part II of the NST following year. He entered the Cavendish in 1888 and stayed there for the next 55 years. For more details of his life and works, see George P. Thomson, "George Frederick Charles Searle, 1864-1954," *Biographical Memoirs of Fellows of the Royal Society 1* (1955): 246-252.

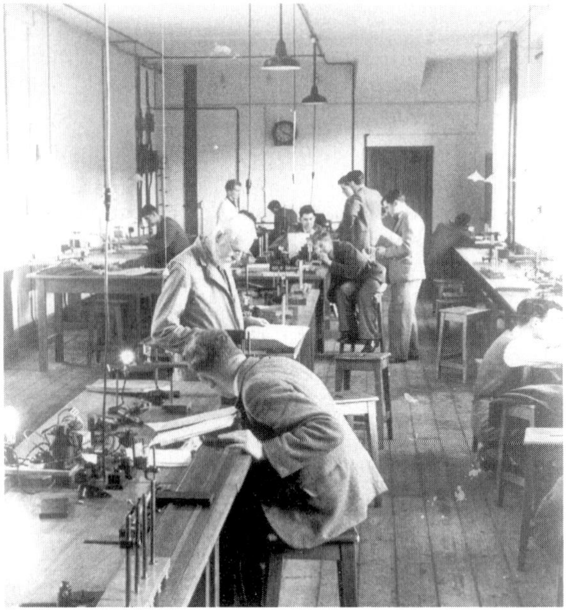

Figure 5.7. Searle examining a student notebook during his Practical Class [Courtesy of the Cavendish Laboratory].

> He showed a combination of the practical and the ideal which seems odd to the normal modern man, and perhaps unusually so to the normal physicist, but which is, one is told, characteristic of the mystic. His experiments aimed at illustrating principles. They did this, but they also showed a good deal of ingenuity in design . . . Certainly Searle's apparatus and methods have been adopted very widely and have influenced the teaching of physics all over the world.[95]

In addition, Searle's classes were always well-organized:

> Searle set a pattern by producing laboratory sheets of instructions, giving sufficient theory to make the objective clear. The experiments were always properly tried out and it was customary for him to include the results obtained by a student who had worked on the experiments for him . . . Searle was very insistent on proper note taking. He expected the student to have a notebook for recording readings and another book for the written up account of the experiment.[96]

Searle's practical class was characterized by short but clear lessons in mathematics and theory, precise measurements, and the use of simple apparatus. These practical elements of his teaching method influenced the research styles of many of the researchers he trained.

Searle's characteristic style of teaching was based on the performance of "real"

---

[95] G. P. Thomson, "G. F. C. Searle," 249.

[96] A. J. Woodall & A. C. Hawkins, "Laboratory Physics and its Debt to G. F. C. Searle," *Physics Today 4* (1969): 283-285.

experimentation, as is vividly illustrated in his laboratory manuals for the study of elasticity, harmonic motion, and optics. In 1908 he published his first textbook, *Experimental Elasticity*, which served as the model for those that followed. In the preface to this book, Searle discussed the evolution of his teaching methods:

> When, in 1890, I was appointed to my present post of Demonstrator in Experimental Physics, I found that, the then existing text-books of practical physics did not entirely meet the needs of the students, partly because they did not, as a rule, show how the formulae required in the experimental work are derived from the principles of the subject. The students themselves added to the difficulty, for their ideas as to those principles were often indistinct. I was thus led to devise some experiments intended to illustrate principles as simply and directly as possible. I also wrote notes explaining how the necessary formulae are obtained from the principles involved in those experiments and describing in detail how the practical work is to be conducted. The students showed very kind appreciation of these earlier notes and thus I was encouraged to prepare others; this work has proved so interesting that I have continued it, as opportunities have occurred, with the result that at the present time the students attending my practical class rely mainly upon these manuscript notes for the necessary instructions.[97]

The structure Searle developed for *Experimental Elasticity* was ingenious. He devoted its first chapter to elementary principles of elasticity, including Hooke's law, stress, strain, expansion and compression, shear, rigidity, Young's modulus, Poisson's ratio, isothermal and adiabatic elasticities, etc. In its second chapter, he provided readers with "the mathematical solutions of some important problems, since these solutions are required in connexion with several experiments." The purpose of the second chapter was to "endeavour to indicate the point at which assumptions are made and to consider the difficulties which arise."[98] In the third chapter, Searle provided fourteen experiments, each of which contained sections for description of apparatus, experimental procedure, mathematical discussion if necessary, and practical example. "Though, in nearly every case, the apparatus is so simple that it may be constructed by any person who is moderately skilled in the use of the tools," Searle commented, "yet the experimental methods, when carried out with care, are capable of yielding definite results."[99] Searle ended his textbook of fewer than 200 pages with a few mathematical notes and "Hints on Practical Work in Physics."

A simple comparison of Searle's *Experimental Elasticity* with Glazebrook and Shaw's legendary *Practical Physics* illustrates the superiority of Searle's book in terms of practicality. Whereas Glazebrook and Shaw's text had few diagrams, Searle's text provided many detailed figures of apparatus. Searle's "Practical examples" with real experimental data also made his textbook much more serviceable. For example, for the determination of Young's modulus, Glazebrook and Shaw offered a method but no figures or actual data, while Searle offered two

---

[97] G. F. C. Searle, *Experimental Elasticity: A Manual for the Laboratory* (Cambridge: Cambridge University Press, 1908), v.

[98] Ibid., 30.

[99] Ibid., 71.

Experiment.—Determine the modulus of elasticity for the material of the given wire

Enter results thus :—

Length of unstretched wire . . . 718·53 cm.
Extension per kilogramme (mean of 4 observations) . . . . . . ·0575 „
Cross-section . . . . . . ·01 sq. cm.
Value of E 1,250,000 kilogrammes per. sq. cm.

54. **Practical example.** The observations may be entered as in the following record of an experiment made on a brass wire.

Length of wire from support to clamp of vernier $= L = 745$ cm.

Readings of screw-gauge on wire

0·0944 0·0943 0·0947 0·0945
0·0943 0·0943 0·0945 0·0945 } mean reading 0·0944 cm.

Correction for zero 0·0002 cm., to be added.

Hence diameter $= 2a = 0{\cdot}0946$ cm.

Cross section $= A = \pi a^2 = \pi \times (0{\cdot}0473)^2 = 0{\cdot}00703$ square cm.

Readings of vernier for increasing and diminishing loads :—

Load grammes	Loading cm.	Unloading cm.	Mean cm.	Extension cm.	$\frac{\text{Extension}}{\text{Load}}$ cm. grm.$^{-1}$
0	1·510	1·515	1·513	0	—
1000	1·625	1·625	1·625	·112	$1{\cdot}12 \times 10^{-4}$
2000	1·730	1·735	1·733	·220	1·10
3000	1·835	1·840	1·838	·325	1·08
4000	1·945	1·945	1·945	·432	1·08
5000	2·045	2·050	2·048	·535	1·07
6000	2·150	2·155	2·153	·640	1·07
7000	2·260	2·260	2·260	·747	1·07
8000	2·355	2·360	2·358	·845	1·06
9000	2·465	—	2·465	·952	1·06

Mean $1{\cdot}079 \times 10^{-4}$

When the extension was plotted against the load, it was found that the points lay very nearly on the straight line cutting the line $M = 0$ at 0·006 cm. and the line $M = 5000$ at 0·533 cm. Hence, for $M = 5000$ grm., $l = 0{\cdot}527$ cm. The corresponding value of l/M, viz. 0·527/5000 or $1{\cdot}054 \times 10^{-4}$ differs by 2·5 per cent. from the mean value of l/M derived from the table. The zero reading 1·513 cm. in the table is clearly abnormal; possibly the weight of the empty pan is insufficient to ensure that the wire is straight. If we treat the reading 1·733 cm., which was found for a load of 2000 grammes, as the zero reading, we find the following values of l/M :—

1·050, 1·060, 1·050, 1·050, 1·054, 1·042, $1{\cdot}046 \times 10^{-4}$.

The mean, $1{\cdot}050 \times 10^{-4}$, now agrees closely with the value derived from the diagram. Using the value $1{\cdot}054 \times 10^{-4}$ for l/M, we have, by § 53,

$$\text{Young's modulus} = E = \frac{MgL}{lA} = \frac{981 \times 745}{1{\cdot}054 \times 10^{-4} \times 0{\cdot}00703}$$
$$= 9{\cdot}86 \times 10^{11} \text{ dynes per square cm.}$$

Figure 5.8. Experiment on Young's modulus in Glazebrook-Shaw's Practical Physics, p. 144 (left); and in Searle's Experimental Elasticity, pp. 85-86 (right).

different methods, together with figures of the apparatus employed, a detailed list of procedures, and actual data from the record of "an experiment made by Mr. T. G. Bedford upon a brass wire" (Figure 5.8).[100] Searle's next two textbooks, *Harmonic Motion* (1915), and *Optics* (1925) also provided superior structure and clarity.[101]

C. T. R. Wilson, who became a Cavendish demonstrator and University lecturer in 1900, was another influential source of inspiration to students. W. Lawrence Bragg, who came to Cambridge in 1909, recalled that C. T. R's lectures on optics were:

> the best, and the delivery was the worst, of any lectures to which I have ever been. He mumbled facing the board, he was very hesitant and jerky in his delivery, and yet the way he presented the subject was quite brilliant. I think his lectures on optics set the standard for similar lectures all over the country when later his pupils got chairs . . . It

[100] Glazebrook and Shaw, *Practical Physics*, 141-144; Searle, *Experimental Elasticity*, 80-90.

[101] The publication of *Experimental Optics* was originally intended to follow *Experimental Elasticity* "in a few months." However, Searle became ill in 1910 and recovered only in 1915. To make a fresh start, he decided "to take the rather easier course of publishing the work done in my class in Experimental Harmonic Motion." See *Experimental Harmonic Motion* (Cambridge: Cambridge University Press, 1915), vi, and *Experimental Optics* (Cambridge: Cambridge University Press, 1925), vi.

> was C. T. R.'s treatment of a grating diffracting white light, which he had given us in his lectures, which set me thinking on the right lines when I gave a simpler treatment in November 1912 of Laue's diffraction experiments. You will see how much I owe to him.[102]

P. M. S. Blackett, who would receive the 1948 Nobel Prize in Physics for "his development of the Wilson cloud chamber method, and his discoveries therewith in the fields of nuclear physics and cosmic radiation," gave a similar account of C. T. R., the teacher:

> I first met C. T. R. Wilson just after World War I when I attended his lectures on light. His voice was not easy to follow and his writing on the blackboard was difficult to read, but somehow I took adequate notes and these are almost the only lecture notes of my student days to which I have repeatedly returned. He had a penetrating but very simple approach to wave phenomena, in particular to interference and diffraction effects, which he treated by the elegant use of amplitude phase diagrams. By these methods he conveyed a deep physical understanding of the Abbé theory of image formation and the closely related general relationship between the Fourier expansion of the light from a periodic object and the intensity of the resulting spectra . . .
>
> C. T. R.'s influence on the teaching of experimental physics in Cambridge was also profound: for many years he was in charge of the Third Year experimental class in the Cavendish Laboratory and early introduced the method giving the students a few simple but searching experimental tasks, which demanded considerable experimental skill and had the character of minor research problems each requiring many weeks of work. This method was then in considerable contrast to the more usual practice of making the students, even in their final year, performed numerous rather stereotyped experiments. For many years he gave devoted attention to the students who passed through the senior laboratory of the Cavendish and inspired them with a deep and lasting affection.[103]

As these recollections indicate, C. T. R. Wilson's influence on students was somewhat different from Searle's. C. T. R. was a powerful researcher pioneering a new area of science with a new instrument he developed. His teaching therefore interacted closely with his research, which attracted a small number of talented students. In this respect, he was very similar to his mentor, J.J. Thomson. C. T. R.'s stimulating classes complemented Searle's well-organized ones. Despite their different teaching styles, they shared a common love of simple apparatus, a hallmark of the Cavendish under J.J.'s directorship.

5.3.2. Finance

As in the previous century, finance remained a serious problem at the Cavendish, although the Laboratory's total revenue grew from about £2200 in 1900 to about £3500 in 1912.[104] Nearly 90% of the Cavendish's total receipts during the first years of the new century came from fees paid by students for attending lectures and demonstrations, which, owing to the ever-increasing number of students attending

[102] Crowther, *The Cavendish Laboratory,* 167.
[103] P. M. S. Blackett, "Charles Thomas Rees Wilson, 1869-1959," *Biographical Memoirs of Fellows of the Royal Society 6* (1960): 268-295 on 292-293.
[104] *CUR* (18 March 1901): 93; (18 March 1913): 114-115.

demonstration classes, rose from £1864 in 1900 to £2125 in 1902, £2578 in 1906, and £3325 in 1911. The University's annual grant for wages and maintenance, however, remained nearly unchanged from the first days of the Cavendish and, of the total receipts, only about 2% were contributed by interest from the Lord Rayleigh's Apparatus Fund and other stocks and by examination fees. The dependence of teaching and research at the Cavendish on lecture and demonstration fees is evident in its expenditures at that time. The lion's share of the total went to lecturers and demonstrators. Nevertheless, the Cavendish could devote, on average, one fourth of its total expenditure to the purchase of apparatus every year. J.J. wisely saved each year's surplus of £100 to £500 for future construction of new buildings for teaching and research.

In short, the Laboratory's finances depended almost exclusively on lecture and demonstration fees, and it had no reliable independent financial sources for research purposes. A few lucky researchers, including Whetham, H. A. Wilson, Richardson, and Kleeman, obtained grants from the Royal Society, but such cases were very rare and the amount received often was very small. Before World War I, J.J. later recalled, the British Government annually provided £4000 for research "which was administered by the Royal Society and had to suffice for the needs of all the sciences, so that there was not much available for any particular science."[105]

In the new century, the Clerk Maxwell Scholarship continued to play an important role in supporting research students. Recipients of this scholarship included H. A. Wilson (1901), O. W. Richardson (1904), F. Horton (1906), E. M. Wellisch (1909), R. T. Beatty (1910), R. D. Kleeman (1910), and F. W. Aston (1913). Except for Richardson, these scholarship recipients were advanced students from outside: H. A. Wilson, Horton, Kleeman, and Beatty had been 1851 Research Scholars; Wellisch came to the Cavendish from Australia as an advanced student; and Aston entered the Cavendish as an assistant to J.J. and only later became an advanced student.

Considering the Cavendish's slim financial resources, providing more room to accommodate more research was a major challenge. In 1903, as part of his annual report, J.J. noted that "the difficulty of finding sufficient accommodation for those desirous of participating in original work has become very acute," and he asked for the conversion of two rooms in the Porter's Lodge for that purpose.[106] In the 1904 annual report he wrote, "The need for more space for research is now urgent, and, if the usefulness of the Laboratory is to be maintained additional accommodation will have to be provided at no distant date."[107]

When the University failed to provide the requested space, the Cavendish's chronic space problem was relieved by a generous gift from Lord Rayleigh. In October of 1905, J.J. reported to Rutherford, "I hardly know how we shall get the people in but we shall have more rooms soon as Lord Rayleigh has given us £5000

---

[105] J.J. Thomson, *Recollections,* 126.
[106] *CUR* (6 June 1903): 931.
[107] *CUR* (22 June 22 1905): 1176.

to build a new research Laboratory."[108] In May 1906, the Museum and Lecture Rooms Syndicate officially recommended the extension of the Cavendish Laboratory as initiated by "Lord Rayleigh's gift of £5000, part of the Nobel Prize."[109] To this generous gift, J.J. added £2000 of accumulated fees. The £8000 total cost of the new building, the Rayleigh Wing, was slightly higher than originally expected, but J.J. cleverly managed the difficulties and won University approval for additional support of £738. 8s. 7d.[110] Lord Rayleigh, then Chancellor of the University, opened the Wing named after himself on June 16, 1908, the 34th anniversary of the Laboratory's opening. The Rayleigh Wing included "a large room for students engaged in original research," two photographic darkrooms, a large lecture room with seating for 120, and "a number of small rooms for private research."[111] The new building not only provided for "the comfort of those engaged in original research" but also enabled "Mr. Whetham's Lectures on Physics," previously conducted at Trinity College, "to be given at the Laboratory."[112] Thus, the Cavendish's entire lecture series could be delivered within the Laboratory "for the first time for many years."[113]

5.3.3. Instruments

In the new century, the Cavendish continued its tradition of using *simple* but *efficient* apparatus. Most research instruments were constructed in the Laboratory either by researchers or workmen. In his annual reports, J.J. often recorded that "a large quantity of apparatus has been made in the workshops of the Laboratory."[114] Thus, the Cavendish's lecture assistants and workmen played the important role of supplying most of the necessary instruments and apparatus used in the facility for teaching and research. In 1899, the Cavendish's chief assistant, W. G. Pye, left the Laboratory to establish his own company for manufacturing scientific instruments. F. Lincoln, who had joined the Cavendish as "an exceedingly small boy in 1892," succeeded Pye in this position.[115] Two lecture assistants, W. H. Hayles and J. Rolfe, continued their service to the Cavendish in the new century, as did J.J.'s private assistant, Everett. These four men worked in the Cavendish for many years and, as J. G. Crowther correctly indicated, "the continuity of the workshop characteristics was one of the strands in the Cavendish tradition."[116]

J.J.'s appointment as Professor of Natural Philosophy at the Royal Institution included the services of an assistant, and some services of this assistant benefited the Cavendish. With the consent of the Institution's manager, this assistant not only

---

108 CUL MSS ADD 7653 T27 (1 October 1905): J.J. to Rutherford.
109 *CUR* (22 May 1906): 944-945.
110 *CUR* (26 February 1907): 606; (19 January 1909): 480-481.
111 *CUR* (26 February 1907): 606.
112 *CUR* (17 June 1909): 1238-1239.
113 *A History of the Cavendish Laboratory*, 13.
114 *CUR* (22 June 1905): 1176.
115 *A History of the Cavendish Laboratory*, 82.
116 Crowther, *The Cavendish Laboratory*, 111.

Figure 5.9. F. W. Aston working with his mass spectroscope [Courtesy of the Cavendish Laboratory.

helped J.J. prepare lectures for the Institution, he also helped J.J. with his research at the Cavendish. J.J.'s first Institution assistant was G. W. C. Kaye, future superintendent of the physics department of the National Physical Laboratory. Kaye frequently assisted the Professor in his private research, which J.J. acknowledged in several of his papers.

In 1910, Kaye was succeeded by F. W. Aston, who had been recommended by J.J.'s old friend Poynting, Aston's former teacher at Mason College. The decision to hire Aston was, as G. Hevesy pointed out:

> not only most fortunate for Aston, but for the great physicist [J.J.] himself, for the Cavendish Laboratory, for Trinity College which Aston entered when coming to Cambridge, and for the speedy development of natural sciences which we have witnessed in the last thirty years.[117]

Aston had very dexterous hands and he made several ingenious improvements to apparatus used for positive ray research that were crucial to J.J.'s research on positive rays after 1910. After the end of World War I, Aston made further

---

[117] G. C. de Hevesy, "Francis William Aston, 1877-1945," *Obituary Notices of Fellows of the Royal Society 5* (1945-1948): 634-650 on 637. See also G. P. Thomson, "Dr. Francis William Aston, F. R. S.," *Nature 157* (1946): 290-292.

developments to the parabola technique that he and J.J. had cultivated before the War. The result was the invention of the mass spectrograph and the Nobel Prize in Chemistry for Aston in 1922.

In addition to devoting funds to the in-house construction of teaching and research apparatus, each year J.J. devoted a few hundred pounds of lecture fees to the purchase of apparatus manufactured outside the Cavendish, generally more complex than the in-house variety. Grants and gifts also helped supply the Cavendish with apparatus. The Gordon Wigan Fund made frequent grants to the Cavendish for this purpose, funding, for example, "an induction coil and an interferometer," "a large automatic air pump," "a gaede molecular pump," and "a constant deviation spectrometer."[118] In 1902, J.J. reported that the Council of Newnham College loaned the Laboratory "a large and valuable collection of instruments" which, J.J. predicted, would be "of great service to the Department."[119] In 1904, Rev. T. C. Fitzpatrick contributed the Laboratory's liquid air plant, and, in 1908, G. F. C. Searle and W. G. Pye donated "one of their new type of spectrometers."[120] One instrument even came from the Japanese Commission, which, J.J. reported, presented the Laboratory with "a valuable Heliostat made by Noguchi" in 1910.[121]

Searle and C. T. R. Wilson, as the principal teachers of the Cavendish's practical classes, were two major figures in the design and manufacture of the Laboratory's apparatus. Searle designed a variety of simple instruments, including an effective rheostat "for accurate adjustment," vibration magnetometers and Robison magnets, an apparatus for determining the curvature of a spherical surface for use "in one of the classes at the Cavendish Laboratory," an apparatus for measuring the surface tension of soap film, a goniometer "devised . . . in conjunction with W. G. Pye and Co.," and others.[122] Many of Searle's designs were manufactured, under his authorization, by Pye's instrument company at Cambridge (see Figure 5.10).

This represented a serious challenge to the Cambridge Scientific Instrument Company which produced many of the necessary instruments for the practical classes at the Cavendish and elsewhere. The company however secured other teachers, most notably C.T.R. Wilson, for the competition with Pye's company. In the early 1900s, C. T. R. constructed a sensitive gold-leaf electrometer, a tilted electrometer, and a micro-electroscope, which soon became the most frequently

---

[118] *CUR* (15 June 1906): 1161; (8 June 1908): 1088; (28 July 1913): 1416; (5 March 1915): 649.

[119] *CUR* (28 May 1902): 955.

[120] *CUR* (10 June 1904): 949; (17 June 1909): 1239.

[121] *CUR* (27 June 1911): 1338.

[122] G. F. C. Searle, "On a simple Rheostat," *Phil. Mag. 6* (1903): 173-175; "Notes on a Vibration Magnetometer and on the Ball-ended Magnets of Robison," *Proc. Camb. Phil. Soc. 12* (1902): 27-33; — & A. C. W. Aldis, & G. M. B. Dobson, "On a Revolving Table Method of determining the Curvature of Spherical Surface," *Phil. Mag. 21* (1911): 218-224; "Some Methods of measuring the Surface Tension of Soap Film," *Proc. Camb. Phil. Soc. 17* (1913): 285-299; "The Determination of the Focal Length of a thick Mirror," *Proc. Camb. Phil. Soc, 18* (1915): 115-126.

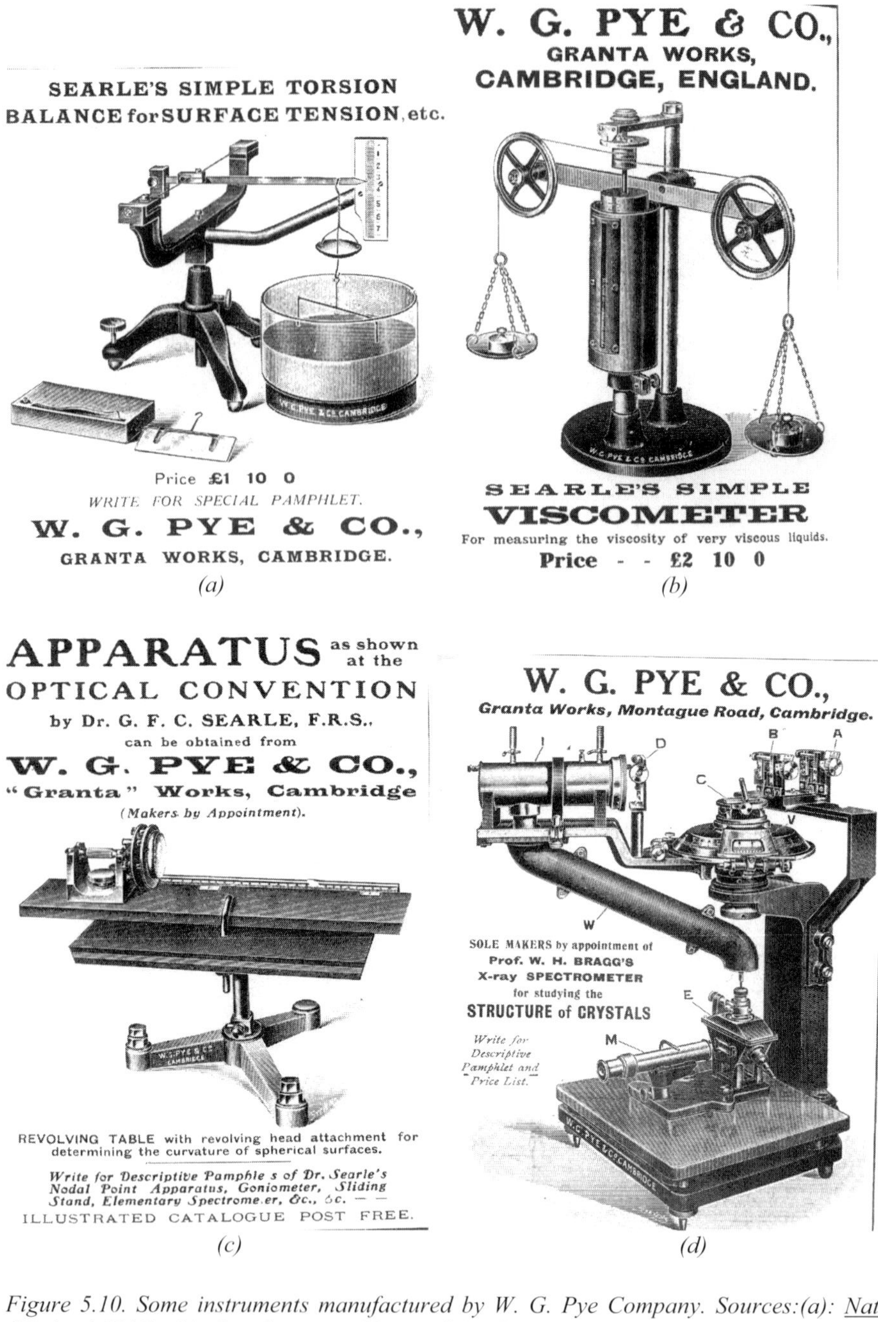

Figure 5.10. Some instruments manufactured by W. G. Pye Company. Sources:(a): Nature (25 April 1912); (b): Supplement to Nature (31 July 1913); (c): Nature (27 June 1912); (d): Supplement to Nature (19 November 1914)].

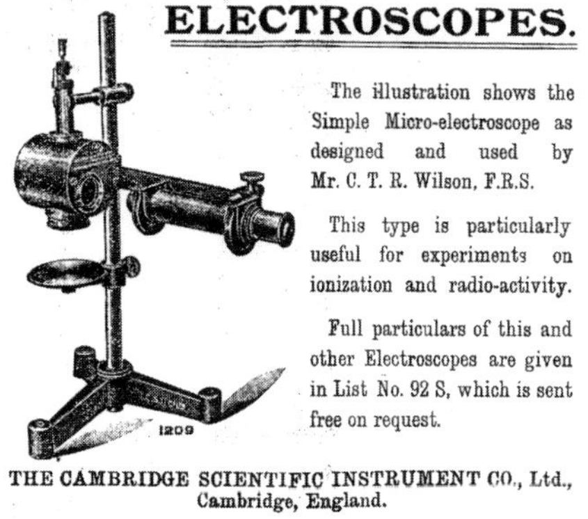

(a)

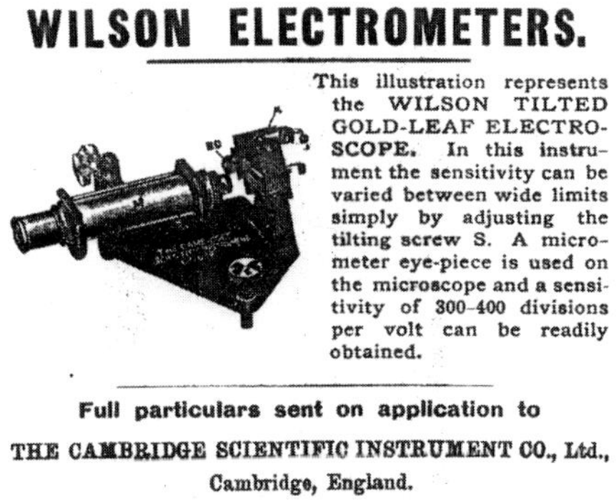

(b)

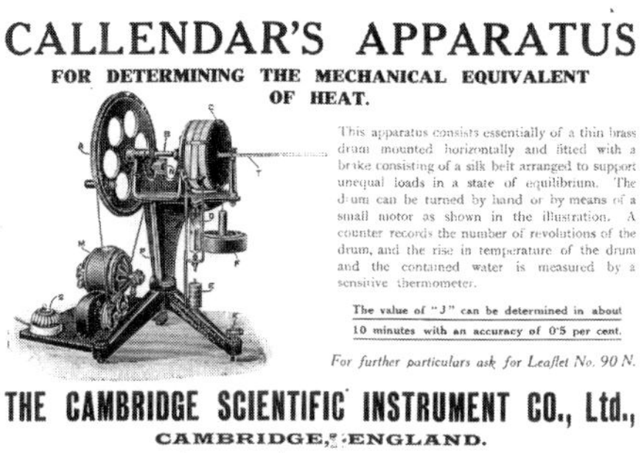

(c)

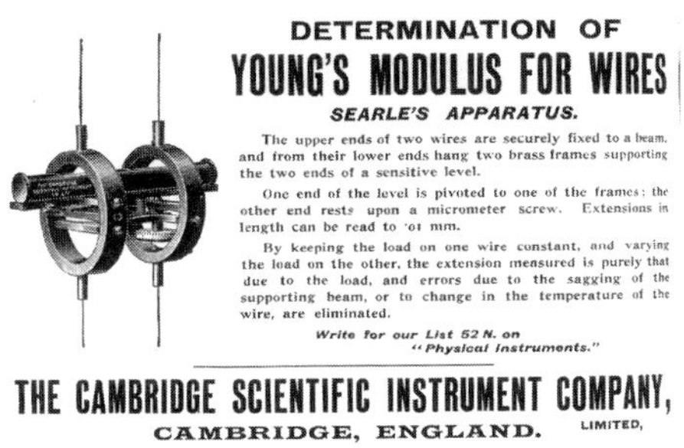

(d)

(e)

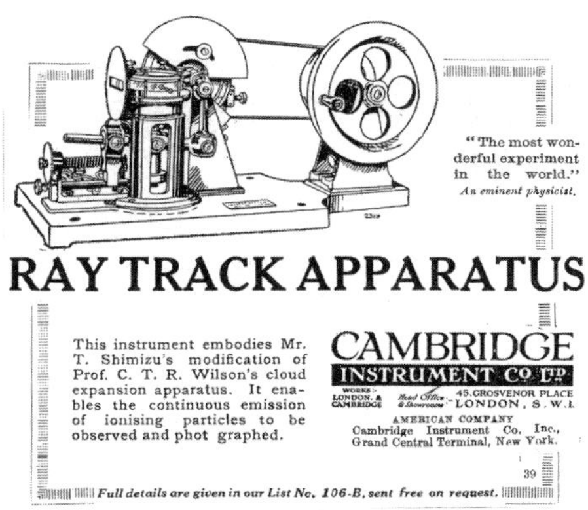

(f)

Figure 5.11. Instruments manufactured by the Cambridge Scientific Instrument Company. Sourcse: (a): Phil. Mag. 25 (1913); (b): Phil. Mag. 26 (1913); (c): Supplement to Nature (May 1913); (d): Supplement to Nature (19 May 1910); (e): Supplement to Nature (15 May 1913); (f): Phil. Mag. 1 (1926).

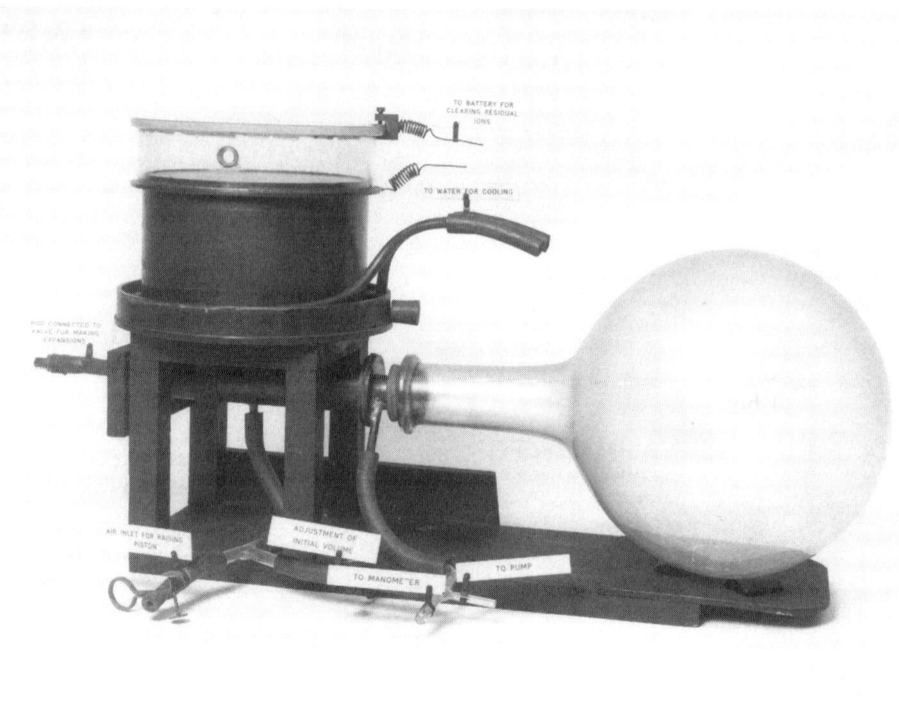

Figure 5.12. C.T.R.Wilson's original Cloud Chamber, the magic instrument [Courtesy of the Cavendish Laboratory].

used instruments at the Laboratory.[123] The Cambridge Scientific Instrument Company quickly commercialized these instruments, and, later manufactured Wilson's Cloud Chamber and T. Shimizu's modified version of the cloud chamber for commercial distribution (Figure 5.11.f).

Other researchers also designed apparatus, and some investigated theory concerning apparatus. G. W. Walker wrote on the theory of the quadrant electrometer.[124] T. H. Laby made a string electrometer, and E. M. Wellisch devised a detector for electromagnetic waves.[125] Aston designed a simple micro-balance to determine the densities of small quantities of gases.[126] A. E. Oxley devised an instrument to produce circularly polarized light "with the idea of combining as far as possible the advantages of the Fresnel rhomb and the quarter-wave plate in one

---

[123] C. T. R. Wilson, "On a Sensitive Gold-Leaf Electrometer," *Proc. Camb. Phil. Soc. 12* (1903): 135-139. See also *A History of the Cavendish Laboratory*, 235-236.

[124] G. W. Walker, "On the Theory of the Quadrant Electrometer," *Phil. Mag. 6* (1903): 238-250.

[125] T. H. Laby, "A String Electrometer," *Proc. Camb. Phil. Soc. 15* (1909): 106-113; E. M. Wellisch, "An Electric Detector for Electromagnetic Waves," *Proc. Camb. Phil. Soc, 15* (1909): 337-339.

[126] F. W. Aston, "A Simple Form of Micro-Balance for determining the Densities of Small Quantities of Gases," *Proc. Roy. Soc. 89* (1914): 439-446.

piece of apparatus."[127] The instruments designed at the Cavendish embodied that institution's hallmark of simple efficiency at low cost. Aston's remarks about his micro-balance clearly delivered the essence of this hallmark: "In conclusion, I may state that the instrument herein described may be assembled in a comparatively short time by any skilled glass-worker, and, thanks to the very small quantity of material used, at a cost of a few shillings."[128]

The primary example of Cavendish-style apparatus of the period was C. T. R. Wilson's Cloud Chamber. Rutherford once called it "the most original and wonderful instrument in scientific history."[129] With this instrument, researchers could photograph the tracks of ions, meaning that they could "see" the traces of these particles. Despite its almost magical power, Wilson's Cloud Chamber was comprised by very simple elements: a glass cylinder for the cloud chamber, a big glass ball for evacuation, several glass tubes, a few India-rubber components, wooden and brass cylinders, and some gelatin (Figure 5.12).[130] The cost of this magical chamber was "only about £5."[131] Wilson's 1911 Cloud Chamber was the modified version of an earlier model that he had used since 1895. The new model was larger, "wide enough to give ample room for the longest α-ray, and high enough to admit of a horizontal beam of X-rays being sent through it without any risk of complications due to the proximity of the roof and floor."[132] However, the most crucial addition to the instrument was a method for photographing the tracks of "even the fastest β-particles, the individual ions being rendered visible."[133] This photographic method had been developed in 1885-86 by A. M. Worthington, a former Cavendish researcher.[134] The elegant simplicity of the Cloud Chamber well reflected both Wilson's artistic skill and the Cavendish's financial stringency.

The invention of the cloud chamber and the resulting photographs of tracks of α-, β-, γ-, and X-rays through gases soon impacted theoretical studies of radiation

---

[127] A. E. Oxley, "On Apparatus for the Production of Circularly Polarized Light," *Phil. Mag. 21* (1911): 517-532 on 517.

[128] F. W. Aston, "A Simple Form of Micro-Balance for determining the Densities of Small Quantities of Gases," 446.

[129] "Professor C. T. R. Wilson (Obituary)," *The Times* (16 November 1959): 16. For more about Wilson's cloud chamber, see David G. Tomas, *Tradition, Context of Use, Style and Function; Expansion Apparatus used at the Cavendish Laboratory during the period 1895-1912,* unpublished master thesis (University of Montreal, 1979); P. Galison & A. Assmus, "Artificial Clouds, Real Particles," in D. Gooding, T. Pinch & S. Schaffer (eds.), *The Use of Experiment: Studies in the Natural Sciences* (Cambridge: Cambridge University Press, 1989), 225-274; Clinton Chaloner, "The Most Wonderful Experiment in the World: a History of the Cloud Chamber," *BJHS 30* (1997): 357-374; and P. Galison, *Image and Logic*, chapter 2: Cloud Chambers: The Peculiar Genius of British Physics.

[130] For the theory and construction of the chamber, see C. T. R. Wilson, "On a Method of making Visible the Paths of Ionising Particles through a Gas," *Proc. Roy. Soc. 85* (1911): 285-288; "On an Expansion Apparatus for making Visible the Tracks of Ionising Particles in Gases and Some Results obtained by its Use," *Proc. Roy. Soc. 87* (1912): 277-292.

[131] Crowther, *The Cavendish Laboratory,* 165.

[132] C. T. R. Wilson, "On an Expansion Apparatus," *Proc. Roy. Soc. 87* (1912): 277.

[133] Ibid.

[134] For the method of illuminating and photographing the tracks, see ibid., 279-282. For Worthington's work of fast photograph, see A. M. Worthington, *A Study of Splashes* (London: Longmans, Green, 1908).

and atomic structure. When, in 1911, C. T. R. showed W. H. Bragg his first photographs of "the cloud formed on ions due to X-rays," Bragg immediately realized that they presented long-awaited evidence supporting his neutral-pairs hypothesis of X-rays over Thomson-Barkla's impulse hypotheses.[135] In his 1911 paper, Wilson noted that X-rays, when ionizing gases, made short but distinctive tracks, "small tread-like objects not more than a few millimetres in length, and many of them being considerably less than 1/10 mm. in breadth," and he concluded that "the results are in agreement with Bragg's view that the whole of the ionisation by X-rays may be regarded as being due to β- or cathode-rays arising from the X-rays" (see "Fig. 2. Cloud formed on Ions due to X-Rays," in Figure 5.13).[136]

Wilson's 1912 photographs of α-ray tracks also lent strong support to Rutherford's nucleus model of the atom (Figure 5.14). In 1911, Rutherford had proposed his nucleus atomic model based on experiments investigating the scattering of α-particles.[137] Based on Rutherford's calculations, Bragg constructed a theoretical speculation of the path of α-particles.[138] Surprisingly, Bragg's hypothetical lines coincided with a photograph of α tracks that C. T. R. provided in his 1912 paper, persuading many to accept Rutherford's nucleus model of the atom (See illustrations 2 and 3 of Figure 5.14). C. T. R. himself recognized the importance of this photograph and embraced the nucleus hypothesis. "As Rutherford has contended," he wrote, "the scattering of large amounts is in the case of α-particles mainly due to the former process, that is to say, it is *the result of the single deflections through considerable angles and not a cumulative effect due to a very large number of minute deviations*."[139] Rutherford, in his 1914 paper on the structure of the atom, rebuked the idea that the large deflection of α-particles resulted from the accumulation of many encounters within the atom:

> It is of interest to note that C. T. R. Wilson, by photographing the trails of the α particle, later showed that the α particle occasionally suffers a sudden deflexion through a large angle. This affords convincing evidence of the correctness of the view that large deflexions do occasionally occur as a result of an encounter with a single atom.[140]

After the end of World War I, use of Wilson's magical chamber rapidly spread to physics laboratories throughout the world but, as Clinton Chaloner pointed out, Cavendish researchers were the first and major group to employ Wilson's photographs and cloud chamber research, perhaps because these researchers

---

[135] Wheaton, *Tiger and Shark*, 165.

[136] C. T. R. Wilson, "On a Method of a making Visible the Paths of Ionising Particles through a Gas," 287-288.

[137] E. Rutherford, "The Scattering of α and β Particles by Matter and the Structure of the Atom," *Phil. Mag. 21* (1911): 669-688.

[138] W. H. Bragg, "Radioactivity as a Kinetic Theory of a Fourth State of Matter," *Archives of the Roentgen Ray* (April, 1911): 402-415.

[139] C. T. R. Wilson, "On an Expansion Apparatus," 284. Emphasis added.

[140] E. Rutherford, "The Structure of the Atom," *Phil. Mag. 27* (1914): 488-498 on 490.

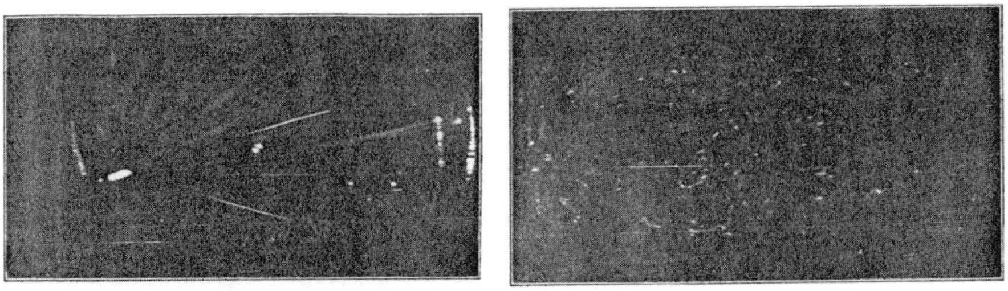

Fig. 1.
Cloud formed on Ions due to α-Rays.

Fig. 2.
Cloud formed on Ions due to X-Rays.

Figure 5.13. Pictures from the Cloud Chamber. From "On a Method of making Visible the Paths of Ionising Particles through a Gas," <u>Proc. Roy. Soc. 85</u> (1911): 258-288.

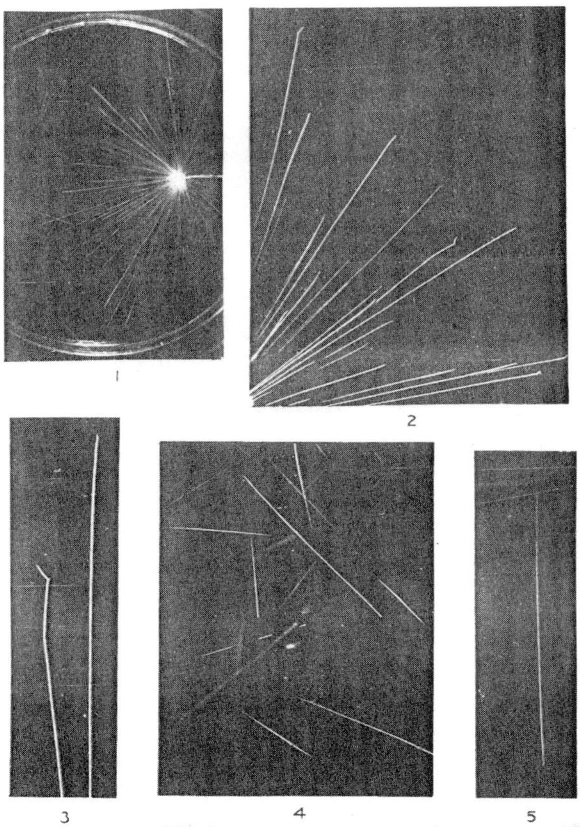

Figure 5.14. Ionisation by α-rays. From "On an Expansion Apparatus for making Visible the Tracks of Ionising Particles in Gases and Some Results obtained by its Use," <u>Proc. Roy. Soc. 87</u> (1912): 277-292.

"shared a familiarity with Cavendish methods . . . and would certainly have been acquainted with the early form of the cloud chamber."[141] At the Cavendish, T. Shimizu modified the apparatus in which the expansion took place by mechanically moving the piston with a reciprocating motion, and he also added a special cinematographic camera to "take a great number of photographs of α-ray tracks . . . to find some evidence to indicate the disruption of atoms by the α-particles."[142] After Shimizu returned to Japan, Blackett took charge of further development of the apparatus, providing yet another example of the workings of semi-independent research subgroups at the Cavendish.

5.4. Researchers and Researches

During the early years of the new century, the Cavendish Laboratory was the true mecca of experimental physics, attracting talented researchers from all over the world. Cavendish researchers produced important studies on ionization, radiation, and the structure of the atom, and the quip that "at the Cavendish there was more physics done to the square centimetre than at any other place in the world" had become more truth than joke.[143]

The demography of the Cavendish researchers changed rapidly, ultimately completing the Cavendish's transformation from a university laboratory to a world research center. The number of researchers "engaged in Original Research in the Laboratory" increased steadily, climbing from 27 in 1900-01 to 35 in 1905-06, and reaching "about 40" in 1912-13, according to J.J.'s annual reports.[144] The Cavendish annual photographs, usually taken in May or June, recorded increasingly larger groups: 15 researchers in 1900, 26 in 1905, and 31 in 1912.[145] In the fifteen years between 1895, when the University first opened its doors to the outside world, and 1910, the number of Cavendish researchers tripled; in the ten years between 1900 and 1910, the number almost doubled (see Figure 5.15).

Significantly, among the Cambridge graduates working at the Laboratory, the NST and the MT-NST graduates became the absolute majority. In the first fourteen years of the new century, only one Cavendish researcher, J. C. M. Garnett, took both Parts I and II of the MT. About ten researchers, including W. L. Bragg, G. I. Taylor, and G. P. Thomson, chose to combine Part I of the MT and Part II of the

---

[141] Clinton Chaloner, "The Most Wonderful Experiment in the World," 371-372.

[142] Takeo Shimizu, "A Preliminary Note on Branched α-Ray Tracks," *Proc. Roy. Soc. 99* (1921): 432-435 on 432.

[143] *A History of the Cavendish Laboratory*, 270.

[144] *CUR* (5 June 1901): 1030; (15 June 1906): 1161; (28 July 1913): 1416. There are three ways to count the number of Cavendish researchers during this period: the first one is to rely on J.J.'s annual reports; the second is to count the number of researchers represented in the annual photographs; and the last is to count the number in the List of the Cavendish researchers in *A History of the Cavendish Laboratory, 1871-1910*.

[145] The number of researchers represented in the annual photos is smaller than the number reported in J.J.'s annual reports mainly because Cambridge graduates at the Laboratory seldom appeared in the annual photos. C. T. R. Wilson and Searle also did not appear in most annual photos taken in the 1900s.

NST. One indicator of the decline in popularity of the MT at the Cavendish was the number of wranglers there. Fewer than five new wranglers came to the Laboratory, and A. S. Eddington who took the MT in 1904 was the *last* senior wrangler to enter the Cavendish before the 1910 regulation abolished the traditional classification of wranglers, senior optimes, and junior optimes. Most Cambridge students at the Cavendish in the new century took both parts of the NST. Among them were J. A. Crowther, Miss J. M. W. Slater, G. Stead, H. Thirkill, and R. Whiddington. However, the Cambridge students generally were less productive than the non-Cambridge students; fourteen of the Cambridge students produced nothing during their time at the Laboratory.

The role of non-Cambridge graduates (including visitors) was becoming more and more important, and the increase in the number of researchers at the Laboratory reflected the presence of these outsiders. The annual photographs recorded a continuous increase in the number of outsiders at the Laboratory, from 11 in 1900, to 19 in 1905 and 24 in 1912, while the number of Cambridge graduates remained flat. The non-Cambridge students—as the major research group—often competed with Cambridge students for scholarships, teaching posts, and the Professor's ear: six of seven Clerk Maxwell Scholarships went to outsiders; H. A. Wilson, Barkla, Kleeman, and others distinguished themselves with successful researches, and A. Wood and T. G. Bedford served as demonstrators for several years.

The core group of non-Cambridge researchers at the Cavendish were the advanced students, many of whom were 1851 Research Scholars from various parts of the British Empire, following the example of Rutherford in the previous century.[146] These researchers came to the Cavendish to work on electrons, radiation, photoelectricity, gaseous discharge, and thermionics, among other subjects; wrote research papers under J.J.'s supervision; and received research certificates in addition to the Cambridge B. A. degree. Barkla came from Liverpool to study electric waves along wires and secondary X-rays. Horton (Birmingham) studied "the effect of temperature on the modulus of torsional rigidity." A. Wood (Glasgow) researched the "radioactivity of ordinary metals and of salts of the alkali metals" and "diurnal periodicity of spontaneous ionisation in closed metal vessels." John Patterson (Toronto) researched the electrical properties of thin metallic films, ionization in air, and "the change of resistance of metals when placed in a magnetic field." H. L. Cooke (Montreal) performed "experiments on the penetrating radiation and on the velocity of light" and thermionics. Jeremiah J. E. Durack (Australia) studied "the connection between the velocity and the mean free path of electrons moving through a gas at some standard temperature and pressure." Kleeman (Australia) researched the nature and characteristics of α, β, γ and X-rays. W. H. Longeman (South Africa) researched "the production of secondary rays by α rays from polonium."[147]

---

[146] For the complete list of advanced students during the period, see *The Historical Register of the University of Cambridge, Supplement, 1911-20* (Cambridge: Cambridge University Press, 1922), 49-52.

[147] *Record of the Science Research Scholars for the Exhibition of 1851*, 24-50.

J.A.Vincent. W.Craig-Henderson. J.E.Almy. Hon. R.J.Strutt. R.L.Wills. H.A.Wilson. A.A.Robb. Dr.Nabl.

J.McClelland. C.G.Barkla. C.T.R.Wilson. Prof. J.J.Thomson. R.S.Willows. J.S.Townsend. J.Burke. H.S.Allen.

J.L.Glasson. A.E.Oxley. Pr.Slocum. S.G.Lusby. G.W.Todd.

D.M.Bose. G.Stead. R.T.Beatty. R.Whiddington. T.Harris. L.Southerns. L.J.P.Jolly.

J.A.Orange. C.S.Allott. J.Stead. F.W.Aston. H.H.Paine. H.Thurkill. R.C.Goudy. A.Ll.Hughes.

R.D.Kleeman. N.R.Campbell. F.Horton. Miss Saltmarsh. Prof. Sir J.J.Thomson. Miss Maltby. J.A.Crowther. J.Satterly. E.M.Wellisch.

Figure 5.15. Annual photographs of 1900 & 1910 [Courtesy of the Cavendish Laboratory].

Most visiting researchers in the Laboratory were either American professors on sabbatical leave or American graduate students spending a year or two at the Cavendish to learn something new. Among the professors were H. A. Bumstead (Yale University), E. C. Bingham (Richmond College), C. D. Child (Colgate University), H. A. Erikson (University of Minnesota), W. B. Huff (Bryn Mawr College), L. T. More (University of Cincinnati), and G. F. Hull (Dartmouth College). The graduate students included E. P. Adams, J. M. Adams, and T. Lyman (Harvard University); J. E. Almy (University of Nebraska); R. C. Gowdy (University of Cincinnati); G. F. Hull (Dartmouth College); E. Rhoads (Johns Hopkins University); and others. From Europe, Canada, Australia, and Japan came K. Przibram and E. Lohr (University of Vienna), S. T. Smoluchowski (University of Lemberg, Austria), L. Vegard (University of Christiania), A. Bestelmeyer and M. Levin (University of Göttingen), G. Jaffé (University of Liepzig), J. Kunz (University of Zürich), W. A. Borodowsky (University of Dorpat, Russia), V. Carlheim-Gyllensköld (University of Stockholm), J. C. McLennan and H. F. Dawes (University of Toronto), F. C. Gates (McGill University), S. G. Lusby (University of Sydney), and T. Noda (Higher Normal School, Tokyo). The presence of these visitors at the Cavendish, together with that of the advanced students from British dominions, transformed the Laboratory into a true international center.

One result of the presence of foreign researchers at the Cavendish was the establishment of a cycle in which researchers returned to their native countries to teach and then sent their students to the Cavendish for further training. J.J. was kept busy replying to letters of inquiry from such former students. The Canadian connection was particularly strong. Rutherford at McGill University sent his students R. K. McClung (1901-04), H. Brooks (1902-03), F. C. Gates (1902, 1905), and H. L. Cooke (1903-06). Rutherford's letters contain many references to students he was sending to the Laboratory, for example, "I hope to get the 1851 Scholarship for [McClung] next year to send him once to you in a little beyond the preparatory stage of research," and "I hope Miss Brooks and McClung are doing well."[148] Typical replies of J.J. were: "McClung is doing very well"; "I shall be very glad to give Miss Brooks permission to work in the Laboratory & to attend lectures"; and "Cooke is at present working at the question of the emanation from uranium."[149] J. C. McLennan of the University of Toronto sent researchers J. Patterson (1900-02), E. F. Burton (1904-06), H. F. Dawes (1906-08), and C. S. Wright (1908-1910). Together with researchers from Queen's University of Kingston, Ontario, notably W. C. Baker (1900-02), these Canadians flourished at the Cavendish and received fair appraisals from J.J., who said, "We have a great many men working here, among them two Canadians, Patterson and Baker, who are both doing well."[150] The majority of them returned to Canada to complete the cycle, and later sent students of their own to the Cavendish.

---

[148] CUL MSS ADD 7654 R66 (9 January 1900) & R67 (26 December 1902): Rutherford to J.J.
[149] CUL MSS ADD 7653 T18 (2 May 1902), T19 (13 May 1902), T23 (18 February 1904): J.J. to Rutherford.
[150] CUL MSS ADD 7653 T14 (15 February 1901): J.J. to Rutherford.

Prior to World War I, an Australian connection also was apparent, but it was weaker than the Canadian connection. Although both W. H. Bragg (University of Adelaide) and R. Threlfall (University of Sydney) began teaching physics in Australia in 1886, Australian researchers appeared at the Cavendish only after 1900, with the single exception of Miss F. Martin who came from the University of Sydney in 1894 to work on electrical discharge. However, in the early decades of the twentieth century, the University of Sydney sent J. J. E. Durack (1900-04), T. H. Laby (1905-09), E. M. Wellisch (1907-10) and S. G. Lusby (1909-11); the University of Adelaide sent Kleeman (1905-13) and J. L. Glasson (1909-11); Queen's College sent R. Hosking (1902-04).[151] Of the Australians, Kleeman, formerly Bragg's assistant, was the most successful at the time, publishing more papers than any other Cavendish researcher in the years around 1910. After the end of the World War I, when Rutherford became Cavendish Professor, the Australian connection would grow much stronger.

The cycling of students from abroad into the Cavendish indicated the maturation of the Cavendish School and also the changing sources of its human resources. Cambridge University no longer was the only reliable source of researchers for the Cavendish because former Cavendish researchers, now working in different parts of the world, could supply the Laboratory with researchers of at least equal quality. Cavendish alumni, by selecting and training young researchers who wished to study at the Laboratory, bore much of the responsibility for the quality of the students heading to the Laboratory. The links forged between the Cavendish and other institutions, particularly institutions outside the British Isles, indicate that the Cavendish had established and now maintained a truly international network.

The changing milieu of the Laboratory in the new century also was demonstrated by the Cavendish researcher's publications. Almost 200 researchers who had worked at the Cavendish between 1901 and 1914 published about 400 papers during the first two decades of the twentieth century. The research subjects covered by these papers are summarized in Table 5.2, which reveals some significant trends at the Cavendish during this period.

The majority of Cavendish researchers were concentrating on fewer subjects, notably ionization, radioactivity, radiation, and thermionics. J.J.'s research preferences certainly influenced this trend. Because this concentration of research around a professor's interests was a new phenomenon at the Cavendish, it naturally caused concern and attracted criticism. In his survey article in *A History of the Cavendish Laboratory 1871-1910*, N. R. Campbell noted anxiety in the Cambridge community concerning the Cavendish's concentration on a particular cluster of problems in physics:

[151] See R. W. Home, *Physics in Australia to 1945: Bibliography and Biographical Register* (Melbourne: University of Melbourne, 1990).

Table 5.2. Research Subjects and Number of Papers at the Cavendish, 1901-1914

Subject	*Number of Papers*
electrons/ionization/ electric discharge & conduction	85
thermionics	28
radioactivity	46
radiation[A]	50
condensation nuclei	7
positive rays	9
structure of the atom	6
electricity & magnetism	30
design of instruments (& experiments)[B]	15
various theoretical research[C]	10
chemistry	25
spectroscopy	13
metal	12
electrolysis	7
osmosis	5
photo-electricity	8
others	35

A: including a large number of studies on the nature of X-rays and γ-rays
B: except 2 papers on the design of the cloud chamber
C: including 2 papers on relativity theory

> The idea seems to be widely current that the Cavendish Laboratory is a place of narrow specialization, that all the researches carried out in it are directed to the solution of a small branch of physical problems, which have little connection with the general body of the science.[152]

The Cavendish Laboratory was rapidly changing from a university laboratory responsible for teaching Cambridge students (and providing them with research facilities) to a rather specialized research center.

Despite this trend toward specialization, the Laboratory continued its laissez-faire tradition. As seen in Table 5.2, various lines of research were still flourishing at the Cavendish in the new century, although they seemed less impressive and less organized in comparison to the major lines of research undertaken there. For example, Searle and T. G. Bedford worked on the measurement of magnetic hysteresis. C. T. R. Wilson began the research on atmospheric electricity and thunderstorms that gave him special satisfaction, "when my work led to the fulfillment of my dream of isolating a portion of the earth's surface and measuring the charge upon it and the current flowing into it from the atmosphere."[153] P. S. Barlow published a series of papers on osmosis.[154] J. C. M. Garnett investigated the colors in metal glasses and in metallic film.[155] D. F. Comstock from Massachusetts Institute of Technology made a theoretical study of "whether the inertia of matter has or has not a complete electromagnetic explanation."[156] In 1908, G. I. Taylor published a paper on the structures of shock waves in gases, which signaled his "debut in fluid mechanics, the field in which he was to publish over 150 papers during the subsequent 60 years."[157] G. Stead and H. Donaldson published two short papers on "the problem of the rotation of a solid cylinder about its axis, in connexion with the Principle of Relativity."[158] Also flourishing in the Laboratory at this time were chemical researches by S. Skinner, T. H. Laby, G. A. Carse, C. Chittock, L. Vegard, E. F. Burton, and others. J. B. B. Burke even conducted research on the origin of life, which "caused a little amusement" in the Laboratory.[159] None of these pursuits related directly either to J.J.'s major research or to the main research lines at the Cavendish.

---

[152] *A History of the Cavendish Laboratory*, 248.

[153] C. T. R. Wilson, "Reminiscences of My Early Years," 170.

[154] P. S. Barlow, "Osmotic Experiments on Mixtures of Alcohol and Water," *Phil. Mag. 10* (1905): 1-12; "On the Osmotic Pressures of Dilute Aqueous Solutions," *Proc. Camb. Phil. Soc. 13* (1905): 229-238; "The Osmotic Pressures of Alcoholic Solutions," *Phil. Mag. 11* (1906): 595-604.

[155] J. C. M. Garnett, "Colours in Metal Glasses and in Metallic Films," *Phil. Trans. 203* (1904): 385-420.

[156] D. F. Comstock, "The Relation of Mass to Energy," *Phil. Mag. 15* (1908): 1-21 on 1.

[157] G. Batchelor, *The Life and Legacy of G. I. Taylor* (Cambridge: Cambridge University Press, 1996), 43.

[158] G. Stead and H. Donaldson, "The Problem of Uniform Rotation treated on the Principle of Relativity," *Phil. Mag. 20* (1910): 92-95; *Phil. Mag. 21* (1911): 319-324. The authors concluded (in the second paper) that "this investigation shows that the relativity theory involves no contractions when applied to the case of uniform rotation." (p. 324)

[159] CUL MSS ADD 7653 S50 (14 September 1905): Searle to Rutherford.

The vigor of the Cavendish's laissez-fair tradition at this time could be seen in the sometimes casual, sometimes dramatic, changes that Cavendish researchers freely made in their research directions. C. T. R. Wilson, for example, moved from investigations of condensation nuclei to researches on atmospheric electricity during the first decade of the new century, but returned to construction of the Cloud Chamber in 1911. Kleeman worked mainly on the nature of radiation until 1910, and then turned his attention to capillarity, chemical attraction, and liquids. Horton worked on a wide area covering torsional rigidity, electrical discharge, and positive electricity. A. E. Oxley worked mostly on magnetic susceptibility, but also performed research on "the detection of small amounts of polarization in light from a dull sky" and the Hall effect in liquid electrolytes.[160] Wellisch published many papers on the movement of ions but also one on the "electric detector for electromagnetic waves."[161] Other researchers at the Cavendish also dug into several different fields. In short, a variety of research subjects were pursued at the Cavendish in the new century, and free movement from one research topic to another was very much in evidence.

Among the research subjects pursued less extensively at the Cavendish, two are worthy of special attention because they reveal both the strong and the weak points of the laissez-fair tradition. These subjects are photoelectricity and chemical research. In the 1900s and 1910s, photoelectricity attracted the notice of preeminent physicists such as P. Lenard, A. Einstein, H. A. Lorentz, M. Planck, A. Sommerfeld, Peter Debye, Karl Compton, Richardson, and Robert Millikan. Some of these physicists investigated photoelectricity in consideration of Einstein's new hypothesis of light quanta, while others were more interested in the nature of light or in the relation between light and electrons. At the Cavendish, where photoelectricity remained a minor subject of research involving only a few researchers, research efforts on this topic were not systematic and lacked a clear leader. A. L. Hughes became the major researcher of photoelectricity at the Cavendish by producing six papers on the subject, but his work failed to provide either new theory or better data for confirming existing theory.[162] His major concerns were the effect of photoelectricity on salts and other compounds, the development of a more sensitive photoelectric cell, and the velocity of photoelectrons from matter. W. M. Varley's work on photoelectric discharge from

---

[160] A. E. Oxley, "The Detection of Small Amount of Polarisation in Light from a Dull Sky," *Proc. Camb. Phil. Soc. 16* (1912): 561-579 on 570; "The Hall Effect in Liquid Electrolytes," *Proc. Roy. Soc. 88* (1913): 588-604.

[161] E. M. Wellisch, "An Electric Detector for Electromagnetic Waves," *Proc. Camb. Phil. Soc. 15* (1909): 337-339.

[162] A. L. Hughes, "On the Velocities of the Electrons produced by Ultra-Violet Light," *Proc. Camb. Phil. Soc. 16* (1911): 167-174; "The Photo-Electric Effects of Certain Compounds," *Proc. Camb. Phil. Soc. 16* (1911): 376-382; "A Note on Short Wave Lengths in the Mercury Arc Spectrum," *Proc. Camb. Phil. Soc. 16* (1911): 428-429; "The Photo-Electric Effect of Some Compounds," *Phil. Mag. 24* (1912): 380-390; "A Sensitive Photo-Electric Cell," *Phil. Mag. 25* (1913): 679-682; "On the Velocities with which Photo-Electrons are emitted from Matter," *Phil. Mag. 25* (1913): 683-686.

metallic surfaces and on the absorption of ultra-violet light in different gases also were functional rather than theoretical.[163] In 1910, however, Kleeman produced two experimental works that confirmed the similarity between X-rays and ordinary light by showing the asymmetry in photo-emission.[164] The relative neglect of this important research subject at the Cavendish, as well as the lack of any serious attempt within the Laboratory to connect photoelectricity with quantum theory, suggests a serious drawback of the Cavendish's laissez-faire tradition, no doubt resulting from J.J.'s less attention on the subject.

If the subject of photoelectricity illustrates a shortcoming of the Cavendish laissez-fair tradition, research related to chemistry proves its benefits. Chemical research did not appear either to attract much attention at the Cavendish or to be well orchestrated with the central research themes. P. V. Bevan wrote a long paper on the combination of hydrogen and chlorine under the influence of light; W. C. D. Whetham and H. A. Wilson worked briefly on electrolysis; Richardson published an article on solubility and diffusion in solutions of dissociated gases; P. S. Barlow and L. Vegard each published a series of papers on osmotic pressure in various solutions; E. F. Burton worked extensively on the colloid; Kleeman researched the law of chemical attraction, the relation between molecular attraction and the properties of liquids, and "the Nature of the Internal Work done during the Evaporation of a Liquid."[165] However, this short list provides only a sampling of the chemical research performed at the Cavendish. Considered in term of total number of papers published (Table 5.2), chemical studies comprised a very large group of studies performed at the Cavendish during the first two decades of the twentieth century, and this work made large, although indirect, contributions to the major lines of research at the Cavendish (such as electrical discharge, electric conduction, and radioactivity) by providing valuable information about the chemical properties of various gases and other substances. In other words, during this period, the Cavendish had its own in-house chemists. Although these chemical researchers

[163] W. M. Varley, "On the Photo-Electric Discharge from Metallic Surface in Different Gases," *Phil. Trans. 202* (1904): 439-458; "On the Absorption of Ultra-Violet Light in different Gases," *Proc. Camb. Phil. Soc. 12* (1904): 510-516.

[164] R. D. Kleeman, "A Difference in the Photoelectric Effect caused by Incident and Divergent Light," *Nature 83* (1910): 339; "On the Direction of Motion of an Electron ejected from an Atom by Ultra-Violet Light," *Proc. Roy. Soc. 84* (1911): 92-99.

[165] Some examples are as follows: P. V. Bevan, "The Combination of Hydrogen and Chlorine under the Influence of Light," *Phil. Trans. 202* (1904): 71-121; W. C. D. Whetham, "The Theory of Electrolytic Dissociation," *Phil. Mag. 5* (1903): 279-290; H. A. Wilson, "The Laws of Electrolysis of Alkali Salt-Vapours," *Phil. Mag. 4* (1902): 207-214; O. W. Richardson, "The Solubility and Diffusion in Solution of Dissociated Gases," *Phil. Mag. 7* (1904): 266-274; P. S. Barlow, "Osmotic Experiments on Mixtures of Alcohol and Water," *Phil. Mag, 10* (1905): 1-12; L. Vegard, "Researches upon Osmosis and Osmotic Pressure," *Phil. Mag. 16* (1908): 247-271 & 396-419; E. F. Burton, "The Action of Electrolytes on Colloidal Solutions," *Phil. Mag. 12* (1906): 472-478; R. D. Kleeman, "An Investigation of the Determination of the Law of Chemical Attraction between Atoms from Physical Data," *Phil. Mag. 21* (1911): 83-102; Kleeman, "Molecular Attraction and the Properties of Liquids," *Phil. Mag. 22* (1911): 566-586; Kleeman, "On the Nature of the Internal Work done during the Evaporation of a Liquid," *Proc. Camb. Phil. Soc. 17* (1914): 402-408.

frequently shifted their interests to other topics or even left the Laboratory, Cavendish researchers investigating ionization or radioactivity did not need to venture outside their own institution for expert help with the chemical aspects of their studies. This extraordinary condition was due not only to the Cavendish's laissez-faire tradition, but also to the careful attention paid by J.J. to physical chemistry since his student days. J.J. frequently made direct suggestions to Cavendish researchers that they investigate specific chemical subjects and encouraged them to carry out chemical studies.[166] At the Cavendish Laboratory under J.J. Thomson's leadership, there was no clear boundary between physics and chemistry, which undoubtedly enriched the various physics researches carried on under its roof.

5.5. The Decline of J.J. Thomson's Leadership

The year 1910 marked J.J. Thomson's twenty-fifth anniversary as Professor of Experimental Physics at Cambridge. To celebrate the event, a volume of collected memoirs of former Cavendish researchers, *A History of the Cavendish Laboratory 1871-1910,* was planned (Figure 5.16). J.J. himself contributed an essay, "Survey of the Last Twenty-Five Years," in which he reviewed his tenure as director of the Cavendish, a task that he appraised as "very difficult as well as very pleasant."[167] On November 12, 1910, as scheduled, the completed volume was presented to J.J. by Glazebrook, now head of the National Physical Laboratory, at a meeting commemorating the anniversary of J.J.'s professorship. In reply, J.J. avowed, "I have not the power to find words to express adequately my gratitude to those who have initiated and carried through this commemoration of my tenure of the Cavendish Professorship for 25 years."[168] That evening a dinner party "of old and present researchers" was attended by "about sixty in all."[169] Many physicists and former Cavendish researchers from around the world sent congratulatory letters praising J.J.'s achievements and wishing him "another quarter [century] in good health on every progress."[170]

J.J. was at the peak of his career, and a decline in his leadership was inevitable. Quantitative evidence of this decline was provided by a decrease in the number of 1851 Exhibition Scholars who chose to study at the Cavendish. Before 1910, the absolute majority of Exhibition Scholars who studied physics in Britain chose the Cavendish as their sole destination. H. A. Wilson, Barkla, Wood, Horton, and Kleeman are just few examples. Starting in about 1910, however, some Exhibition

---

166 For example, Bevan thanked J.J. "for having suggested this investigation, and for his advice during the course of the experiment." See Bevan, "The Combination of Hydrogen and Chlorine under the Influence of Light," *Phil. Trans. 202* (1904): 121.

167 *A History of the Cavendish Laboratory*, 75.

168 CUL MSS ADD 7654, Cam. 23.

169 L. Badash, *Rutherford and Boltwood: Letters on Radioactivity* (New Haven: Yale University Press, 1969), 237.

170 CUL MSS ADD 7654 K31 (4 December1910): J. Kranich (Moravia) to J.J. Brackets added.

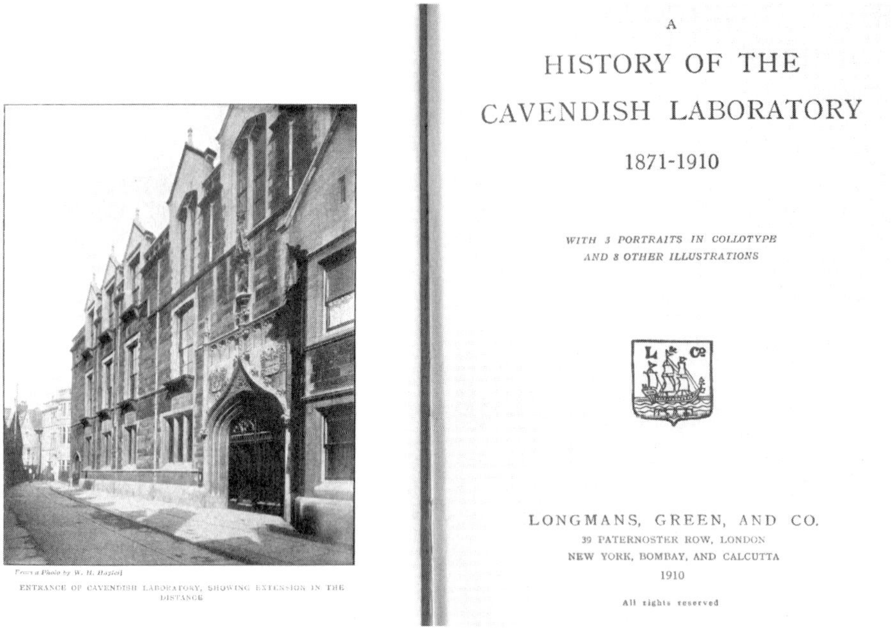

Figure 5.16. Frontispiece and Title page of A History of the Cavendish Laboratory, 1871-1910.

Scholars, such as T. Royds, R. W. Boyle, and J. A. Gray, chose to study radioactivity under Rutherford at Manchester. Others, like G. E. Jauncey and S. E. Peirce, chose to pursue X-ray research under W. H. Bragg at University of Leeds. Other Scholars, including James Chadwick, elected to study at German universities. [171] Although Exhibition Scholars who chose the Cavendish still outnumbered those who did not, and although some who did not originally choose the Cavendish moved there after their foreign studies, a sea-change had begun.

The studies performed at the Cavendish after 1910 were less exciting than those undertaken at the turn of the century. As the century opened, professor and researchers had concentrated their energy on developing an understanding of the newly discovered electron. In 1910, the professor and researchers did not share a central research topic through which J.J. might exert his intellectual leadership: positive rays never played the important role that corpuscles did in sparking the imaginations of Cavendish researchers. Worse, at the Cavendish the newly emerging theoretical tools of the quantum and relativity theories were not fully cultivated.[172] Moreover, other research centers, notably the Curie laboratory in Paris

---

[171] For more information, see *Record of the Science Research Scholars of the Royal Commission for the Exhibition of 1851; 1891-1929*, 39-48.

[172] For the reception of Einstein's relativity theory at Cambridge, see Warwick, "Cambridge Mathematics and Cavendish Physics: Cunningham, Campbell and Einstein's Relativity 1905-1911. Part I: The Use of Theory," & "Part II: Comparing Traditions in Cambridge Physics."

and the Rutherford laboratory in Manchester, seemed to surpass the Cavendish in experimental work on radioactivity.[173] Between 1911 and 1914 few Cavendish researchers paid attention to the newly discovered nucleus, while Rutherford's team at Manchester pursued it with vigor and confidence: in a letter to Boltwood in March of 1914, Rutherford commented:

> I then go down to London on Thursday to introduce a discussion on the structure of the atom. I am speculating whether J.J.T. will turn up, because he knows that I think his atom is only fitted for a museum of scientific curiosities. The ideas of a nucleus atom are really working out exceedingly well. You will have seen the work of Bohr and Moseley.[174]

Although J.J., C. T. R. Wilson, Aston, and others still carried out some important work between 1910 and 1914, and although the number of published research papers by Cavendish researchers did not decline, the Cavendish was losing vitality.

This decline was demonstrated by Niels Bohr's decision to leave the Cavendish for Victoria University of Manchester. In late September of 1911, an enthusiastic Bohr had arrived at Cambridge hoping to perform postdoctoral research under J.J. at the mecca of atomic physics.[175] Bohr wrote to his fiancée: "After my arrival the first thing I did was to pay my respects to Thomson. He was extremely pleasant, and we had a little talk during which he said he would be interested in seeing my treatise when it was finished. You can imagine how happy I was when I left."[176] Bohr's elation, however, did not last long. He soon realized that life at Cambridge was not what he had expected. He was puzzled by the "state of molecular chaos," as he termed it, prevailing at the Laboratory, and he got nothing from an experiment on cathode rays that J.J. recommended he perform.[177] His greatest disappointment, however, was in J.J., who seemed to pay little attention to Bohr's doctoral thesis on the electron theory of metals. The intellectual milieu of the Cavendish fell far short of what Bohr had expected.

For Bohr, relief came from Manchester. In December, he had the opportunity to meet and talk with Rutherford, who had come to Cambridge to attend the annual Cavendish Laboratory Dinner. By January, Bohr had decided to move to Manchester to work under Rutherford, where he spent four very productive months advancing his theory of atomic structure which extensively utilized both J.J.'s electron-based model and Rutherford's new nucleus model. Unlike J.J., Rutherford paid attention to Bohr's work, as he later recalled, with "characteristic caution

---

[173] See J. L. Davis, "The Research School of Marie Curie in the Paris Faculty, 1907-14," *Annals of Science 52* (1995): 321-355. For more about Rutherford's years at Manchester, see J. B. Birks (ed.), *Rutherford at Manchester* (London: Heywood & Co., 1962).

[174] L. Badash, *Rutherford and Boltwood*, 292.

[175] For J.J.'s influence on Bohr's early atomic model, see J. L. Heilbron and T. S. Kuhn, "The Genesis of the Bohr Atom," *HSPS 1* (1969): 211-290.

[176] S. Rozental (ed.), *Niels Bohr: His Life and Works as seen by His Friends and Colleagues* (Amsterdam: North-Holland Personal Library, 1967): 40

[177] Ibid., 41.

against overstating the bearing of the atomic model and extrapolating from comparatively meagre experimental evidence." [178] At Manchester, Bohr and Rutherford formed a lifelong friendship, and Bohr also became acquainted with G. Hevesy and C. G. Darwin, who would greatly influence the birth of his famous trilogy. Certainly Rutherford and his laboratory at Manchester were what Bohr had originally expected from J.J. and the Cavendish Laboratory.

What began the decline of the Cavendish Laboratory? I believe the chief cause of its descent was the waning of J.J.'s leadership. His research interests (notably positive rays) were no longer at the cutting edge and, as a result, his intellectual leadership was slowly eroding. Unlike Aston, most of the talented youngsters had turned their attention to the nature of radiation or to radioactivity. By 1910, more productive and more reliable mentors than J.J. were available in these fields, particularly Rutherford, Barkla, and W. H. Bragg, all of whom worked outside the Cavendish. Worse, J.J. had been unsympathetic to the quantum and relativity theories and even to Rutherford's nucleus atomic model. J.J.'s adherence to the old mechanical models, like the vortex-ring and Faraday's tube, continued until the outbreak of World War I. His conservative view is well documented in his 1907 book, *The Corpuscular Theory of Matter*, in which he stated, "The corpuscular theory of matter with its assumptions of electrical charges and the forces between them is *not nearly so fundamental as the vortex atom theory of matter*." [179] Even as late as 1914, J.J. considered his vortex atom to be a useful model.[180] By this time almost no Cavendish researchers were following the Professor's old ideas, nor were they paying much attention to the quantum, relativity, and nucleus theories.

J.J. also was experiencing difficulty concentrating his energies on managing the Laboratory. Now a public figure, he increasingly was called upon to perform duties away from the Cavendish. In 1908, he was knighted as a Knight Bachelor. In 1909, he traveled to Winnipeg, Canada, to preside over a meeting of an organization of which he was president, the British Association. In 1911, he was elected President of the Physical Society of London. After attending several meetings and dinners of the Royal Society, in 1913 he was approached to accept its presidency, an offer which he first declined, but later accepted, serving in that position from 1915 to 1920. J.J.'s most serious diversion, however, arose from his 1905 appointment as Professor of Natural Philosophy at the Royal Institution. Among his responsibilities there was the delivery of a series of public lectures, including the famous Friday-

---

[178] N. Bohr, "Reminiscences of the Founder of Nuclear Science and of Some Developments based on His Work," *Proceedings of the Physical Society of London 78* (1961): 1085. See also AIP MSS OH48 (Margrethe & Aage Bohr) & OH50 (Niels Bohr). Bohr went to Manchester to take an experimental course on radioactivity that was prepared by Geiger, Makower, and Marsden, not to research on the structure of the atom. However, while he waited for radioactive materials, he read Darwin's new paper on the absorption of alpha particles and thereafter entered the research on atomic structure.

[179] J. J. Thomson, *The Corpuscular Theory of Matter*, 2. Emphasis added. See also his *Electricity and Matter*, chapter 2, especially pp. 50-51.

[180] J. J. Thomson, *The Atomic Theory*, 25. J.J. subscribed to this mechanical world-view until his death, as demonstrated in a letter written by him in 1937. See Strutt, *Life of J.J. Thomson*, 202-203.

evening lectures. J.J. prepared these lectures in advance, mostly at the Cavendish.[181]

The Cavendish lacked an organizational instrument to make up for the absences of its director, such as deputies to substitute for J.J. when he was not at the Laboratory or go-betweens to intermediate between himself and the Cavendish researchers. J.J. always had maintained direct connections with the researchers, and his absences therefore proved quite troublesome. Searle and C. T. R. were highly capable senior demonstrators, but they were less effective than J.J. at directing junior researchers. When Rutherford succeeded J.J. as Cavendish Professor, he secured the help of Chadwick, who acted as his lieutenant, but J.J. had no such assistance. Nor did J.J. secure such help as had so generously been provided to him by Glazebrook and Shaw in the previous century.

The Cavendish also was beginning to face competition from research centers founded by some of J.J.'s former students. Rutherford, Henry Bragg, Townsend, Barkla, and R. J. Strutt all were energetically carrying out research programs in their own research centers on British soil. Rutherford had created not only his own research center but also his own school at Victoria University of Manchester, where in 1907 he had been appointed Langworthy Professor of Physics. Rutherford had surrounded himself with talented researchers, particularly H. Geiger, H. Moseley, E. Marsden, Chadwick, Darwin, Hevesy, and Bohr. They concentrated on the study of radioactivity, the nature of radiation, and the structure of the atom, achieving during the short period from 1910 to 1914 such brilliant successes as α-scattering experiments by Rutherford, Geiger and E. Marsden; Rutherford's hypothesis of the nucleus model of the atom; Bohr's trilogy; and Mosley's law. The Rutherford School had fewer researchers than the Cavendish School, but these researchers were more homogeneous in their interests and more productive. Above all, the Rutherford School was better organized and better financed, and it enjoyed the use of 350 mg of radium loaned by the Austrian Academy of Science. In Rutherford, the School had an energetic, ambitious, and self-confident leader capable of declaring to his American friend Boltwood in December of 1910:

> I can devise an atom much superior to J.J.'s, for the explanation of and stoppage of α and β particles, and at the same time I think it will fit in extraordinarily well with the experimental numbers. It will account for the reflected α particles observed by Geiger, and generally, I think, will make a fine working hypothesis.[182]

Although Henry Bragg at the University of Leeds, Townsend at Oxford University, Barkla at the University of Liverpool and at King's College in London, and Strutt at the Imperial College in London were less successful than Rutherford at Manchester, the existence of their small-scale research centers meant that Cavendish was no longer *the* center in all fields of physics, even in Britain.

J.J.'s timely resignation, and the fortuitous election of the right successor, saved the Cavendish before its decline became apparent. However, this decline, it must be said, along with the decline of J.J.'s leadership, were relative. During the

[181] Strutt, *Life of J.J. Thomson*, 149.

[182] L. Badash, *Rutherford to Boltwood*, 235.

years between the beginning of the twentieth century and the outbreak of World War I, the Cavendish Laboratory was the world's most important research center for experimental physics. The fact that former Cavendish researchers were able to compete with their mother school from their own research centers can certainly be regarded among the Cavendish School's great successes. Especially considering the very close and friendly research network that had been formed among the Cavendish researchers, it is quite possible to regard these daughter institutions as *extensions* of the Cavendish School.

CHAPTER 6

THE END OF AN ERA, 1914-1919

> Your own work in theoretical and experimental science has been monumental; but your giving to the world so many leading physicists seems a work as unique and far-reaching.
>
> L. L. Campbell[1]

> No one knows better than I how second-rate any man must feel who succeeds to your chair. In fact I think that will be its greatest drawback; your successor will by this circumstance be kept in a very humble frame of mind, if not an unhappy one.
>
> C. G Barkla[2]

6.1. World War I and the Cavendish Laboratory

"The Great War" of 1914-1918 had a significant impact on the usual activities of the Cavendish Laboratory. During that war, the Cavendish virtually ceased its two basic functions of research and teaching. Military personnel were billeted in some areas of the Cavendish, and its workshops and workmen were employed in the production of various gauges. Most researchers left the Cavendish early in the war to fight Germans. Those who remained left the Cavendish later to participate in war-related scientific research, principally at naval research centers and the National Physical Laboratory. The Cavendish then was maintained by few crippled researchers, senior demonstrators, and some foreign visitors.

Early in the war, the British government indicated almost no intention of utilizing scientific talent for the war effort. Cavendish researchers, including J.J.'s own son, George P. Thomson, were enlisted as ordinary soldiers and then were sent to the front. As is well illustrated in his correspondence, J.J. was deeply anxious about George's safety during the later half of 1914 when George was stationed at the front.[3] W. Lawrence Bragg, the twenty-four-year-old Cavendish researcher who already was a Nobel Laureate:

> was posted to the Leicestershire Royal Horse Artillery where he was a fish out of water among a lot of hunting men. But after a year's curious experience of men and horses the War Office plucked him out and sent him to France to take over the French method of locating enemy guns by sound, and start sound-ranging for the British Forces.[4]

---

[1] CUL MSS ADD 7654 C7 (27 November 1926): L.L. Campbell to J.J. for celebrating his seventieth birthday.

[2] CUL MSS ADD 7654 B18 (11 March 1919): Barkla to J.J.

[3] Strutt, *Life of J.J. Thomson*, 176-178.

[4] G. M. Caroe, *William Henry Bragg, 1862-1942: Man and Scientist* (Cambridge: Cambridge University Press, 1978), 79.

A few British physicists were killed in action. H. G. J. Moseley, Rutherford's promising protégé, was the most famous figure among them.[5] Moseley, who just before the war had discovered the very important rule indicating the close relationship between an atomic number and the characteristic radiation of that element, was killed in August of 1915 at Gallipoli in what was considered by many military experts to be one of the most futile military campaigns of World War I. His death in battle, lamented by scientists around the world as a "striking example of the misuse of scientific talent," was widely publicized and effectively employed to withdraw scientists from the front.[6] The Cavendish Laboratory remained relatively unscathed in this first high-casualty war: Edouard Jacot, an 1851 Research Scholar who had worked at the Cavendish from 1910 to 1912, was killed in 1917; G. Owen, another former Cavendish worker, was wounded while serving in the New Zealand Expeditionary Force. Most Cavendish men, however, survived and none of the Laboratory's great names perished.

By mid-1915, it was obvious that the war would continue longer than had been expected. German submarine attacks on ships headed to the British Isles, air raids by Zeppelins, and the demand of improved wireless telegraphy and aviation technologies drove the British government to mobilize scientists. In July of that year, Arthur J. Balfour, then First Lord of the Admiralty, set up the Board of Invention and Research (B. I. R.) "to initiate, investigate and advise generally upon proposals in respect to the application of science and engineering to naval warfare."[7] The B. I. R.'s Central Committee, its governing body, was comprised by George Beilby, Charles Parsons, and J.J. and was chaired by Lord Fisher. Its scientific advisory panel consisted of H. B. Baker, W. H. Bragg, H. C. H. Carpenter, William Crookes, W. Duddell, P. F. Frankland, B. Hopkinson, Oliver J. Lodge, W. J. Pope, E. Rutherford, G. G. Stoney, and R. J. Strutt. Richard Threlfall and others were added later. One of the B. I. R.'s first missions was to withdraw scientists from the front in order to "apply to such persons [to] assisting in the experimental work being carried out for the B. I. R."[8]

J.J. actively participated in B. I. R. business from its inception. In an August 1915 letter, J.J. made suggestions to Captain Thomas E. Crease, the first secretary of the B. I. R., concerning "how the Scientific Experts on the Board could deal most expeditiously and efficiently with the various suggestions and inventions sent in."[9] He examined the huge number of suggestions, which varied from worthy ideas like an effective way of destroying Zeppelin airships and an acoustic method of detecting submarines to rather ridiculous ideas like training sea-lions equipped with

---

[5] John L. Heilbron, *H. G. J. Moseley: The Life and Letters of an English Physicist, 1887-1915* (Berkeley: University of California Press, 1974).
[6] Ibid., 124-125.
[7] For the activities of B. I. R., see J.J. Thomson, *Recollections*, 206-224.
[8] CUL MSS ADD 7654 Gov. 13 (or C40: 2 December 1915): the First Lord of Admiralty to the Secretary of the B. I. R. Brackets added.
[9] CUL MSS ADD 7654 Gov. 1 (or C34: 5 August 1915): J.J. to T. E. Crease.

torpedoes to attack to German submarines (just like Japanese kamikajes during WWII) and a perpetual motion machine. J.J. himself proposed some useful ideas for war-related equipment, for example, "a non-contact firing device for unmoored mines" and a piezo-electric device for recording pressures caused by explosions.[10] He also offered to help in the effort to develop new vacuum tubes for wireless telegraphy, an effort that had been carried out at the Cavendish by G. Stead, S. F. Prasad, and J.J.'s private assistant, Everett (Figure 6-1).[11] As a senior representative of Britain's scientific community, J.J. acted as chairman of the B. I. R. when Lord Fisher was absent.

One of J.J.'s most important wartime roles, however, was to recommend, and sometimes select, physicists for positions appropriate to their individual talents. For example, William H. Eccles of the War Office at Burlington House asked J.J.'s

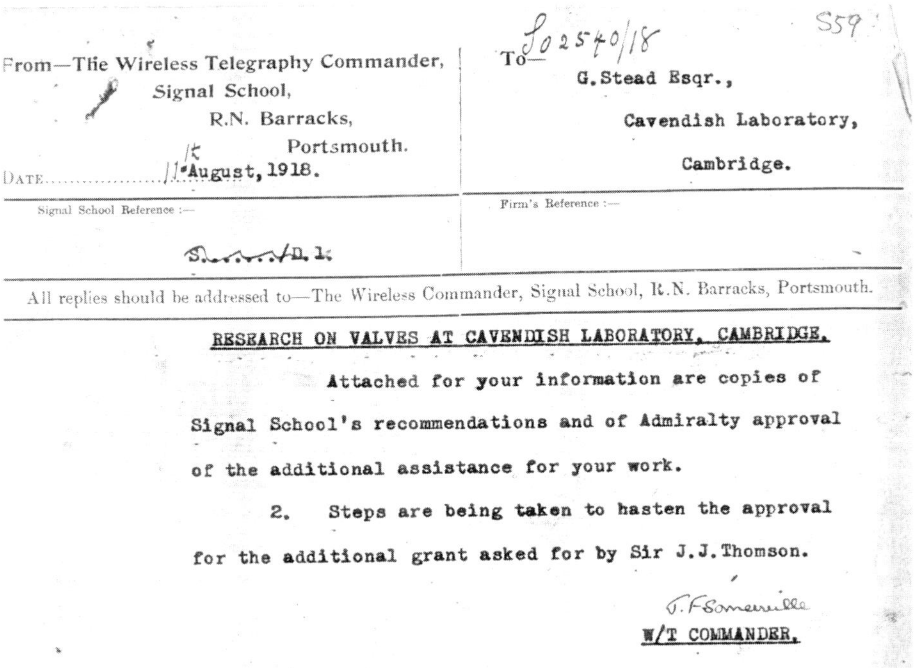

S02540/18 S59

From—The Wireless Telegraphy Commander,
Signal School,
R.N. Barracks,
Portsmouth.
DATE 11 August, 1918.

To— G.Stead Esqr.,
Cavendish Laboratory,
Cambridge.

Signal School Reference :—
Firm's Reference :—

All replies should be addressed to—The Wireless Commander, Signal School, R.N. Barracks, Portsmouth.

RESEARCH ON VALVES AT CAVENDISH LABORATORY, CAMBRIDGE.

Attached for your information are copies of Signal School's recommendations and of Admiralty approval of the additional assistance for your work.

2. Steps are being taken to hasten the approval for the additional grant asked for by Sir J.J.Thomson.

G. F. Somerville
W/T COMMANDER.

Figure 6.1. A letter from the Signal Corps to G. Stead of the Cavendish Laboratory. During the Great War, a few Cavendish workers participated in the project to improve vacuum tubes for wireless telegraphy (CUL MSS ADD 7654 Gov. 37) [Courtesy of Cambridge University Library].

[10] For the method of sweeping unmoored mines, see CUL MSS ADD 7654 Gov. 5.

[11] The research on vacuum tubes for more effective wireless telegraphy was carried out first at the National Physical Laboratory, and later at the Cavendish. See CUL MSS ADD 7654 Gov. 4 (13 September 1915), Gov. 10 (23 October 1915), Gov. 16 (19 December 1915), Gov. 22 (29 April 1916), Gov. 23 (2 June 1916), and Gov. 37 (11 August 1918).

"advice about getting a complete list of (physicists and chemists) connected with, or trained at, the University of Cambridge."[12] Captain Crease asked J.J. to recommend two scientists to perform experimental work with a naval officer on acoustic detection of submarines.[13] With thirty years of experience as Director of the Cavendish Laboratory, J.J. surpassed his contemporaries in the function of matching the talents of individuals to various assignments. He knew most of the British physicists and was able to quickly appraise their abilities. When J.J. did not personally know a young physicist, he relied on appraisals of that individual by his former students. J.J.'s position as President of the Royal Society between 1915 and 1920 added weight to his recommendations.

J.J.'s former students often were anxious to utilize him in their attempts to find suitable jobs in the war effort. J. J. E. Durack, an 1851 Research Scholar who between 1900 and 1904 had worked at the Cavendish on the connection between velocity and the mean free path of electrons moving through a gas, wrote J.J. from Bombay to ask for "more useful employment for me than on which I am engaged at present." A second lieutenant, Durack reported that he had spent a month as "superintendent of manure carts" and then was sent to India to bring back "a shipload of horses." "Is there any scientific work connected with the War," Durack asked, "for which you could recommend me?[14] Durack ultimately was unsuccessful in his search for scientific defense work, and ended up performing regular military service in France, Egypt, and Mesopotamia until the end of the war.

The B. I. R.'s most successful research during the war centered on detection of German submarines. W. H. Bragg was in charge of B. I. R.'s section II, which was intended to deal with "submarines, mines, searchlights, wireless telegraph and general electrical, electromagnetic, optical and acoustical subjects" but soon concentrated on anti-submarine research. This was a new kind of research that demanded not only scientific knowledge but also engineering skill.[15] An acoustic method of submarine detection was quickly adopted as the most promising, but developing reliable hydrophones proved very difficult. Worse, as Rutherford indicated, few physicists were "particularly expert in sound."[16] However, physicists soon "got a pretty good grip of the subject," proving how useful and versatile physicists could be. At the naval research center at Hawkcraig and later at Harwich, Bragg "marvelously" managed naval officers and important "visitors" who had less knowledge of modern science. Bragg was the right person to manage this new type of organization, which was large, complicated, and intended to produce a specific

[12] CUL MSS ADD 7654 E2 (20 August 1915): W. H. Eccles to J.J.

[13] CUL MSS ADD 7654 Gov. 9 (or C39: 9 October ny): T. E. Crease to J.J.

[14] CUL MSS ADD 7654 D33 (25 November 1915): J. J. E. Durack to J.J.

[15] For Henry Bragg's activities during the war, see Caroe, *William Henry Bragg*, 79-92. The early report on the anti-submarine project indicates many difficulties in developing effective equipment. See CUL MSS ADD 7654 Gov. 20 (C42A), a classified document stamped of "secret."

[16] CUL MSS ADD 7654 R70 (13 October 1915): Rutherford to J.J.

result. Bragg was assisted by A. B. Wood, another protégé of Rutherford. J.J. played no active role in the anti-submarine project.

The absence of direct contributions by J.J. to war-related researches, as well as his role as senior representative of the British scientific community, indicated the advent of new leadership for this community. The key figures of this new leadership were Rutherford and William Henry Bragg. It was Rutherford who went to the United States to represent British delegates in joint scientific efforts to defeat Germany. At the end of the war, Rutherford worked hard for establishment of a naval physics research laboratory, and he circulated the memorandum which eventually persuaded the government to establish that facility.[17] Bragg's success with the anti-submarine project convinced both the military and the general public of the utility of scientific talent in the war effort. The hydrophone was such an impressive development that it "would remain a most useful equipment both for anti-submarine forces and for U-boats themselves throughout World War II."[18] Recognition of this contribution made it possible for Bragg to attract generous post-war financial support from the newly established Department of Scientific and Industrial Research to run a large research team on X-ray crystallography at University College in London.

Lastly, J.J. made an important but rather unknown contribution during the war to both Cambridge University and the Cavendish. That was his strong support for the establishment of the Cambridge doctoral degree, which prepared for further development of the Cavendish after the War's end. In 1916, he already had raised in the Senate the issue of establishment of new degrees for research students:

> I regret that I cannot be present to-morrow at the discussion of the Report on the new degrees in Science. Had I been able to attend I should have raised the question whether it might not be possible to give to students who after six terms' residence have completed a satisfactory piece of research the titular degree of Doctor. I think it is probable that after the War many students from neutral countries will be unwilling to go to Germany for their post-graduate studies, and would much prefer to come to England. I have reason to believe that such students attach considerable importance to the attainment of the degree of Doctor (a titular degree is all they require) and unless some such degrees were obtainable they would not be likely to come in any considerable number.[19]

J.J.'s argument certainly was based on his long experience with foreign students at the Cavendish. Still there was strong hesitation to establish the degree of "Doctor of Science." The conservative faction of the University had successfully defeated similar schemes for the establishment of a doctoral degree at Cambridge in the early 1870s and again in 1895. But the time had changed. Although conservatism and anti-German feelings ran strong, the war so dramatically weakened opposition to the institution of a Cambridge doctorate that conservatives were able to delay its development only for few years. In 1920, the Ph.D. finally arrived at Cambridge.

[17] CUL MSS ADD 7654 F12, R74, R76, Gov. 36, and M31.
[18] John Terraine, *The U-Boat Wars, 1916-1945* (New York: G. P. Putnam's Sons, 1989), 29.
[19] *CUR* (14 November 1916): 219. For the report on the degrees for research, see *CUR* (6 October 1916): 50-51.

The next Cavendish Professor, Rutherford, would enjoy the benefits of the new degree system.

6.2. The End of the Thomson Era

In early 1918, Prime Minister Lloyd George offered J.J. the Mastership of Trinity College, because, as the Prime Minister confessed, J.J.'s "super-eminence as a scientist was known even to barbarians like myself who never had the advantage of any university training."[20] This offer was a great honor for any scientist, but especially for a graduate of Trinity like J.J. The position came with a huge salary of approximately £3200, a generous expense account, a nice lodge, few duties, and tenure for life. The traditional ceremonies, which were held on March 5, 1918, attracted a large number of spectators including "a considerable number of army cadets then stationed" at Cambridge.[21]

His new office at Trinity and the end of the war eight months later, in November of 1918, prompted J.J. to take a step that he had long contemplated: resignation of his Cavendish professorship. J.J. did not intend this step to cut off his relationship with the Cavendish or to end his research career. Instead, he looked to Cambridge to provide him with a new professorship and perhaps a new laboratory. He also expected to amicably share with his successor, who certainly would be one of his former students, the Cavendish space and personnel. With characteristic optimism, J.J. neglected to consider the difficulties that his successor might confront when this scheme was put into practice. Ironically, a decade previously, in 1906, he had cautioned Rutherford on this very point:

> I was very glad to hear from Schuster that there was a chance of your coming to England. I hope it is true . . . The laboratory is a very good one. The only spot I see on the prospect, is the possibility of a dual management of the laboratory by Schuster & yourself proving a little inconvenient. I think it would conduce to a smooth working in the future to have the respective parts to be played by Schuster & yourself defined as clearly as possible at the outset. It will be very delightful to have you back again. We are just about to begin to build an extension of the Laboratory. We shall pull down the houses adjoining it in Free School Lane & build a new laboratory close to the old one, arranged so that if need be it can be used as a separate laboratory so that if at any time there are two Professor of Physics each can have a laboratory.[22]

Schuster had freely relinquished the laboratory at Manchester that he had spent almost three decades developing in order to attract Rutherford. Now J.J., who had raised the Cavendish from a small university laboratory to a world-renowned research center, wished to share his laboratory with his successor as, this letter made clear, he had been planning for some time.

The vacancy of the Cavendish Professorship of Experimental Physics was officially declared on March 3, 1919, and candidates were asked to submit their

---

[20] Strutt, *Life of J.J. Thomson*, 205.
[21] Ibid., 208. For the details of the ceremonies, the congratulatory notes, and other stories about the Mastership, see also ibid., 205-214.
[22] CUL MSS ADD 7653 T28 (18 December 1906): J.J. to Rutherford.

applications by March 26.[23] Rutherford was widely regarded as the most suitable candidate, and those who were concerned about the future of the Cavendish, including Joseph Larmor, Cambridge's Lucasian Professor of Mathematics, pressed Rutherford to stand for the election.[24] Rutherford hesitated. He had no clear reason to leave his own flourishing school at Manchester, a school whose fame had eclipsed even that of the Cavendish. Moreover, Rutherford had only recently returned to his laboratory following the end of the war to begin his revolutionary experiments on the bombardment of α-particles with nitrogen, which later would be called "artificial disintegration." Before deciding whether to submit his application for the Cavendish professorship, Rutherford wrote to J.J. to clear up "several important points."

> My dear Professor,
>
> I have been thinking over the Cavendish matter seriously but before coming to any decision in the matter, there are several important points on which I would like your views & frank opinion. If I decided to stand & were elected for the post, I feel that no advantages of the post could possibly compensate for any disturbance of our long continued friendship or for any possible friction whether open or latent that might possibly arise if we did not have a clear mutual understanding with regard to the Laboratory and research arrangements.
>
> In the first place, I should say that I would welcome your presence in the Laboratory and its work & would be only too glad if you would help as far as you feel able in training researchers in the Lab. I am confident in the near future there will be more advanced students than the Laboratory can take or that [sic] two of us can look after properly. Under these conditions, it might prove awkward for both of us to have students make a decision in which part of the Lab. or under whose direction they wished to work.
>
> To avoid such undesirable complications, there are two alternatives either for you to work solus with your Laboratory assistants, or for the Director of the Lab to have charge of all advanced students and to assign them their line of work & their supervisor. Under the latter, one would naturally place under your supervision those students who worked along lines in which you were specially interested or on topics in which you wished further investigation. If we kept closely in touch on projected lines of research, I would know your wants & we would as occasional arose turn men over for your supervision; but of course, such students would be students of the Laboratory first and of their supervision supervisor second.
>
> How does the above appeal to you as a good working arrangement for both sides? I should welcome any other suggestions of your own. Another point. The new Director might feel it desirable and necessary to make changes in the organization of the teaching and research in the Lab. & possibly . . . in the personnel—with which changes you might not altogether concur. It would be very unfortunate that any trouble in difference should arise in that score, for the Director, it seems to me, must take the personal responsibility for the efficiency of the Lab. & its teaching.
>
> I have spoken quite frankly of possible series of misunderstanding & I feel if I were elected it would be of the greatest importance for both of us to have such a clear understanding of the situation & its difficulties that we would work in complete harmony . . .
>
> I am afraid I am troubling you on a number of questions but I should be very glad of your views as soon as possible as I will have to make a decision whether I will become a candidate during the next few days . . . [25]

[23] *CUR* (4 March 1919): 495.

[24] For Larmor's role in the election, see David Wilson, *Rutherford: Simple Genius* (Cambridge, Mass.: MIT Press, 1983), 408-412.

Rutherford's letter was characteristically frank. He recognized that his management style, which was to maintain direct control over researchers and staff, was different from J.J.'s traditional "laissez-faire" style. J.J. put forward his suggestions and assured Rutherford that problems would not be forthcoming.

> Dear Rutherford,
>
> I am very glad to find that you are still entertaining the possibility of coming to Cambridge as Professor. If you do you will find that I shall leave you an absolutely free hand in the management of the Laboratory. The scheme which has been adopted was not the one originally proposed: the first idea was to appoint a Director of the Laboratory and leave me as Cavendish Professor. I felt however that this would not do, and that it was better to appoint a New Professor with exactly the same powers as I had had during my tenure. Under the new scheme I have nothing to do with the management of the Laboratory or its Finance. A few rooms in the North wing are assigned to me, one of these will be filled up as workshop and I shall be allowed the services of a skilled mechanic. This man will be entirely under my orders as will of course my private assistant but I shall have nothing to do with any other assistants or with the teaching staff . . .
>
> I think everything can be made quite definite before we begin except the assignment of the research students. I do not see at present how this can be made a matter of formal regulation but I should think that it would solve itself quite naturally from the consideration of the subjects the students were working at. You would naturally take those who were doing what specially interested you while I should take those working at things which specially interested me. If any took subjects on which we were neither of us very keen we might tore up . . .
>
> Yours ever
>
> (signed) [26]

From Rutherford's point of view, J.J.'s reply did not clear up the "difficulties" that Rutherford feared might lead to a deterioration in their friendship, but instead highlighted them. J.J. had revealed his desire to share the Laboratory's space and, worse, its research students. On March 15 Rutherford confided his indecision to Schuster.

> Dear Schuster,
>
> Larmor, I believe told you they are wanting me to be a candidate for the Cavendish Chair and I expect you know the arrangements whereby J.J.T. has to have his own rooms, mechanic, etc., and to have research people to work with him. Of course the post has many attractions but has some obvious disadvantages and difficulties. As the time approaches for a definite decision, I find myself oscillating a good deal between the relative certainties of Manchester (including a fine Lab and a show of my own) and the advantages of Cambridge as a place to live in and opportunities for advancing Physics. At the same time, the presence of J.J.T., working in a part of the Lab even without technical power has certain obvious difficulties which are not diminished by the fact that our lines of work are very parallel . . . Well, if you feel like expressing an opinion, please do so soon as I shall have to make the irreversible decision by the end of the week. Larmor is doing everything in his power to make the post as comfortable as possible and is very pressing as also are other Cambridge people.[27]

---

[25] CUL MSS ADD 7653 T43 (7 March 1919): Rutherford to J.J.

[26] CUL MSS ADD 7653 T44 (10 March 1919): J.J. to Rutherford.

A week later, Rutherford sent another letter to Schuster, indicating that he was inclined to remain at Manchester: "It has been a worrying problem what to do but I think Manchester has first call under the conditions."[28] Informed by Larmor that Rutherford might not apply for the Cavendish professorship, J.J. again wrote to Rutherford to persuade his hesitating "pupil" to change his mind.

> Dear Rutherford,
>
> I am anxious to make the position with regard to the Cavendish Laboratory as clear as possible even at the risk of repeating what I said before. The intention is to make the two Professors as independent as if their Laboratories were in separate buildings and as soon as the necessary funds can be got the University will, I am sure, take steps to obtain a new Laboratory. In the meantime a few rooms in the new wing of the present Laboratory are assigned for the use of the new Professorship, & these and these only will be subject to his control. Speaking for myself I should never dream of interfering in any way with the rest of the Laboratory or expressing any opinion about matters of policy. I should treat it just as I would a separate Laboratory a mile away. The details as to the rooms to be assigned & etc. are not yet settled. They await the appointment of the new Professor and I shall be most anxious to do all I can to fall in with his views. I regard the new Professorship which the University has asked me to take, as being a distant and separate post and having—if I may so express it, no continuity with the Cavendish Professorship which I have resigned. When I sent in my resignation I had thought out the question and determined to sever myself entirely from any connection with the management or policy of the Cavendish Laboratory, so that my successor might have a perfectly free hand to carry out any policy he might see fit to adopt.
>
> I shall confine myself to my own rooms and do all I can to get a new Laboratory built. I said most of this in my former letter but I think it cannot be too often repeated that as far as I am concerned the new Cavendish Professor will be as independent as if he were in an entirely separate building.
>
> Yours Sincerely
> (signed)
>
> I am sure you can rely on everyone in the University doing every thing in his power to meet your views. There is a very keen hope that you may see your way to come to Cambridge. Personally nothing would give me so much pleasure as to have for my successor my most distinguished pupil.[29]

Having humbly asked his middle-aged former student to succeed to him, the old professor declared that he would yield everything to please his "most distinguished pupil." It is noteworthy that J.J., for the first time, closed his letter to Rutherford with "Yours Sincerely" rather than "Yours ever," as he was accustomed to do. Rutherford's reticence to share the Cavendish with him would always remain an unpleasant memory for J.J., and he refrained from mentioning it in his autobiography of 1936. J.J.'s biographers, Robert Strutt and George P. Thomson,

---

[27] CUL MSS ADD 7653 S43 (15 March 1919): Rutherford to Schuster. This letter was published in Mark Oliphant, *Rutherford: Recollections of the Cambridge Days* (Amsterdam: Elsevier, 1972), 16.

[28] Oliphant, ibid., 17.

[29] CUL MSS ADD 7653 T46 (23 March 1919): J.J. to Rutherford.

also did their best to soften this episode in their reportage. J.J.'s greatness of spirit and characteristic optimism enabled him to swallow this bitter pill, however, and the old man even acquiesced to Rutherford's request to take charge of the Laboratory during the Easter Term.[30] Rutherford, assured of his absolute status as both Cavendish Professor and Director of the Laboratory if elected, sent his application by telegraph on the day of the election, April 2, 1919, and was elected as the fourth Cavendish Professor of Experimental Physics. The Cavendish Laboratory's long, prosperous era under J. J. Thomson's directorship had come to a close.

As planned, J.J. was elected to the newly established position of Professor of Physics on July 15, 1919.[31] Cambridge's Vice-Chancellor, A. E. Shipley, sent J.J.

Figure 6.2. Two giants, talking perhaps about the results of recent cricket matches [Courtesy of the Cavendish Laboratory].

[30] CUL MSS ADD 7654 R78 (13 April 1919): Rutherford to J.J.
[31] *CUR* (14 June 1919): 919; (18 July 1919): 1036.

Figure 6.3. The annual photograph of 1932. J.J. Thomson and Rutherford sat at the center of the front line, surrounded by many present and future Nobel Laureates [Courtesy of the Cavendish Laboratory].

notification of his election: "We duly elected you in less than a minute today. It seems absurd to congratulate you so I will congratulate the University."[32] The promised new laboratory for J.J., however, did not materialize, as Rutherford had realistically anticipated. Instead, J.J. worked in the "Garage" in the basement of the new building, concentrating on investigations of positive rays and various aspects of gaseous discharges. Aston, who shared the space in the "Garage," continued as J.J.'s sole collaborator. Often young physicists like Mark Oliphant, following Rutherford's advice, visited this place to find the two Nobel Prize winners working together.

J.J. kept his promise to regard his beloved Cavendish as being "a mile away,"

[32] CUL MSS ADD 7654 S50 (15 July 1919): A. E. Shipley to J.J.

and did not interfere with Laboratory business. Rutherford treated his old mentor graciously, and there seems to have been no more conflict between these two great men. In his letters to Boltwood, Rutherford expressed his satisfaction with the new post at Cambridge as well as his relationship with J.J.: "I enjoy life very well here & find I keep pretty lively for an old man. J.J.T. is still energetic but his time is much occupied with Trinity & the Royal Society" (4 December 1919); "J.J.T. is very flourishing and I think enjoys his job as Master of Trinity. He is looking after a number of research students and thus relieves me of what would be an intolerable burden" (2 November 1920); "Cambridge suits me well alround [sic] & even the presence of the Master of Trinity in the Lab adds a fillip to life. He is as keen as a youngster still & is still searching for an 'bahn-brechende' scientific event as in the old days (30 May 1922)."[33] Cavendish researchers often saw the two white-haired physicists, J.J. and Rutherford, deep in conversation, perhaps about the results of recent cricket matches.

The Cavendish Laboratory under Rutherford developed into a global center for the study of nuclear physics and other experimental subjects, and it attracted a large number of talented researchers from around the world.[34] Chadwick, P. M. S. Blackett, C. D. Ellis, J. D. Cockcroft, E. T. S. Walton, Oliphant, and P. Kapitza were some of the researchers who worked in the Cavendish during his tenure. Although quite different from his former mentor in many respects, Rutherford too received a great deal of respect and love from his students, who often called him "papa" or "crocodile (a nickname given him by Kapitza)." A young Italian physicist, Emilio Segrè, remembered Rutherford in 1934 as follows:

> When he went around the laboratory, he would sit on one of the laboratory stools, extract the butt of a pencil from his waistcoat, and check the results of the experiment in progress. The research workers, when addressed, nearly stood at attention and ran rather than walked. This response certainly did not arise out of formal discipline but from the intrinsic respect that Rutherford elicited. A comment from Rutherford, whether good or bad, was not taken lightly. In other places I have seen famous laboratory directors treated almost condescendingly by young scientists, but this certainly did not happen to Rutherford.[35]

As we have seen, the history of the Cavendish Laboratory is not only the history of a great physics laboratory but also the history of great men. There may be no better explanation for the remarkable success of the Cavendish during the late nineteenth and early twentieth centuries than the following Chinese proverb: "Although blue originates from indigo-blue, the former is bluer than the latter."

---

[33] L. Badash, *Rutherford and Boltwood*, 322, 336, 351.

[34] For the history of the Cavendish Laboratory under Rutherford, see Oliphant, *Rutherford: Recollections*; Crowther, *The Cavendish Laboratory*, 176-268; J. Hendry (ed.), *Cambridge Physics in the Thirties* (Bristol: Adam Hilger, 1984); and Jeffrey A. Hughes, *The Radioactivists: Community, Controversy and the Rise of Nuclear Physics* unpublished Ph.D. dissertation (Cambridge: University of Cambridge, 1993).

35 E. Segrè, *From X-rays to Quarks: Modern Physicists and Their Discoveries* (New York: W. H. Freeman and Co., 1980), 113.

REFERENCES

Adie, R. H. "On the Osmotic Pressure of Salts in Solution," *Journal of the Chemical Society* 59 (1891): 344-372.

________. "A Direct Comparison of the Physical Constants involved in the Determination of Molecular Weights by Raoult's Method," *Proceedings of the Chemical Society 7* (1891): 26.

Adams, E. P. "Water Radioactivity," *Phil. Mag. 6* (1903): 563-569.

Almy, J. E. "On a Diminution of the Potential Difference between the Electrodes of a Vacuum Tube, produced by a Magnetic Field at the Cathode," *Proc. Camb. Phil. Soc. 11* (1901): 183-190.

Anderson, D. L. *The Discovery of the Electron* (Princeton: Van Nostrand, 1964).

[Anonymous Editorial]. "The New Physical Laboratory of the University of Cambridge," *Nature 10* (1874): 139-142.

Aston, F. W. "The Distribution of Electric Force in the Crookes Dark Space," *Proc. Roy. Soc. 84* (1910): 526-535.

________. "Sir J. J. Thomson's New Method of Chemical Analysis," *Scientific Progress* (July, 1912): 48.

________ & H. E. Watson. "On the Relation between Current, Voltage, Pressure, and the Length of the Dark Space in different Gases," *Proc. Roy. Soc. 86* (1911): 168-180.

________. "On the Discharge between Concentric Cylinders in Gases at Low Pressures," *Proc. Roy. Soc. 87* (1912): 428-436.

________. "On the Influence of the Nature of the Cathode on the Length of the Crookes Dark Space," *Proc. Roy. Soc. 87* (1912): 437-451.

________. "A Simple Form of Micro-Balance for determining the Densities of small Quantities of Gases," *Proc. Roy. Soc. 89* (1913): 439-446.

________. "J.J. Thomson," *The Times* (4 September 1940).

Badash, L. *Rutherford and Boltwood: Letters on Radioactivity* (New Haven: Yale University Press, 1969).

Baker, W. C. "Note on the Influence of Ultra Violet Radiation on the Discharge in a Vacuum Tube having a polished Zinc Electrode," *Proc. Camb. Phil. Soc. 11* (1902): 472-476.

Ball, W. W. R. *History of the Mathematical Tripos* (Cambridge, 1880).

________. *History of the Study of Mathematics at Cambridge* (Cambridge, 1889).

________. *Cambridge Papers* (London: Macmillan and Co., 1918).

Barkla, C. G. "The Velocity of Electric Waves along Wires," *Phil. Mag. 1* (1901): 652-666.

________. "Secondary Radiation from Gases subject to X-rays," *Phil. Mag. 5* (1903): 685-698.

Barlow, P. S. "Osmotic Experiments on Mixtures of Alcohol and Water," *Phil. Mag. 10* (1905): 1-12.

________. "On the Osmotic Pressures of Dilute Aqueous Solutions," *Proc. Camb. Phil. Soc. 13* (1905): 229-238.

________. "Osmotic Experiments on Mixtures of Alcohol and Water," *Phil. Mag. 10* (1905): 1-12.

________. "The Osmotic Pressures of Alcoholic Solutions," *Phil. Mag. 11* (1906): 595-604.

Barrell, H. "The Rayleighs and the National Physical Laboratory," *Applied Optics 3* (1964): 1125-1128.

Bartholomew, J. R. *The Formation of Science in Japan* (New Haven: Yale University Press, 1989).

Basalla, G., W. Coleman & R. Kargon (ed.). *Victorian Science: A Self Portrait from the Presidential Addresses of the British Association for the Advancement of Science* (Garden City, N.Y.: Doubleday, Anchor Books, 1970).

Batchelor, G. *The Life and Legacy of G. I. Taylor* (Cambridge: Cambridge University Press, 1996).

Beatty, R. T. "The Production of Cathode Particles by Homogeneous Röntgen Radiations, and their Absorption by Hydrogen and Air," *Phil. Mag. 20* (1910): 320-330.

________. "The Production of Cathode Particles by Homogeneous Röntgen Radiations," *Proc. Camb. Phil. Soc. 15* (1910): 416-422.

________. "On a Dissymetry in the Emission of the Cathode Particles which are produced by Homogeneous Röntgen Radiations," *Proc. Camb. Phil. Soc. 15* (1910): 492-494.

________. "The Ionisation of Heavy Gases by X-Rays," *Proc. Roy. Soc. 85* (1911): 230-239.

________. "The direct Production of Characteristic Röntgen Radiation by Cathode Particles," *Proc. Roy. Soc. 87* (1912): 511-518.

________. "The Energy of Röntgen Rays," *Proc. Roy. Soc. 89* (1913): 314-327.

________. "On the Energy required to ionize an Atom," *Phil. Mag. 26* (1913): 183-186.

Becher, H. W. "Voluntary Science in Nineteenth Century Cambridge University to the 1850's," *BJHS 19* (1986): 57-87.

Bennett, J. et al. *Empires of Physics: A Guide to the Exhibition* (Cambridge: Whipple Museum of the History of Science, 1993).

Bevan, P. V. "On some Phenomena connected with the Combination of Hydrogen and Chlorine under the Influence of Light," *Proc. Camb. Phil. Soc. 11* (1901): 264-266.

________. "Reflection and Transmission of Light by a charged Metal Surface," *Proc. Camb. Phil. Soc. 11* (1902): 438-444.

________. "On the Influence on Light reflected from and transmitted through a Metal of a Current through the Metal," *Proc. Camb. Phil. Soc. 11* (1902): 380-390.

________. "The Combination of Hydrogen and Chlorine under the Influence of Light," *Phil. Trans. 202* (1903): 71-121.

________. "On the Joule-Thomson Effect," *Proc. Camb. Phil. Soc. 12* (1903): 127-134.

________. "A Lecture Experiment to illustrate the Effect of a straight Current on a Magnetic Pole" *Proc. Camb. Phil. Soc. 12* (1903): 212-214.

________. "The Temperature Effect in the Combination of Hydrogen and Chlorine under the Influence of Light," *Proc. Camb. Phil. Soc. 12* (1904): 398-405.

________. "Note on some Physical Properties of Sodium Vapour," *Proc. Camb. Phil. Soc. 13* (1905): 129-131.

________. "A Method of following the course of certain Chemical Actions, and a Period of Induction in the Action of Excess of Water on Monochloracetic Acid," *Proc. Camb. Phil. Soc. 13* (1906): 269-281.

________. "Lloyd's Fringes for internal Reflexion, and the Change of Phase of ordinarily reflected Light," *Phil. Mag. 14* (1907): 503-508.

Birk, J. B. (ed.) *Rutherford at Manchester* (London: Heywood & Co., 1962).

Bispham, J. S. "Potential Gradient in Glow Discharge from a Point to a Plane," *Proc. Roy. Soc. 81* (1908): 477-495.

Blackett, P. M. S. "The Craft of Experimental Physics," in Harold Wright (ed.) *University Studies, Cambridge 1933* (London: Ivory Nicholson & Watson, 1933): 67-96.

________. "Charles Thomas Rees Wilson, 1869-1959," *Biographical Memoirs of Fellows of the Royal Society 6* (1960): 268-295.

Bohr, N. "Note on the Electron Theory of Thermoelectric Phenomena," *Phil. Mag. 23* (1912): 984-986.

________."Reminiscences of the Founder of Nuclear Science and of Some Developments based on His Work," *Proceedings of the Physical Society of London 78* (1961): 1085.

Bragg, W. L. "The Diffraction of Short Electromagnetic Waves by a Crystal," *Proc. Camb. Phil. Soc. 17* (1912): 43-57.

________. "The Crystalline Structure of Copper," *Phil. Mag. 28* (1914): 355-360.

________. "The Austin Wing of the Cavendish Laboratory," *Nature 158* (1946): 326-327.

________. "Organization and Work of the Cavendish Laboratory," *Nature 161* (1948): 627-628.

________. "Famous Experimental Apparatus in the Cavendish Laboratory, Cambridge," *Nature 166* (1950): 7-9.

________. "The Cavendish Laboratory Archives," *Nature 169* (1952): 684-686.

Bragg, W. H. "Radioactivity as a Kinetic Theory of a Fourth State of Matter," *Archives of the Roentgen Ray* (April, 1911): 402-415.

Breul, K. *Students' Life and Work in the University of Cambridge: Two Lectures*, revised edition (Cambridge: Bowes and Bowes, 1910).

[British Association Committee on Standards of Electrical Resistance]. "Report of the Committee appointed by the British Association on Standards of Electrical Resistance," *B. A. Report* (1862): 125-163; *B. A. Report* (1863): 111-176; *B. A. Report* (1864): 345-367; *B. A. Report* (1865): 308-313.

[British Association Committee for Testing Experimentally Ohm's Law]. "Report of the Committee for Testing Experimentally Ohm's Law," *B. A. Report* (1876): 36-63.

Bryan, G. B. "On the Determination of the Conductivity of Liquids in Thin Layers," *Phil. Mag. 45* (1898): 253-273.

Buchwald, Jed Z. *From Maxwell to Microphysics* (Chicago: University of Chicago Press, 1985).

________ & A. Warwick (eds.) *Histories of the Electron* (Cambridge, Mass.: MIT Press, forthcoming).

Bumstead, H. A. "The Heating Effects produced by Röntgen Rays in different Metals, and their Relation to the Question of Change in the Atom," *Phil. Mag. 11* (1906): 292-317.

________. "On the Absence of excited Radio-activity due to temporary Exposure to γ Rays," *Proc. Camb. Phil. Soc. 13* (1905): 125-128.

Bunge, M. & W. R. Shea (ed.) *Rutherford and Physics at the Turn of the Century* (New York: Dawson & S. H. P., 1979).

Burke, J. B. B. "On the Phosphorescent Glow in Gases," *Phil. Mag. 1* (1901): 342-356 & 455-464.

________. "On Luminosity and the Kinetic Theory," *B. A. Report* (1902): 549.

________. "On a Modification of FitzGerald's Model of the Ether," *B. A. Report* (1904): 478-478.

Burton, C. V. "Experiments on Colour-perception and on a Photo-voltaic Theory of Vision," *Proc. Camb. Phil. Soc. 6* (1888): 308-313.

Burton, E. F. "On the Properties of Electrically prepared Colloidal Solutions," *Phil. Mag. 11* (1906): 425-447.

________. "The Action of Electrolytes on Colloidal Solutions," *Phil. Mag. 12* (1906): 472-478.

Burton, W. "The Refractive Index and Dispersion of Light in Argon and Helium," *Proc. Roy. Soc. 80* (1908): 390-405.

Cahan, D. "The Institutional Revolution in German Physics, 1865-1914," *HSPS 15:2* (1985): 1-65.

________. *An Institute for an Empire: The Physikalisch-Technisch Reichsanstalt, 1871-1918* (Cambridge: Cambridge University Press, 1989).

Callendar, H. L. "On the Practical Measurements of Temperature; Experiments made at the Cavendish Laboratory," *Phil. Trans. 178* (1887): 161-230.

________. "On the Method of measuring Surface Tension by Observations of the Form of Liquid Drops," *Proc. Camb. Phil. Soc. 6* (1888): 179.

[Cambridge University Physics Society]. *A Hundred Years and More of Cambridge Physics* (Cambridge, 1974, 1980, 1995).

Campbell, L. & W. Garnett. *The Life of James Clerk Maxwell* (London: MacMillan & Co., 1882).

Campbell, L. L. "Disintegration of the Aluminum Cathode," *Phil. Mag. 28* (1914): 347-354.

Campbell, N. R. "Some Experiments on the Electrical Discharge from a Point to a Plane," *Phil. Mag. 6* (1903): 618-627.

________. "A Null Method of measuring small Ionisations," *Proc. Camb. Phil. Soc. 13* (1905): 132.

________. "The Radiation from Ordinary Materials," *Phil. Mag. 9* (1905): 531-544; *Phil. Mag. 11* (1906): 206-224.

________. "Radioactivity and Chemical Change," *Phil. Mag. 9* (1905): 545-549.

________. "The Radio-activity of Metals and their Salts," *Proc. Camb. Phil. Soc. 13* (1906): 282-287.

________ & A. Wood, "The Radio-activity of the Alkali Metals," *Proc. Camb. Phil. Soc. 14* (1906): 15-21.

________. "The Effect of Change of Temperature on Spontaneous Ionisation," *Phil. Mag. 13* (1907): 614-622.

________. "The β Rays from Potassium," *Proc. Camb. Phil. Soc. 14* (1907): 211-216.

________. "The Number of Electrons in an Atom," *Proc. Camb. Phil. Soc. 14* (1907): 287-291.

________. "The Radioactivity of Potassium, with special reference to Solutions of its Salts," *Proc. Camb. Phil. Soc. 14* (1908): 557-567.

________. "The Absorption of β Rays by Liquids," *Phil. Mag. 17* (1909): 180-190.

________. "The Radioactivity of Rubidium," *Proc. Camb. Phil. Soc. 15* (1909): 11-12.

________. "The Study of Discontinuous Phenomena," *Proc. Camb. Phil. Soc. 15* (1909): 117-136.

________. "Discontinuity in Light Emission," *Proc. Camb. Phil. Soc. 15* (1909): 310-328 & (1910): 513-525.

Capstick, J. W. "On the Ratio of the Specific Heats of the Paraffins, and their Monohalogen Derivatives," *Phil. Trans. 185* (1894): 1-36.

________. "On the Ratio of the Specific Heats of Some Compound Gases," *Phil. Trans. 186* (1895): 567-592.

________. "Exhibition of Goldstein's Experiments on Kathode Rays," *Proc. Camb. Phil. Soc. 8* (1895): 305.

________. "On the Kathode Fall of Potential in Gases," *Proc. Roy. Soc. 63* (1898): 356-365.

Cardwell, D. S. L. *The Organization of Science in England,* revised edition (London: Heinemann, 1972).

Caroe, G. M. *William Henry Bragg, 1862-1942: Man and Scientist* (Cambridge: Cambridge University Press, 1978).

Carse, G. A. & T. H. Laby. "On a Relation between the Velocity and the Volume of the Ions of certain Organic Acids and Bases," *Proc. Camb. Phil. Soc. 13* (1906): 288-295.

________ & ________. "A Relation between the Velocity and Volume of Organic Ions in Aqueous Solutions," *Proc. Camb. Phil. Soc. 14* (1906): 1-11.

Cassie, W. "On the Effect of Temperature on the Specific Inductive Capacity of a Dielectric," *Phil. Trans. 181* (1890): 1-17.

Cattermole, M. J. G. & A. F. Wolfe. *Horace Darwin's Shop: A History of the Cambridge Scientific Instrument Company, 1878 to 1968* (Bristol: Adam Hilger, 1987).

Chaloner, C. "The Most Wonderful Experiment in the World: a History of the Cloud Chamber," *BJHS 30* (1997): 357-374.

Chapman, J. C. "An Attempt to detect a Fatigue Effect in the production of Secondary Röntgen Radiations," *Proc. Camb. Phil. Soc. 16* (1911): 142-147.

________ & E. D. Guest. "The Intensity of Secondary Homogeneous Röntgen Radiation from Compounds," *Proc. Camb. Phil. Soc. 16* (1911): 136-141.

________. "Preliminary Note on the Characteristic Homogeneous Röntgen Radiation from Elements of High Atomic Weight," *Proc. Camb. Phil. Soc. 16* (1911): 399-404.

________. "Fluorescent Röntgen Radiation from Elements of High Atomic Weight," *Proc. Roy. Soc. 86* (1912): 439-451.

________. "A Comparison of the Spectra of Fluorescent Röntgen Radiations," *Proc. Roy. Soc. 88* (1912): 24-37.

________. "A Negative Result on the Refraction of X-rays," *Proc. Camb. Phil. Soc. 16* (1912): 574-579.

Child, C. D. "Anode Drop with Hot CaO Cathode," *Physical Review 29* (1909): 351-368.

Chittock, C. "The Migration Constants of Dilute Solutions of Hydrochloric Acid," *Proc. Camb. Phil. Soc. 15* (1909): 55-63.

Chree, C. "Conduction of Heat in Liquids," *Proc. Roy. Soc. 43* (1887): 30-48.

________. "Conduction of Heat in Liquids, Historical Treatment," *Phil. Mag. 24* (1887): 1-27.

________. "On Æolotropic Elastic Solids," *Proc. Roy. Soc. 44* (1888): 214-218.

________. "On the Effect of an Electric Current on Saturated Solutions," *Proc. Camb. Phil. Soc. 6* (1888): 223-229.

________. "On the Stresses in Rotating Spherical Shells," *Proc. Camb. Phil. Soc. 6* (1889): 332-334.

________. "On the Effects of Pressure on the Magnetisation of Cobalt," *Phil. Trans. 181* (1890): 329-387; also *Proc. Roy. Soc. 47* (1890): 41-44.

________. "Some Experiments on Liquid Electrodes in Vacuum Tubes," *Proc. Camb. Phil. Soc. 7* (1891): 222-46.

Chrystal, G. "On Bi- and Unilateral Galvanometer Deflection," *Phil. Mag. 2* (1876): 401-414.

________ & S. A. Saunder. "Results of a Comparison of the British-Association Units of Electrical Resistance," *B. A. Report* (1876): 13-19.

Clayden, A. W. & C. T Heycock. "The Spectra of Indium," *Phil. Mag. 2* (1876): 387-389.

Coldridge, W. "On the Electrical and Chemical Properties of Stannic Chloride; together with the Bearing of the Results therein obtained on the Problems of Electrolytic Conduction and Chemical Action," *Phil. Mag. 29* (1890): 383-490.

Comstock, D. F. "The Relation of Mass to Energy," *Phil. Mag. 15* (1908): 1-21.

Cook, H. L. "Experiments on Penetrating Radiation," *Proc. Camb. Phil. Soc. 13* (1905): 160-164.

Crosland, M. P. (ed.) *The Emergence of Science in Western Europe* (London: MacMillan, 1975).

Crowther, J. A. "On the Coefficient of Absorption of the β Rays from Uranium," *Phil. Mag. 12* (1906): 379-392.

________. "On the Secondary Röntgen Radiation from Gases and Vapours," *Phil. Mag. 14* (1907): 653-675.

________. "On the Scattering of the β-Rays from Uranium by Matter," *Proc. Roy. Soc. 80* (1907): 186-206.
________. "On the Fatigue of Metals subjected to Radium Rays," *Proc. Camb. Phil. Soc. 14* (1907): 340-350.
________. "On the Passage of Röntgen Rays through Gases and Vapours," *Proc. Roy. Soc. 82* (1909): 103-127.
________. "On the Effect of Temperature on Ionisation," *Proc. Roy. Soc. 82* (1909): 351-357.
________. "On the Relation between Ionisation and Pressure for Röntgen Rays in different Gases," *Proc. Camb. Phil. Soc. 15* (1909): 34-37.
________. "On the Relative Ionisation produced by Röntgen Rays in different Gases," *Proc. Camb. Phil. Soc. 15* (1909): 38-39.
________. "On the Secondary Röntgen Radiation from Air and Ethyl Bromide," *Proc. Camb. Phil. Soc. 15* (1909): 101-105.
________. "On the Scattering of the β-Rays from Radium by Air," *Proc. Camb. Phil. Soc. 15* (1909): 273-274.
________. "On the Scattering of Homogeneous β-Rays and the Number of Electrons in the Atom," *Proc. Roy. Soc. 84* (1910): 226-247.
________. "On the Energy and Distribution of Scattered Röntgen Radiation," *Proc. Roy. Soc. 85* (1910): 29-43.
________. "On the Transmission of β-Rays," *Proc. Camb. Phil. Soc. 15* (1910): 442-458.
________. "On the Distribution of the Secondary Röntgen Radiations round the Radiator," *Proc. Camb. Phil. Soc. 16* (1910): 112-120.
________. "On an attempt to detect Diffusion in a Pencil of Röntgen Rays," *Proc. Camb. Phil. Soc. 16* (1911): 177-187.
________. "Further Experiments on Scattered Röntgen Radiation," *Proc. Camb. Phil. Soc. 16* (1911): 188.
________. "On the Scattered Radiation from different Radiators," *Proc. Camb. Phil. Soc. 16* (1911): 365-369.
________. "On the Distribution of the Scattered Röntgen Radiation," *Proc. Roy. Soc. 86* (1912): 478-494.
________. "On a Theory of the disymmetrically distributed Secondary Röntgen Radiation," *Proc. Camb. Phil. Soc. 16* (1912): 534-539.
________. "Research Work in the Cavendish Laboratory in 1900-1918," *Supplement to Nature 118* (1926): 58-60.
Crowther, J. G. *British Scientists of the Nineteenth Century* (London: Kegan, Paul, Trench Trubner & Co., 1935).
________. *British Scientists of the Twentieth Century* (London: Routledge & Kegan Paul, 1952).
________. *The Cavendish Laboratory, 1874-1974* (New York: Science History Publications, 1974).
Cunningham, J. A. "The Discharge of Electricity through Gases and the Temperature of the Electrodes," *Phil. Mag. 4* (1902): 684-703.
________. "On an Attempt to detect the Ionisation of Solutions by the action of Light and Röntgen Rays," *Proc. Camb. Phil. Soc. 11* (1902): 431-433.
________. "The Cathode Fall from Hot Carbon," *Phil. Mag. 9* (1905): 193-207.
Dahl, P. F. *Flash of the Cathode Rays: A History of J.J. Thomson's Electron* (Bristol: Institute of Physics, 1997).
Davis, E. A. & I. J. Falconer. *J.J. Thomson and the Discovery of the Electron* (London: Taylor & Francis, 1997).
Davis, J. L. "The Research School of Marie Curie in the Paris Faculty, 1907-14," *Annals of Science 52* (1995): 321-355.
Darwin, G. H. "On the Formation of Ripple-Mark in Sand," *Proc. Roy. Soc. 36* (1884): 18-43.
________ & H. Darwin. "First Report of British Association Committee for the Measurement of the Lunar Disturbance of Gravity," *B. A. Report* (1880): 25-26.
________ et al. "Second Report of British Association Committee for the Measurement of the Lunar Disturbance of Gravity," *B. A. Report* (1882): 95.
Domb, C. "James Clerk Maxwell—100 Years later," *Nature 282* (1979): 235-39.

________. "James Clerk Maxwell in London, 1859-1865," *Notes and Records of the Royal Society of London 35* (1980): 67-103.

Donaldson, H. "On the Spectra of the Electrodeless Ring Discharge in certain Gases," *Phil. Mag. 22* (1911): 720-727.

Douglas, M. *How Institutions Think* (Syracuse, N. Y.: Syracuse University Press, 1986).

Durack, J. J. E. "Lenard Rays," *Phil. Mag. 4* (1902): 29-45.

________. "On the Specific Ionisation produced by the Corpuscles given out by Radium," *Phil. Mag. 5* (1903): 550-561.

Edmonds, S. A. "Quasi-Radio-activity produced by the Point Discharge," *Proc. Camb. Phil. Soc. 12* (1904): 469-473.

Emery, G. F. "Thermo-electric Properties of Salt Solutions," *Proc. Roy. Soc. 55* (1894): 356-373.

Erikson, H. A. "The Recombination of the Ions in Air at different Temperature," *Phil. Mag. 18* (1909): 328-336.

Erskine-Murray, J. R. "On the Effect of the Röntgen X Rays on the Contact Electricity of Metals," *Proc. Roy. Soc. 59* (1896): 333-339.

________. "On Contact Electricity of Metals," *Proc. Roy. Soc. 63* (1898): 113-146.

Evans, Ivor B. N. *Man of Power: The Life Story of Baron Rutherford of Nelson* (London: Stanley Paul & Co., 1930).

Eve, A. S. *Rutherford, being the Life and Letters of the Rt Hon. Lord Rutherford, O. M.* (Cambridge: Cambridge University Press, 1939).

Everitt, C. W. F. *James Clerk Maxwell: Physicist and Natural Philosopher* (New York: Charles Scribner's Sons, 1975).

Ewing, J. A. & H. G. Klaassen. "Magnetic Qualities of Iron," *Phil. Trans. 184* (1893): 985-1039.

Falconer, I. *The Cavendish Laboratory: An Outline Guide to the Museum*, a pamphlet (Cambridge, 1980).

________. "Corpuscles, Electrons and Cathode Rays: J.J. Thomson and the 'Discovery of the Electron," *BJHS 20* (1987): 241-276.

________. "J.J. Thomson's Work on Positive Rays, 1906-1914," *HSPBS 18:2* (1988): 265-310.

________. "J.J. Thomson and 'Cavendish Physics,'" in Frank A. J. L. James (ed.), *The Development of the Laboratory* (London: MacMillan Press, 1989), 104-117.

________. "Transfer of Concept & Practice between J.J. Thomson and His Research Students," Papers for the Meeting on the Transfer of Concepts and Practice in the Physical and Mathematical Sciences (Maison Francaise, Oxford, 23 May 1996).

Fawcett, P. G. "The Electric Strength of Mixtures of Nitrogen and Hydrogen." *Proc. Roy. Soc. 56* (1894): 263-271.

Feather, N. *Lord Rutherford*, new edition (London: Priory Press, 1973).

________. "Rutherford's Cavendish," *New Scientist 7* (1960): 598-600.

Feffer, S. M. "Arthur Schuster, J.J. Thomson, and the Discovery of the Electron," *HSPS 20* (1989): 33-61.

Fitzpatrick, T. C. "On the Application of Alternating Currents to the Determination of the Conductivity of Electrolytes," *B. A. Report* (1886): 328-334.

________. "On the Action of the Solvent in Electrolytic Conduction," *Phil. Mag. 24* (1887): 379-391.

________. "On the Specific Resistance of Copper," *B. A. Report* (1890): 120-130.

________. "Table of Electro-chemical Properties of Aqueous Solutions," *B. A. Report* (1893): 146-213.

Fleming, J. A. "On a New Form of Resistance-Balance adapted for comparing Standard Coils," *Phil. Mag. 9* (1880): 109-117.

________. "Problems on the Distribution of Electric Currents in Network of Conductors treated by the Method of Maxwell," *Phil. Mag. 20* (1885): 221-258.

Forman, P., J. L. Heilbron & S. Weart. "Physics circa 1900: Personnel, Funding and Productivity of the Academic Establishments," *HSPS 5* (1975): 1-185.

Fruton, J. S. *Contrasts in Scientific Style: Research Groups in the Chemical and Biochemical Sciences* (Philadelphia: American Philosophical Society, 1990).

Galison, P. & A. Assmus. "Artificial Clouds, Real Particles," in D. Gooding, T. Pinch & S. Schaffer (eds.) *The Use of Experiment* (Cambridge: Cambridge University Press, 1989): 224-274.

Galison, P. *Image and Logic: A Material Culture of Microphysics* (Chicago: University of Chicago Press, 1997).

Garnett, J. C. M. "Colours in Metal Glasses and in Metallic Films," *Phil. Trans. 203* (1904): 385-420; *Phil. Trans. 205* (1905): 237-288.

Gascoigne, J. "Mathematics and Meritocracy: The Emergence of the Cambridge Mathematical Tripos," *Social Studies of Science 14* (1984): 547-584.

Geison, G. L. *Michael Foster and the Cambridge School of Physiology* (Princeton: Princeton University Press, 1978).

________ & Frederic L. Holmes (ed.) *Research Schools: Historical Reappraisals, Osiris 8* (1993).

Gill, H. V. "On a New Kind of Glow from Palladium in Vacuum Tubes," *Proc. Roy. Soc. 82* (1909): 464-476.

Glaisher, J. W. L. "The Mathematical Tripos," *Proceedings of the London Mathematical Society 18* (1886): 4-38.

Glasson, J. L. "Secondary Röntgen Rays from Metallic Salts," *Proc. Camb. Phil. Soc. 15* (1910): 437-441.

________. "The Variation of Ionizing Power with the Velocity of Cathode Rays," *Phil. Mag. 22* (1911): 647-656.

Glazebrook, R. T. "An Experimental Determination of the Values of the Velocities of Normal Propagation of Plane Waves in different directions in a Biaxal Crystal, and a Comparison of the Results with Theory," *Phil. Trans. 170* (1878): 287-377.

________. "Note on Nicol's Prism," *Phil. Mag. 10* (1880): 247-254.

________. "Double Refraction and Dispersion in Iceland Spar: an Experimental Investigation, with a Comparison with Huyghens' Construction for the Extraordinary Waves," *Phil. Trans. 171* (1880): 421-449.

________. "Note on the Reflection and Refraction of Light," *Proc. Camb. Phil. Soc. 3* (1880): 326-339.

________. "On some Equations connected with the Electromagnetic Theory of Light," *Proc. Camb. Phil. Soc. 4* (1881): 155-167.

________. "On the Measurement of Small Resistances," *Phil. Mag. 11* (1881): 291-295.

________. "On a Method of Comparing the Electrical Capacities of two Condensers," *Phil. Mag. 11* (1881): 370-377.

________. "On the Molecular Vortex Theory of Electromagnetic Action," *Phil. Mag. 11* (1881): 397-413.

________. "On the Refraction of Plane Polarised Light at the Surface of a Uniaxal Crystal," *Phil. Trans. 173* (1882): 595-620.

________. "On the Isochromatic Curves of Polarised Light seen in a Uniaxal Crystal cut at Right Angles to the Optic Axis," *Proc. Camb. Phil. Soc. 4* (1883): 292-304.

________. "On a Spectrophotometer," *Proc. Camb. Phil. Soc. 4* (1883): 304-308.

________. "On a Common Defect of Lenses," *Proc. Camb. Phil. Soc. 4* (1883): 308.

________. "On Polarising Prisms," *Phil. Mag. 15* (1883): 352-362.

________. "On Curved Diffraction Gratings," *Phil. Mag. 15* (1883): 414-423.

________, J. M. Dodds & E. B. Sargant. "Experiments on the Value of the British Association Unit of Resistance," *Phil. Trans. 174* (1883): 223-268.

________. "A Comparison of Maxwell's Equations of the Electromagnetic Field with those of Helmholtz and Lorenz," *Proc. Camb. Phil. Soc. 5* (1884): 120-131.

________. "On the General Equations of the Electromagnetic Field," *Proc. Camb. Phil. Soc. 5* (1884): 142-153.

________. "On the Effect of Moisture in modifying the Refraction of Plane Polarised Light by Glass, *Proc. Camb. Phil. Soc. 5* (1884): 169-175.

________. "On a Method of Measuring the Electrical Capacity, and on the Determination by Electrical Observations of the Period of a Tuning- fork," *Phil. Mag. 18* (1884): 98-105.

________. "On a Young's Eriometer," *Proc. Camb. Phil. Soc. 5* (1885): 223.

________. "On the Theory of some Experiments of Fröhlich in the Position of the Plane of Polarization of Light diffracted at Reflexion from a Grating," *Proc. Camb. Phil. Soc. 5* (1885): 254-257.

________. "A Comparison of the Standard Resistance Coils of the British Association, with Mercury Standards constructed by M.J.R. Benoit of Paris and Herr Strecher of Wurzburg," *Phil. Mag. 20* (1885): 343-354.

________ & W. N. Shaw. *Practical Physics*, second edition (London: Longmans, Greens & Co., 1886).

________ & T. C. Fitzpatrick. "On the Specific Resistance of Mercury," *Phil. Trans. 179* (1888): 351-376.

________ & T. C. Fitzpatrick. "On the Permanence of the Original Standards of Resistance of the British Association and of other Standard Coils," *B. A. Report* (1888): 57-69.

________. "On the Air Condensers of the British Association," *B. A. Report* (1890): 102-120.

________. "On the Value of some Mercury Resistance Standards," *Phil. Mag. 32* (1891): 70-75.

________ & S. Skinner. "Some Notes on Clark's Cells," *Proc. Camb. Phil. Soc. 7* (1892): 250.

________ & S. Skinner. "On the Clark Cell as a Standard of Electromotive Force," *Phil. Trans. 183* (1892): 567-628.

________. *Mechanics: An Elementary Text-Book, Theoretical and Practical* (Cambridge: Cambridge University Press, 1895).

________. *James Clerk Maxwell and Modern Physics* (New York: MacMillan & Co., 1896).

________. "The Cavendish Laboratory 1876-1900," *Supplement to Nature 118* (1926): 52-58.

Gooday, G. "Precision Measurement and the Genesis of Physics Teaching Laboratories in Victorian Britain," *BJHS 23* (1990): 25-51.

Gordon, J. E. H. "On the Determination of Verdet's Constants in Absolute Units," *Phil. Trans. 167* (1877): 1-34.

________. "Measurements of Electrical Constants—No. II. On the Specific Inductive Capacities of Certain Dielectrics. Part I," *Phil. Trans. 170* (1879): 417-446.

Gould, P. "Women and the Culture of University Physics in Late Nineteenth-Century Cambridge," *BJHS 30* (1997): 127-149.

Harman, P. M. *Energy, Force, and Matter* (Cambridge: Cambridge University Press, 1982).

________ (ed.) *Wranglers and Physicists: Studies on Cambridge Mathematical Physics in the Nineteenth Century* (Manchester: Manchester University Press, 1985).

Harris, T. "On the Distribution of Electric Force in the Discharge at Low Temperature," *Phil. Mag. 30* (1915): 182-192.

Harrison, E. P. "On the Temperature-Variation of the Coefficient of Expansion of Pure Nickel," *Phil. Mag. 7* (1904): 626-634.

Heilbron, J. L. *H. G. J. Moseley: The Life and Letters of an English Physicist, 1887-1915* (Berkeley: University of California Press, 1974).

________. "Lectures on the History of Atomic Physics, 1900-1922," in C. Weiner (ed.), *History of Twentieth Century Physics: Proceedings of the International School of Physics <<Enrico Fermi>>, Course 57* (New York: Academic Press, 1977), 40-108.

________ & T. S. Kuhn, "The Genesis of the Bohr Atom," *HSPS 1* (1969): 211-290.

________ & Robert W. Seidel, *Lawrence and His Laboratory: A History of the Lawrence Berkeley Laboratory*, vol. 1 (Berkeley: University of California Press, 1989).

Henderson, W. C. "On Electrical Effects due to Evaporation of Sodium in Air and other Gases," *Proc. Roy. Soc. 66* (1900): 183-186.

________ & J. Henry. "Experiments on the Motion of the Æther in an Electromagnetic Field," *Phil. Mag. 44* (1897): 20-26.

Hendry, J. (ed.) *Cambridge Physics in the Thirties* (Bristol: Adam Hilger, 1984).

________. *James Clerk Maxwell and the Theory of the Electromagnetic Field* (Bristol: Adam Hilger, 1986).

Henry, J. "Experiments on the Effect of Ultra-Violet Light on the Conductivity of Iodine Vapour," *Proc. Camb. Phil. Soc. 9* (1895-98): 319-322.

________. "On the Deflexion by Magnetic Force of the Electric Discharge through Gases," *Phil. Mag. 46* (1898): 429-452.

Hevesy, G. C. de. "Francis William Aston, 1877-1945," *Obituary Notices of Fellows of the Royal Society 5* (1945-1948): 634-650.

Hiebert, E. N. "Electric Discharge in Rarefied Gases: The Dominion of Experiment. Faraday, Plücker, Hittorf," in A. J. Fox (ed.), *No Truth Except in the Details: Essays in Hounour of Martin J. Klein* (Dodrecht: Kluwer Academic Publishers, 1995), 95-134.

Hilken, T. J. N. *Engineering at Cambridge University, 1783-1965* (Cambridge: Cambridge University Press, 1967).

Home, R. W. *Physics in Australia to 1945: Bibliography and Biographical Register* (Melbourne: University of Melbourne, 1990).

Hopley, I. B. "Maxwell's Work on Electrical Resistance. 1. The Determination of the Absolute Unit of Resistance," *Annals of Science 13* (1957): 265-272.

________. "2. Proposals for the Re-determination of the B.A. Unit of 1863," *Annals of Science 14* (1958): 197-210.

________. "3. Improvement of Mance's Method for the Measurement of Battery Resistance," *Annals of Science 15* (1959): 51-55.

________. "Maxwell's Determinations of the Number of Electrostatic Units in One Electromagnetic Unit of Electricity," *Annals of Science 15* (1959): 91-108.

Horton, F. "The Effects of Changes of Temperature on the Modulus of Torsional Rigidity of Metal Wires," *Phil. Trans. 204* (1904): 1- 55.

________. "On the Modulus of Torsional Rigidity of Quartz Fibres and its Temperature Coefficient," *Phil. Trans. 204* (1905): 407-431.

________. "The Electrical Conductivity of Metallic Oxides," *Phil. Mag. 11* (1906): 505-531.

________. "The Spectrum of the Discharge from a glowing Lime Cathode in Mercury Vapour," *Proc. Camb. Phil. Soc. 14* (1908): 501-507.

________. "The Emission of Positive Rays from heated Phosphorus Compounds," *Proc. Camb. Phil. Soc. 15* (1909): 329-333.

________. "A Spectroscopic Investigation of the Nature of the Carriers of Positive Electricity from heated Aluminum Phosphate," *Proc. Roy. Soc. 84* (1910): 433-449.

________. "The Vacuum Tube Spectra of Mercury," *Proc. Roy. Soc. 85* (1911): 288-302.

________. "On the Origin of Spectra," *Proc. Camb. Phil. Soc. 16* (1911): 313-317.

________. "The Positive Ionisation produced by Platinum and by certain Salts when heated," *Proc. Roy. Soc. 88* (1912): 117-146.

________. "On the Origin of the Electron Emission from Glowing Solids," *Proc. Roy. Soc. 90* (1914): 548-549.

Hosking, R. "The Electrical Conductivity and Fluidity of Solutions," *Phil. Mag. 7* (1904): 469-484.

Howard, J. N. "John William Strutt, third Baron Rayleigh," *Applied Optics 3* (1964): 1091-1101; "Eleanor Mildred Sidgwick and the Rayleighs," 1120-1122; "The Rayleigh Notebooks," 1129-1132.

Huff, W. B. "The Electrostatic Deviation of α-Rays from Radio-Tellurium," *Proc. Roy. Soc. 78* (1906): 77-79.

Hughes, A. L. "On the Mobility of the Ions produced in Air by Ultra-Violet Light," *Proc. Camb. Phil. Soc. 15* (1910): 483-491.

________. "On the Velocities of the Electrons produced by Ultra-Violet Light," *Proc. Camb. Phil. Soc. 16* (1911): 167-174.

________. "The Photo-Electric Effects of Certain Compounds," *Proc. Camb. Phil. Soc. 16* (1911): 376-382.

________. "A Note on Short Wave Lengths in the Mercury Arc Spectrum," *Proc. Camb. Phil. Soc. 16* (1911): 428-429.

________. "On the Ultra-Violet Light from the Mercury Arc," *Phil. Mag. 21* (1911): 393-404.

________. "The Photo-Electric Effect of Some Compounds," *Phil. Mag. 24* (1912): 380-390.

________. "A Sensitive Photo-Electric Cell," *Phil. Mag. 25* (1913): 679-682.

________. "On the Velocities with which Photo-Electrons are emitted from Matter," *Phil. Mag. 25* (1913): 683-686.

Hughes, J. A. *The Radioactivists: Community, Controversy and the Rise of Nuclear Physics* unpublished Ph.D. thesis (Cambridge: University of Cambridge, 1993).

________. " 'Brains in their fingertips': physics at the Cavendish Laboratory 1880-1940," in R. Mason (ed.) *Cambridge Minds* (Cambridge: Cambridge University Press, 1994), 160-176.

Hunt, B. J. " 'Practice vs. Theory': The British Electrical Debate, 1881-1891," *Isis 74* (1983): 341-355.

________. *The Maxwellians* (Ithaca: Cornell University Press, 1991).

Innes, P. D. "On the Velocity of the Cathode Particles emitted by Various Metals under the Influence of Röntgen Rays, and its bearing on the Theory of Atomic Disintegration," *Proc. Roy. Soc. 79* (1907): 442-462.

Jaffe, G. "On the Conductivity of Gases contained in small Vessels," *Phil. Mag. 8* (1904): 556-567.

________. "Recollections of Three Great Laboratories," *Journal of Chemical Education 29* (1952): 230-238.

Jarausch, K. H. (ed.) *The Transformation of Higher Learning 1860-1930: Expansion, Diversification, Social Opening, and Professionalization in England, Germany, Russia, and the United States* (Chicago: University of Chicago Press, 1983).

Jenkins, W. A. "On the Effect of a Magnetic Field on Metallic Resistance," *Phil. Mag. 27* (1914): 731-739.

Jolly, H. L. P. "The Distribution of Energy in the Spectra of Gases," *Phil. Mag. 26* (1913): 801-827.

Jolly, W. P. *Sir Oliver Lodge* (London: Constable, 1974).

Jones, R. V. "The Evolution of the Laboratory," *Science Journal* (January, 1967): 81-85.

________. "James Clerk Maxwell at Aberdeen, 1851-1860," *Notes and Records of the Royal Society of London 28* (1973): 57-81.

Jungnickel, C. & R. McCormmach. *Intellectual Mastery of Nature: Theoretical Physics from Ohm to Einstein*, 2 vols. (Chicago: University of Chicago Press, 1986).

Kalandyk, S. J. "On the Conductivity of Salt Vapours," *Proc. Roy. Soc. 90* (1914): 634-647.

Kargon, R. H. *Science in Victorian Manchester* (Baltimore: Johns Hopkins University Press, 1977).

________ & P. Achinstein (ed.) *Kelvin's Baltimore Lectures and Modern Theoretical Physics* (Cambridge, Mass.: MIT Press, 1987).

Kaye, G. W. C. "The Emission and Transmission of Röntgen Rays," *Phil. Trans. 209* (1908): 123-151.

________. "On the Distribution of the Röntgen Rays from a Focus Bulb," *Proc. Roy. Soc. 83* (1909): 189-194.

________. "The Selective Absorption of Röntgen Rays," *Proc. Camb. Phil. Soc. 14* (1907): 236-245.

________. "The Emission of Röntgen Rays from Thin Metallic Sheets," *Proc. Camb. Phil. Soc. 15* (1909): 269-272.

Keller, A. *The Infancy of Atomic Physics* (Oxford: Oxford University Press, 1983).

Kim, Dong-Won. *The Emergence of the Cavendish School: An Early History of the Cavendish Laboratory, 1871-1900* unpublished Ph.D. dissertation (Cambridge, Mass.: Harvard University, 1991).

________. "J.J. Thomson and the emergence of the Cavendish School, 1885-1900," *BJHS 28* (1995): 191-226.

Klaassen, H. G. "On the Effect of Temperature on the Conductivity of Solutions of Sulphuric Acid." *Proc. Camb. Phil. Soc. 7* (1889-92): 137-141.

Kleeman, R. D. "On the Recombination of Ions made by α, β, γ, and X-rays," *Phil. Mag. 12* (1906): 273-297.

________. "On the Secondary Cathode Rays emitted by Substances when exposed to the γ Rays," *Phil. Mag. 14* (1907): 618-644.

________. "On the Ionisation of Various Gases by α-, β-, and γ-Rays," *Proc. Roy. Soc. 79* (1907): 220-233.

________. "On the different Kinds of γ Rays of Radium, and the Secondary γ Rays which they produce," *Phil. Mag. 15* (1908): 638-663.

________. "On the Velocity of the Cathode Rays ejected by Substances exposed to the γ Rays of Radium," *Proc. Roy. Soc. 82* (1909): 128-145.

________. "The Ionisation in various Gases by Secondary γ-Rays," *Proc. Roy. Soc. 82* (1909): 358-377.

________. "On the Nature of the Ionisation of a Molecule by an α- Particle," *Proc. Roy. Soc. 83* (1909): 195-199.

________. "The Determination of a Constant in Capillarity," *Phil. Mag. 18* (1909): 39-44.

________. "Some Relations in Capillarity," *Phil. Mag. 18* (1909): 491-510.

________. "Some Relations between the critical Constants and certain Quantities connected with Capillarity," *Phil. Mag. 18* (1909): 901-908.

________. "The Nature of the Ionisation produced in a Gas by γ Rays," *Proc. Camb. Phil. Soc. 15* (1909): 169-177.

________. "A Difference in the Photoelectric Effect caused by Incident and Divergent Light," *Nature 83* (1910): 339.

________. "The Ionisation of Various Gases by the β-Rays of Actinium," *Proc. Roy. Soc. 83* (1910): 530-533.

________. "On the Direction of Motion of an Electron ejected from an Atom by Ultra-Violet Light," *Proc. Roy. Soc. 84* (1910): 92-99.

________. "An Investigation of the Determination of the Law of Chemical Attraction between Atoms from Physical Data," *Phil. Mag. 21* (1911): 83-102.

________. "The Heat of Mixture of Substances and the Relative Distribution of the Molecules in the Mixture," *Phil. Mag. 21* (1911): 535-553.

________. "Molecular Attraction and the Properties of Liquids," *Phil. Mag. 22* (1911): 566-586.

________. "On the Nature and Velocity of an Ion in a Gas," *Proc. Camb. Phil. Soc. 16* (1911): 285-298.

________. "The Heat of Combustion of a Molecule and its Chemical Attraction Constant," *Proc. Camb. Phil. Soc. 16* (1911): 299-312.

________. "The Kinetic Properties of a Molecule in a Substance," *Phil. Mag. 24* (1912): 101-118.

________. "On the different Internal Energies of a Substance I," *Proc. Camb. Phil. Soc. 16* (1912): 540-559; "On the different Internal Energies of a Substance II," 584-599.

________. "On the Ionisation produced by the Collision of Positive Ions in Gaseous Mixtures," *Proc. Camb. Phil. Soc. 16* (1912): 621-630.

________. "On the Properties of Substances connected with the Kinetic Properties of the Molecules, I.," *Proc. Camb. Phil. Soc. 16* (1912): 631-642.

________. "On the Properties of a Liquid connected with its Surface Tension," *Proc. Camb. Phil. Soc. 17* (1912): 149-159.

________. "The Atomic Constants and the Properties of Substances," *Proc. Camb. Phil. Soc. 17* (1913): 175-179.

________. "The Unstable Nature of the Ion in a Gas," *Proc. Camb. Phil. Soc. 17* (1913): 263-279.

________. "On the Dependence of the Relative Ionisation in various Gases by β Rays on their Velocity, and its bearing on the Ionisation produced by γ Rays," *Proc. Camb. Phil. Soc. 17* (1913): 314-320.

________. "On the Nature of the Internal Work done during the Evaporation of a Liquid," *Proc. Camb. Phil. Soc. 17* (1914): 402-408.

Knipp, C. T. "Rays of Positive Electricity from the Wehnelt Cathode," *Phil. Mag. 22* (1911): 926-933.

Knott, C. G. *Life and Scientific Work of P. G. Tait: Supplementing the Two Volumes of Scientific Papers published in 1898 and 1900* (Cambridge: Cambridge University Press, 1911).

Kohler, R. E. Lords of the Fly: Drosophila Genetics and the Experimental Life (Chicago: University of Chicago Press, 1994).

Kohlrausch, F. *An Introduction to Physical Measurements with Appendices on Absolute Electrical Measurement, etc.*, translated from the second German edition by T.H. Waller & H.R. Proctor (London: J & A Churchill, 1873).

Kunz, J. "An Abrupt Limit of Distance in the Power of the Positive Rays to produce Phosphorescence," *Phil. Mag. 14* (1907): 614-617.

________. "On Cathode and Canal Rays from Hollow Cathodes," *Phil. Mag. 16* (1908): 161-183.

Laby, T. H. "The Total Ionisation of various Gases by the α-Rays of Uranium," *Proc. Roy. Soc. 79* (1907): 206-219.

________."The Supersaturation and Nuclear Condensation of Certain Organic Vapour," *Phil. Trans. 208* (1908): 445-474.

________. "A Recalculation of the Vapour Pressures of Mercury," *Phil. Mag. 15* (1908): 789-796.

________ & G. W. C. Kaye. "Gaseous Ionisation and Pressure," *Phil. Mag. 15* (1908): 879-889.

________. "A String Electrometer," *Proc. Camb. Phil. Soc. 15* (1909): 106-113

Laird, E. R. "Entladungsstrahlen," *Physical Review 28* (1909): 225-228.

Larmor, J. "James Clerk Maxwell," *Supplement to Nature 118* (1926): 44-47.

Larsen, E. *The Cavendish Laboratory: Nursery of Genius* (London: Ward, 1962).

Latour, B. and Steve Woolgar, *Laboratory Life: The Social Construction of Scientific Facts* (Beverly Hills, Calif.: Sage, 1979).

Laws, S. C. "Experiments on the Thomson Effect in Alloys of Bismuth and Tin," *Proc. Camb. Phil. Soc. 12* (1903): 179-190.

________. "The Thomson Effect in Alloys of Bismuth and Tin," *Phil. Mag. 7* (1904): 560-578.

________. "The Magnetic Susceptibility of Alloy of Bismuth and Tin," *Phil. Mag. 8* (1904): 49-57.

Leahy, A. H. "On the Movement of Solids through Ether," *Proc. Camb. Phil. Soc. 5* (1886): 374.

Levin, M. "Über einige Eigenschaften des Uranium und Aktinium," *Phisikalische Zeitschrfit 8* (1907): 129-133.

Lindsay, R. B. *Lord Rayleigh: The Man and His Work* (Oxford: Pergamon Press, 1970).

Lodge, O. "Sir J.J. Thomson," *Supplement to Nature 118* (1926): 49-51.

________. "Maxwell and the Cavendish Laboratory," *Nature 119* (1927): 46.

________. *Past Years: An Autobiography* (London: Hodder and Stoughton, 1931).

Logman, W. H. "Note on the Production of Secondary Rays by α Rays from Polonium," *Proc. Roy. Soc. 78* (1906): 212-217.

Lusby, S. G. "The Mobility of the Positive Flame Ion," *Proc. Camb. Phil. Soc. 16* (1910): 26-34.

________. "Some Experiments on Ionisation in Dried Air," *Proc. Camb. Phil. Soc. 15* (1910): 459-464.

________. "The Mobility of the Positive Ion in Flames," *Phil. Mag. 22* (1911): 775-791.

Lyman, T. "Note on the Behaviour of a Potassium Amalgam Cathode in a Vacuum Tube," *Proc. Camb. Phil. Soc. 12* (1902): 45-46.

McClelland, J. A. "Selective Absorption of Röntgen Rays," *Proc. Roy. Soc. 60* (1896): 146-148.

________. "Cathode and Lenard Rays," *Proc. Roy. Soc. 61* (1897): 227-235.

________. "On the Conductivity of the Hot Gases from Flames," *Phil. Mag. 46* (1898): 29-42.

________. "On the Figures produced on Photographic Plates by Electric Discharge," *Proc. Camb. Phil. Soc. 9* (1898): 522-525.

________. "On the Conductivity of Gases from an Arc and from Incandescent Metals," *Proc. Camb. Phil. Soc. 10* (1899): 241-257.

________. "On the Action of Incandescent Metals in producing Electric Conductivity in Gases," *Proc. Camb. Phil. Soc. 11* (1901); 296-305.

McClung, R. K. "The Relation between the Rate of Recombination of Ions in Air and the Temperature of the Air," *Phil. Mag. 6* (1903): 656-666.

________. "The Effect of Temperature on the Ionisation produced in Gases by the Action of Röntgen Rays," *Phil. Mag. 7* (1904): 81-95.

________. "The Relative Amount of Ionisation produced in Gases in Röntgen Rays of different Types," *Phil. Mag. 8* (1904): 357-373.

McCormmach, R. "J.J. Thomson and the Structure of Light," *BJHS 3* (1967): 362-387.

McConnel, J. C. "Measurement of the Dark Rings in Quartz," *Proc. Camb. Phil. Soc. 5* (1883): 53-61.

________. "On the Effect of Self-induction of the Galvanometer in the Determination of the Capacity of a Condenser," *Proc. Camb. Phil. Soc. 5* (1884): 211-217.

________. "Note on the Use of Nicol's Prism," *Phil. Mag. 19* (1885): 317-332.

________. "An Experimental Investigation into the Form of the Wave Surface of Quartz," *Proc. Roy. Soc. 39* (1885): 409-411.

MacKenzie, A. S. "The Deflexion of α Rays from Radium and Polonium," *Phil. Mag. 10* (1905): 538-548.

MacLennan, J. C. "Electrical Conductivity in Gases traversed by Cathode Rays," *Phil. Trans. 195* (1901): 49-77.

________. "On a kind of Radioactivity imparted to certain Salts by Cathode Ray," *Phil. Mag. 3* (1902): 195-203.

MacLeod, R. M. "Resources of Science in Victorian England: The Endowment of Science Movement, 1868-1900," in P. Mathias (ed.) *Science and Society, 1600-1900* (Cambridge: Cambridge University Press, 1972): 111-166.

________ & E. Kay Andrews, "Scientific Careers of 1851 Exhibition Scholars," *Nature 218* (1968): 1011-1016.

________ & R. Moseley. "The 'Naturals' and Victorian Cambridge: Reflections on the Anatomy of an Elite, 1851-1914," *Oxford Review of Education 6* (1980): 177-195.

MacTaggart, H. A. "The Electrification at Liquid-Gas Surface," *Phil. Mag. 27* (1914): 297-314; *Phil. Mag. 28* (1914): 367-378.

Makower, W. "The Molecular Weights of Radium and Thorium Emanation," *Phil. Mag. 9* (1905): 56-77.

Mallik, D. N. "Magnetic Rotation of Electric Discharge," *Phil. Mag. 15* (1908): 531-550.

Martin, F. "Expansion produced by Electric Discharge," *Proc. Camb. Phil. Soc. 9* (1895-98): 11-16.

Maxwell, J. C. *A Treatise on Electricity and Magnetism*, 2 vols., 3rd edition (New York: Dover, 1954).

[Maxwell, J. C.] *The Scientific Papers of James Clerk Maxwell*, edited by W.D. Niven, 2 vols. new edition (New York: Dover Publication, 1965).

Mendelssohn, K. "The Clarenden Laboratory, Oxford," *Cryogenics 6* (1966): 129-140.

Metcalfe, E. P. "On Ionisation in various Gases," *Phil. Mag. 18* (1909): 878-889.

Monckman, J. "On the Effect of Occluded Gases on the Thermo-electric Properties of Bodies, and on their Resistances; also on the Thermo-electric and other Properties of Graphite and Carbon," *Proc. Roy. Soc. 44* (1888): 220-236.

________. "The Specific Resistance and other Properties of Sulphur," *Proc. Roy. Soc. 46* (1889): 136-154.

________. "On the Arrangement of Electrified Cylinders when attracted by an Electrified Sphere," *Proc. Camb. Phil. Soc. 6* (1888): 179-181.

________. "Note on some Experiments on the Creeping of Solutions," *Proc. Camb. Phil. Soc. 6* (1888): 211-212.

________. "On certain Cases of Discharge in Vacuo and on the Zigzag Path of Lightning," *Proc. Camb. Phil. Soc. 9* (1896): 216-221.

More, L. T. "The Fatigue of Metals subjected to Röntgen Radiation," *Phil. Mag. 13* (1907): 708-721.

Morrell, J. B. "The Chemist Breeders: The Research Schools of Liebieg and Thomson," *Ambix 19* (1972): 1-46.

Moseley, R. "Tadpoles and Frogs: Some Aspects of the Professionalization of British Physics, 1870-1939," *Social Studies of Science 7* (1977): 423-446.

________. "The Origin and Early Years of the National Physical Laboratory: A Chapter in the Pre-history of British Science Policy," *Minerva 16* (1978): 222-250.

Mott, N. *A Life in Science* (London: Taylor & Francis, 1986).

Morrow, B. A. "On the Discovery of the Electron," *Journal of Chemical Education 46* (1969): 584-588.

Murray, D. *Memories of the Old College of Glasgow: Some Chapters in the History of the University* (Glasgow: Jackson, Wylie & Co., 1927).

Newall, H. F. "On Peculiarities observed in Iron and Steel at a Bright Red Heat," *Phil. Mag. 24* (1887): 435-439.

________. "On the Recalescence of Steel and Allied Phenomena," *Proc. Camb. Phil. Soc. 6* (1888): 172-179.

Niven, W. D. "On the Calculation of the Trajectories of Shot." *Proc. Roy. Soc. 26* (1876): 268-287.

Noda, T. "On the Ionisation of Gases exposed simultaneously to Röntgen Rays and the Radiation from Radio-active Substances," *Proc. Camb. Phil. Soc. 13* (1906): 356-362.

Novák, V. "Specific Electric Conductivities and Freezing-points of Solutions of Water in Formic Acid." *Phil. Mag. 44* (1897): 9-20.

Nye, M. J. (ed.) *The Question of the Atom: From the Karlsruhe Congress to the First Solvay Conference, 1860-1911* (New York: American Institute of Physics, 1984).

________. *Before Big Science: The Pursuit of Modern Chemistry and Physics, 1800-1940* (New York: Twayne, 1996).

Olearski, C. "Some Experiments on the Dielectric Strength of Mixtures of Gases," *Proc. Camb. Phil. Soc. 5* (1885): 325-330.

Olesko, K. M. *Physics as a Calling: Discipline and Practice in the Königsberg Seminar for Physics* (Ithaca, New York: Cornell University Press, 1991).

Oliphant, M. *Rutherford: Recollections of the Cambridge Days* (Amsterdam: Elsevier, 1972).

Orange, J. A. "On certain Phenomena of the Kathode Region," *Proc. Camb. Phil. Soc. 15* (1909): 217-232.

________. "On the Shape of Beams of Canal-Rays," *Proc. Camb. Phil. Soc. 15* (1909): 334-336.

Owen E. A. "On the Scattering of Röntgen Radiation," *Proc. Camb. Phil. Soc. 16* (1911): 161-166.

________. "The Passage of Homogeneous Röntgen Rays through Gases," *Proc. Roy. Soc. 86* (1912): 426-439.

Owen, G. "On the Condensation Nuclei produced in Air and Hydrogen by heating a Platinum Wire," *Phil. Mag. 6* (1903): 306-315.

________. "On the Discharge of Electricity from a Nernst Filament," *Phil. Mag. 8* (1904): 230-258.

________. "On the Magnetic Deflexion of the Negative Current of Electricity from a Hot Platinum Wire at Low Pressures," *Proc. Camb. Phil. Soc. 12* (1904): 493-509.

Owen, G. E. "The Discovery of the Electron," *Annals of Science 11* (1955): 173-182.

Owens, R. B. "Thorium Radiation," *Phil. Mag. 48* (1899): 360-387.

Oxley, A. E. "On the Magnetic Susceptibilities of Certain Compounds," *Proc. Camb. Phil. Soc. 16* (1910): 102-111.

________. "On Apparatus for the Production of Circularly Polarized Light," *Phil. Mag. 21* (1911): 517-532.
________. "A Magnetic Examination of the Function of the Water Molecule in certain Dilute Solutions," *Proc. Camb. Phil. Soc. 16* (1911): 421-427.
________. "The Variation of Magnetic Susceptibility with Temperature," *Proc. Camb. Phil. Soc. 16* (1912): 486-490.
________. "The Detection of Small Amount of Polarisation in Light from a Dull Sky," *Proc. Camb. Phil. Soc. 16* (1912): 561-579.
________. "The Hall Effect in Liquid Electrolytes," *Proc. Roy. Soc. 88* (1913): 588-604.
Paine, H. H. "The Coagulation of Colloidal Copper," *Proc. Camb. Phil. Soc. 16* (1912): 430-455.
Palmer, B. J. "Some Methods of increasing the Spark-Length of the Wimshurst Machine," *Proc. Camb. Phil. Soc. 13* (1905): 60-68.
Paradis, J. & T. Postlewait (eds.) *Victorian Science and Victorian Values: Literary Perspectives* (New Brunswick, N.J.: Rutgers University Press, 1985).
Patterson, J. "On the Effect of Magnetic Field on the Resistance of Thin Metallic Films," *Proc. Camb. Phil. Soc. 11* (1901): 118-119.
________. "Note on Spontaneous Ionisation in Air at different Temperatures and Pressures," *Proc. Camb. Phil. Soc. 12* (1902): 44.
________. "On the Change of the Electric Resistance of Metals when placed in a Magnetic Field," *Phil. Mag. 3* (1902): 643-656.
________. "On the Electric Properties of Thin Metal Films," *Phil. Mag. 4* (1902): 652-678.
________. "On the Ionisation in Air at different Temperature and Pressures," *Phil. Mag. 6* (1903): 231-238.
Pawlow, W. I. "On the Ionisation of Gases by Collision and the Ionising Potential for Positive Ions and Negative Corpuscles," *Proc. Roy. Soc. 90* (1914): 398-411.
Peace, J. B. "On the Potential Difference required to produce a Spark between two Parallel Plates in Air at different Pressures," *Proc. Roy. Soc. 52* (1892-93): 99-114.
Pearson, D. B. "Note on an Attempt to Detect a Difference in the Magnetic Properties of the Two kinds of Ions of Oxygen," *Proc. Camb. Phil. Soc. 15* (1909): 373-374.
Phillips, P. "Ionic Velocities in Air at different Temperatures," *Proc. Roy. Soc. 78* (1906): 167-191.
Pickering, E. C. *Elements of Physical Manipulation*, 2 vols. (New York: Hurd & Houghton, 1873 (part I), 1876 (part II)).
Pippard, B. "The Cavendish Tradition," *Nature 249* (1974): 602-603.
Porter, R. "The Natural Science Tripos and the 'Cambridge School of Geology,' 1850-1914," *History of Universities 2* (1982): 193-216.
Poynting, J. H. "On the Graduation of the Sonometer," *Phil. Mag. 9* (1880): 59-64.
________. "On a Simple Form of Saccharimeter," *Phil. Mag. 10* (1880): 18-21.
________. "On a Determination of the Mean Density of the Earth," *Phil. Trans. 182* (1891): 565-656.
________ & J. J. Thomson, *A Text-Book of Physics: Sound* (London: Charles Griffin & Co., 1899).
________. *Properties of Matter* (London: Charles Griffin & Co., 1902).
________. *Heat* (London: Charles Griffin & Co., 1904); *Electricity and Magnetism* (London: Charles Griffin & Co., 1914).
Price, D. J. "The Cavendish Laboratory Archives," *Notes and Records of the Royal Society 10* (1953): 139-147.
________. "Sir J.J. Thomson, O.M., F.R.S.: A Centenary Biography," *Discovery 17* (1956): 494-502.
________. "Sir J.J. Thomson, O.M., F.R.S," *Nuovo Cimento 4* (1957): 1609-1629.
________, J. A. Hughes, & S. Schaffer, *The Cavendish Laboratory: Introduction to Prosopography*, unpublished typescript (Cambridge: Whipple Museum for the History of Science, 1991).
Przibram, K. "On the Point-Discharge in Mixtures of Gases," *Phil. Mag. 6* (1903): 176-180.
________. "Über die oszillierende Spitzenentladung bei vermindertem Luftdruck," *Physikalische Zeitschrift 4* (1903): 581-583.
________. "Über die Entladung in Gasgemischen," *Physikalische Zeitschrift 4* (1903): 842-845.
Randall, J. "Aspects of the Life and Work of James Clerk Maxwell," *Nature 195* (1962): 427-434.
Richardson, O. W. "On an Attempt to detect Radiation from the Surface of Wires carrying Alternating Currents of High Frequency," *Proc. Camb. Phil. Soc. 11* (1901): 168-178.
________. "On the Negative Radiation from Hot Platinum," *Proc. Camb. Phil. Soc. 11* (1901): 286-295.

________. "The Electrical Conductivity imparted to a Vacuum by Hot Conductors," *Phil. Trans. 201* (1903): 497-549.
________. "On the Positive Ionisation produced by Hot Platinum in Air at Low Pressures," *Phil. Mag. 6* (1903): 80-96.
________. "The Theory of the Rate of Recombination of Ions in Gases," *Proc. Camb. Phil. Soc. 12* (1903): 144-152.
________. "The Solubility and Diffusion in Solution of Dissociated Gases," *Phil. Mag. 7* (1904): 266-274.
________, J. Nicol & T. Parnell. "The Diffusion of Hydrogen through Hot Platinum," *Phil. Mag. 8* (1904): 1-29.
________. "The Effect of a Luminous Discharge on the Ionisation produced by Hot Platinum in Gases at Low Pressures," *Phil. Mag. 8* (1904): 400-410.
________. "The Electrical Properties of Hot Bodies," *B. A. Report* (1904): 472-473.
________. "Note on the Diffusion of Hydrogen through Palladium," *Proc. Camb. Phil. Soc. 13* (1904): 27-32.
________. "The Discharge of Electricity from Hot Platinum in Phosphorus Vapour," *Phil. Mag. 9* (1905): 407-413.
________. "The Structure of Ions formed in Gases at High Pressure," *Phil. Mag. 10* (1905): 177-179.
________. "The Rate of Recombination of Ions in Gases," *Phil. Mag. 10* (1905): 242-253.
________. "Note on the Positive Leak from Hot Platinum in Air at Low Pressures," *Proc. Camb. Phil. Soc. 13* (1905): 58-59.
________. "The Effect of Hydrogen on the Discharge of Electricity from Hot Platinum," *Proc. Camb. Phil. Soc. 13* (1905): 192-193.
________. "The Ionisation produced by Hot Platinum in different Gases," *Phil. Trans. 207* (1906): 1-64.
Righi, A. "Scientific Worthies, XL. Sir J.J. Thomson, O.M., F.R.S," *Nature 91* (1913): 1-5.
Roberts, G. K. "The Liberally-Educated Chemist: Chemistry in the Cambridge Natural Sciences Tripos, 1851-1914," *HSPS 11* (1980): 157-183.
Robotti, N. "J.J. Thomson at the Cavendish Laboratory: the History of an Electric Charge Measurement," *Annals of Science 52* (1995): 265-284.
Rothblatt, S. *The Revolution of the Dons: Cambridge and Society in Victorian England* (London: Faber & Faber, 1968).
________. "The Student Sub-Culture and the Examination System in Early 19th Century Oxbridge," in Lawrence Stone (ed.), *The University in Society, vol. 1: Oxford and Cambridge from the 14th to the Early 19th Century* (Princeton: Princeton University Press, 1974), 247-303.
[Royal Commission for the Exhibition of 1851]. *Record of the Science Research Scholars of the Royal Commission for the Exhibition of 1851,1891-1929* (London, 1930).
[Royal Commission on Scientific Instruction and the Advancement of Science]. "The Report of the Science Commission on the Old Universities," *Nature 8* (1873): 317-319 & 337-341.
Rozental, S. (ed.) *Niels Bohr: His Life and Works as Seen by His Friends and Colleagues* (Amsterdam: North-Holland Personal Library, 1967).
Rudd, E. & R. Simpson. *The High Education: A Study of Graduate Education in Britain* (London: Routledge & Kegan, 1975).
Rudge, W. A. D. "On the Difference of Potential between the Terminals of a Vacuum Tube," *Proc. Camb. Phil. Soc. 12* (1903): 155-162.
________. "Experiments on the Retention of an Electric Charge by Gases," *Proc. Camb. Phil. Soc. 13* (1905): 194-196.
________. "On the Electrification produced during the raising of a Cloud of Dust," *Proc. Roy. Soc. 90* (1914): 256-272.
________. "On some Sources of Disturbance of the Normal Atmospheric Potential Gradient," *Proc. Roy. Soc. 90* (1914): 571-582.
________. "On the Electrification given to the Air by a Steam Jet," *Proc. Camb. Phil. Soc. 18* (1915): 127-134.
Russell, C. *Science and Social Change in Britain and Europe, 1700-1900* (New York: St. Martin Press, 1983).
Rutherford, E. "A Magnetic Detector of Electrical Waves and some of its Applications," *Phil. Trans. 189* (1897): 1-24.

________. "On the Electrification of Gases exposed to Röntgen Rays, and the Absorption of Röntgen Radiation by Gases and Vapours," *Phil. Mag. 43* (1897): 241-255.
________. "The Velocity and Rate of Recombination of the Ions of Gases exposed to Röntgen Radiation," *Phil. Mag. 44* (1897): 422-440.
________. "The Discharge of Electrification by Ultra-violet Light," *Proc. Camb. Phil. Soc. 9* (1898): 401-416.
________. "Uranium Radiation and the Electrical Conduction produced by it," *Phil. Mag. 47* (1899): 109-163.
________. "The Scattering of α and β Particles by Matter and the Structure of the Atom," *Phil. Mag. 21* (1911): 669-688.
________. "The Structure of the Atom," *Phil. Mag. 27* (1914): 488-498.
Saltzman, M. "J.J. Thomson and the Modern Revival of Dualism," *Journal of Chemical Education 50* (1973): 59-61.
Satterly, J. "The Amount of Radium Emanation in the Atmosphere," *Phil. Mag. 15* (1908): 584-615.
________. "On the Amount of Radium Emanation in the Lower Regions of the Atmosphere and its Variation with the Weather," *Phil. Mag. 20* (1910): 1-36.
________. "Some Experiments on the Absorption of Radium Emanation by Coconut Charcoal," *Phil. Mag. 20* (1910): 778-788.
________. "Note on the Radium-content of the Waters of the Cam, Cambridge Tap Water and some Varieties of Charcoal," *Proc. Camb. Phil. Soc. 15* (1910): 540-544.
________. "The Radium-content of Salts of Potassium," *Proc. Camb. Phil. Soc. 16* (1910): 67-69.
________. "A Study of the Radium Emanation contained in the Air of various Soils," *Proc. Camb. Phil. Soc. 16* (1911): 36-355.
________. "The Radioactivity of Marsh Gas," *Proc. Camb. Phil. Soc. 16* (1911): 356-359.
________. "On the Radium-content of various Fresh and Sea Waters and some other Substances," *Proc. Camb. Phil. Soc. 16* (1911): 360-363.
________. "The Quantities of Radium and Thorium Emanations contained in the Air of certain Soils," *Proc. Camb. Phil. Soc. 16* (1912): 514-533.
________. "On the Number of α- Particles expelled when an Atom of Thorium Emanation disintegrates," *Proc. Camb. Phil. Soc. 16* (1912): 667-673.
________. "The Postprandial Proceedings of the Cavendish Society," *The American Physics Teachers 7* (1939): 179-185 & 244-248.
Saunder, S. A. "On the Variations of the Electromotive Force of a New Form of Leclanché's Cell," *Nature 12* (1875): 564.
Schaffer, S. "Late Victorian Metrology and its Instrumentation: A Manufactory of Ohm," in Robert Bud & Susan E. Cozzens (eds.), *Invisible Connections: Instruments, Institutions, and Science* (Bellingham, WA: Spie Optical Engineering Press, 1992), 23-56.
________. "Rayleigh and the Establishment of Electrical Standards," *European Journal of Physics 15* (1994): 227-285.
________. "Accurate Measurement is an English Science," in M. Norton Wise (ed.), *The Value of Precision*, (Princeton: Princeton University Press, 1995), 135-172.
Schuster, A. "Researches on the Spectra of Metalloids," *Nature 15* (1877): 401-402.
________. "Spectra of Metalloids," *Nature 15* (1877): 447-448.
________. "On the Presence of Oxygen in the Sun," *Nature 17* (1877): 148-149.
________. "On the Passage of Electricity through Gases," *Proc. Camb. Phil. Soc. 3* (1877): 5-61.
________. "On the Spectra of Metalloids. Spectrum of Oxygen," *Phil. Trans. 170* (1879): 37-54.
________. "An easy Method for adjusting the Collimator of a Spectroscope," *Phil. Mag. 7* (1879): 95-98.
________. "On Spectra of Lightning," *Phil. Mag. 7* (1879): 316-321.
________. "Spectra of Metalloids," *B. A. Report* (1880): 258-274.
________. "On the Influence of Temperature and Pressure on the Spectra of Gases," *B. A. Report* (1880): 274-298.
________. "On Harmonic Ratios in the Spectra of Gases." *Proc. Roy. Soc. 31* (1880): 337-347.
________. "On the Dynamical Theory of Radiation." *Phil. Mag. 12* (1881): 261-266.
________. *The Progress of Physics during 33 Years (1875-1908)* (Cambridge: Cambridge University Press, 1911).

________. "Lord Rayleigh," *Supplement to Nature 118* (1926): 47-49.
________. *Biographical Fragments* (London: MacMillan & Co., 1932).
Searle, G. F. C. "On an Experiment of Sir Humphry Davy," *Proc. Camb. Phil. Soc. 7* (1891): 250.
________. "On a Compound Magnetometer for Testing the Magnetic Properties of Iron and Steel," *Proc. Camb. Phil. Soc. 7* (1892): 330-335.
________. "Problems in Electric Convection," *Phil. Trans. 187* (1896): 675-713.
________. "A Method of Measuring the Loss of Energy in Hysteresis," *Proc. Camb. Phil. Soc. 9* (1896): 2-6.
________. "On the Steady Motion of an Electrified Ellipsoid," *Phil. Mag. 44* (1897): 329-341.
________. "Notes on a Vibration Magnetometer and on the Ball-ended Magnets of Robison," *Proc. Camb. Phil. Soc. 12* (1902): 27-33.
________. "The Ballistic Measurement of Hysteresis," *Electrician 49* (1902): 100-103 & 219-222.
________. "On a simple Rheostat," *Phil. Mag. 6* (1903): 173-175.
________. "On the Measurement of Magnetic Force," *Electrician 51* (1903): 319-320.
________. "Notes on a Vibration Magnetometer and on the Ball-ended Magnets of Robison," *Proc. Camb. Phil. Soc. 12* (1903): 27-33.
________. "On the Calculation of Capacities in terms of the Coefficients of Electrostatic Induction," *Proc. Camb. Phil. Soc. 12* (1904): 378-387.
________. "A Method of determining the Thermal Conductivity of Copper," *Phil. Mag. 9* (1905): 125-129.
________ & J. R. Airey. "Standards of Mutual Induction," *Electrician 56* (1905): 318-321.
________. "The Magnetic Inertia of a Charged Sphere in a Field of Electric Force," *Nature 74* (1906): 198-199.
________. "The Expansion of a Gas into a Vacuum and the Determination of the Specific Heat at Constant Pressure for Gases," *Proc. Camb. Phil. Soc. 13* (1906): 241-249.
________. "On the Force required to stop a Moving Electrified Sphere," *Proc. Roy. Soc. 79* (1907): 550-563.
________. "The Impulsive Motion of Electrified Systems," *Phil. Mag. 13* (1907): 118-148; *Phil. Mag. 17* (1909): 43-71.
________. "An Experiment with a pair of Robison Ball-ended Magnets," *Proc. Camb. Phil. Soc. 14* (1907): 185-189.
________. "A Method of determining the Thermal Conductivity of India-rubber," *Proc. Camb. Phil. Soc. 14* (1907): 190-193.
________. "Über die durch eine sprungweise Änderung der Winkel-geschwindigkeit einer electrisierten Kugel hervorgerufene Energiestrahlung," *Phisikalische Zeitschrit 9* (1908): 878-884.
________. "Die Energiestrahlung seitens eines Elektrons, das sich harmonisch oder elliptischen Bahn bewegt," *Phisikalische Zeitschrift 9* (1908): 884-887.
________. *Experimental Elasticity: A Manual for the Laboratory* (Cambridge: Cambridge University Press, 1908).
________, A. C. W. Aldis, & G. M. B. Dobson, "On a Revolving Table Method of determining the Curvature of Spherical Surface," *Phil. Mag. 21* (1911): 218-224.
________. "A Simple Method of determining the Viscosity of Air," *Proc. Camb. Phil. Soc. 17* (1912): 183-192.
________. "Experiments illustrating Flare Spots in Photography," *Proc. Camb. Phil. Soc. 17* (1913): 205-219.
________. "Some Methods of measuring the Surface Tension of Soap Film," *Proc. Camb. Phil. Soc. 17* (1913): 285-299.
________. "Some Methods of measuring the Surface Tension of Soap Film," *Proc. Camb. Phil. Soc. 17* (1913): 340-350.
________. "Experiment on the Harmonic Motion of a Rigid Body," *Proc. Camb. Phil. Soc. 18* (1913): 31-38.
________. "Calculation of the Electrical Resistance of a certain Network of Conductors," *Proc. Camb. Phil. Soc. 18* (1915): 111-114.
________. "The Determination of the Focal Length of a thick Mirror," *Proc. Camb. Phil. Soc. 18* (1915): 115-126.

________. "The Determination of the effective Aperture of the Stop of a Photographic Lens," *Proc. Camb. Phil. Soc. 18* (1915): 195-206.
________. *Experimental Harmonic Motion* (Cambridge: Cambridge University Press, 1915).
________. *Experimental Optics* (Cambridge: Cambridge University Press, 1925).
Segrè, E. *From X-rays to Quarks: Modern Physicists and Their Discoveries* (New York: W. H. Freeman and Co., 1980).
Shakespear, G. A. "The Application of an Interference Method to the Investigation of Young's Modulus for Wires, and its Relation to Changes of Temperature and Magnetization; and a further Application of the same Method to study of the change in Dimensions of Iron and Steel Wires by Magnetization," *Phil. Mag. 47* (1899): 539-556.
Shaw, A. N. "On the Charged Surface Layers in Contact Potential Phenomena between Metals," *Phil. Mag. 25* (1913): 241-256.
________. "Interference Phenomena with Gamma Rays," *Phil. Mag. 26* (1913): 190-194.
Shaw, W. N. "On the Measurement of Temperature by Water Vapour Pressure," *Proc. Camb. Phil. Soc. 5* (1883): 52-53.
________. "On Dimensional Equations and Change of Units," *Proc. Camb. Phil. Soc. 5* (1884): 137-141.
________. "On a Practical point in connection with the Comparison of Resistance." *Phil. Mag. 17* (1884): 398-400.
________. "Focal Lines," *Nature 31* (1884): 185-186.
________. "On the Electrical Resistance of Platinum at High Temperatures," *Proc. Camb. Phil. Soc. 5* (1885): 226.
________. *Practical Work at the Cavendish Laboratory: Heat* (Cambridge: Cambridge University Press, 1886).
________. "On the Verification of Faraday's Law of Electrolysis with reference to Silver and Copper," *B. A Report* (1886): 318-328.
________. "On an Experiment in Ventilation," *Proc. Camb. Phil. Soc. 5* (1886): 410-415.
________. "Some Laboratory Notes," *Proc. Camb. Phil. Soc. 6* (1887): 98-100.
________. "On the Atomic Weights of Silver and Copper," *Phil. Mag. 23* (1887): 138-141.
________ & F. M. Turner. "On Some Measurements of the Frequencies of the Notes of a Whistle of Adjustable Pitch," *Proc. Camb. Phil. Soc. 6* (1887): 90-94.
________. "Report on Hygrometric Methods; First part, including the Saturation Method and the Chemical Method, and Dew-Point Instruments," *Phil. Trans. 179* (1888): 73-149.
________. "On an Apparatus for determining Temperature by the Variation of Electrical Resistance," *B. A. Report* (1888): 590-591.
________. "On Some Accurate Charts of Kew Corrections for Mercury Thermometers," *B. A. Report* (1888): 590.
________. "On the Relation between Viscosity and Conductivity of Electrolytes," *Proc. Camb. Phil. Soc. 7* (1889): 21-31.
________. "Report on the Present State of our Knowledge in Electrolysis and Electrochemistry," *B. A. Report* (1890): 185-223.
________. "On a Pneumatic Analogue of the Wheatstone Bridge," *Proc. Roy. Soc. 47* (1890): 462-469.
________. "On the General Theory of Ventilation," *B. A. Report* (1890): 730-731.
________."The Motion of Clouds considered with reference to their Mode of Formation," *Quarterly Journal of the Royal Meteorological Society 20* (1894): 166-180.
________. "Dalton's Law," *B. A. Report* (1898): 801-803.
________. "On a Pneumatic Analogue of the Potentiometer," *B. A. Report* (1898): 778-779.
________ & T. G. Bedford. "The Measurement of Magnetic Hysteresis," *Phil. Trans. 198* (1902): 33-104.
________. "The Cavendish Laboratory as a Factor in a Counter-Revolution," *Nature 118* (1926): 885-887.
Shimizu, T. "A Preliminary Note on Branched α-Ray Tracks," *Proc. Roy. Soc. 99* (1921): 432-435.
Simpson, T. K. "Maxwell and the Direct Experimental Test of His Electromagnetic Theory," *Isis 57* (1966): 411-432.
Sinclair, S. B. "J.J. Thomson and the Chemical Atom: from Ether Vortex to Atomic Decay," *Ambix 34* (1987): 89-116.

Skinner, S. "On the Relation between the Contraction of Volume and the Heat Developed on mixing certain Liquids," *Proc. Camb. Phil. Soc. 6* (1888): 212.

________. "An Apparatus for Measuring the Compressibility of Liquids," *Phil. Mag. 32*(1891): 79-80.

________. "Physical Properties of Solutions of some Metallic Chlorides," *Journal of the Chemical Society 61* (1892): 339-344.

________. "The Clark Cell when Producing a Current," *Phil. Mag. 38* (1894): 271-279; *Phil. Mag. 39* (1895): 375-376.

________. "The Tin-Chromic Chloride Cell," *Phil. Mag. 39* (1895): 444-447

________. "The Carbon-Consuming Cell of Jacques," *B. A. Report* (1898): 804.

________. "Affinity Constants of Dihydroxymaleic, Dihydroxyfumaric, and Tartronic Acids," *Journal of the Chemical Society 73* (1898): 483-490.

________. "Observations on the minute Structure of the Surface Ice of Glaciers," *Proc. Camb. Phil. Soc. 11* (1900): 33-36.

________. "Note on a Comparison of the Silver deposited in Voltameters containing different Solvents, Water and Pyridine" *B. A. Report* (1901): 32-34.

________. "On Cavitation in Liquids, and its Occurrence in Lubrication," *Proc. Camb. Phil. Soc. 12* (1902): 34-35.

________. "Note on the Slipperiness of Ice," *Proc. Camb. Phil. Soc. 12* (1903): 86.

________. "Note on the Action of Radium Rays on Mecurous Salts," *Proc. Camb. Phil. Soc. 12* (1903): 260-261.

Slater, J. M. W. "On the Exited Activity of Thorium," *Phil. Mag. 9* (1905): 628-644.

________. "On the Emission of Negative Electricity by Radium and Thorium Emanations," *Phil. Mag. 10* (1905): 460-466.

Smith, C. & M. N. Wise. *Energy and Empire: A Biographical Study of Lord Kelvin* (Cambridge: Cambridge University Press, 1989).

Smith, H. "The Spectroscopy of the Electric Brush Discharge in weak Acids and Solutions," *Phil. Mag. 27* (1914): 801-823.

________. "A Comparison of the Positive Rays with the Spectrum of the Positive Column in a Mixture of Helium and Hydrogen," *Phil. Mag. 30* (1915): 805-811.

Southerns, L. "Experimental Investigation as to Dependency of the Weight of a Body on its state of Electrification," *Proc. Camb. Phil. Soc. 15* (1909): 352-372.

________. "A Determination of the Ratio of Mass to Weight for a Radioactive Substance," *Proc. Roy. Soc. 84* (1910): 325-344.

Spitzer, P. G. *Joseph John Thomson: An Unfinished Social and Intellectual Biography* unpublished Ph.D. dissertation (Baltimore: Johns Hopkins University, 1970).

Spurge, C. "On the Curves of Constant Intensity of Homogeneous Polarized Light seen in a Uniaxal Crystal cut at Right Angles to the Optic Axis," *Proc. Camb. Phil. Soc. 5* (1884): 74-86.

Stead, G. "On the Separation of Spectra in Compound Gases," *Phil. Mag. 22* (1911): 727-733.

________ & H. Donaldson, "The Problem of Uniform Rotation treated on the Principle of Relativity," *Phil. Mag. 20* (1910): 92-95; *Phil. Mag. 21* (1911): 319-324.

________. "On the Anode and Cathode Spectra of various Gases and Vapours," *Proc. Roy. Soc. 85* (1911): 393-401.

________. "Note on the Spectrum of Argon," *Proc. Camb. Phil. Soc. 16* (1912): 607-609.

Stewart, B. & A. W. Ward (ed.) *Essays and Addresses by Professors and Lecturers of the Owens College, Manchester* (London: MacMillan & Co., 1874).

Strutt, G. R. "Robert John Strutt, Fourth Baron Rayleigh," *Applied Optics 3* (1964): 1105-1112.

[Strutt, J. W. (the third Baron Rayleigh)] *Scientific Papers by Lord Rayleigh*, 6 vols., reprinted (New York: Dover Publication, 1964).

Strutt, R. J. (the fourth Baron Rayleigh). "The Dispersion of the Cathode Rays by Magnetic Force," *Phil. Mag. 48* (1899): 478-480.

________. "On the Behaviour of the Becquerel and Röntgen Rays in Magnetic Field," *Proc. Roy. Soc. 66* (1900): 75-79.

________. "On the Tendency of the Atomic Weights to approximate to Whole," *Phil. Mag. 1* (1901): 311-314.

________. "On the Conductivity of Gases under the Becquerel Rays," *Phil. Trans. 196* (1901): 507-527.

________. *Life of John William Strutt: Third Baron Rayleigh O.M. F.R.S.*, an augmented edition (Madison: University of Wisconsin Press, 1963).
________. *The Life of Sir J.J. Thomson, O.M.* (Cambridge: Cambridge University Press, 1942).
Sviedrys, R. "The Rise of Physical Science at Victorian Cambridge," *HSPS 2* (1970): 127-151.
________. "An Analysis of the Cavendish Laboratory, 1874-1914," in *Actes XIIe Cong. Int. Hist. of Sci. in 1968*, vol. 2 (1971): 123-127.
________. "The Rise of Physics Laboratories in Britain," *HSPS 7* (1976): 405-436.
Taylor, F. S. "The Teaching of Science at Oxford in the Nineteenth Century," *Annals of Science 8* (1952): 82-112.
Taylor, G. I. "Interference Fringes with Feeble Light," *Proc. Camb. Phil. Soc. 15* (1909): 114-115.
________. "The Conditions necessary for discontinuous Motion in Gases," *Proc. Roy. Soc. 84* (1910): 371-377.
Terraine, J. *The U-Boat Wars, 1916-1945* (New York: G. P. Putnam's Sons, 1989).
Thirkill, H. "On the Recombination of the Ions produced by Röntgen Rays in Gases and Vapours," *Proc. Roy. Soc. 88* (1913): 477-494.
Thompson, S. P. *The Life of William Thomson, Baron Kelvin of Largs*, 2 vols. (London: MacMillan & Co., 1910).
Thomson, G. P. "Dr. Francis William Aston, F. R. S.," *Nature 157* (1946): 290-292.
________. "George Frederick Charles Searle, 1864-1954," *Biographical Memoirs of Fellows of the Royal Society 1* (1955): 246-252.
________. "J.J. Thomson," *Nature 178* (1956): 1317-1319.
________. "J.J. Thomson and the Discovery of the Electron," *Physics Today 9* (Aug., 1956): 19-23.
________, & J. Thomson. "J.J. Thomson as We Remember Him," *Notes and Records of Royal Society of London 12* (1957): 201-210.
________. *J.J. Thomson and the Cavendish Laboratory* (London: Thomas Nelson & Sons, 1964).
________. "Rutherford in 19th Century Cambridge," *Proc. Roy. Soc. 283* (1965): 481-490.
________. *J.J. Thomson, Discoverer of the Electron* (Garden City, NY: Anchor, 1966).
________. "The Septuagenarian Electron," *Physics Today 20* (May, 1967): 55-61.
Thomson, J. J. "Experiments on Contact Electricity between Non-Conductors," *Proc. Roy. Soc. 25* (1876): 169-171.
________."On the Resolution of the Product of Two Sums of Eight Squares into the Sum of Eight Squares," *Messenger of Mathematics 7* (1877): 73-74.
________. "An Extension of Arbogast's Method of Derivations," *Messenger of Mathematics 7* (1878): 142-143.
________. "Vortex Motion in a Viscous Incompressible Fluid," *Messenger of Mathematics 8* (1879): 174-175.
________. "On Maxwell's Theory of Light," *Phil. Mag. 9* (1880): 284-291.
________. "On Some Electromagnetic Experiments with Open Circuits," *Phil. Mag. 12* (1881): 49-60.
________. "On the Electric and Magnetic Effects produced by the Motion of Electrified Bodies," *Phil. Mag. 11* (1881): 229-249.
________. "On the Dimensions of a Magnetic Pole in the Electrostatic System of Units," *Phil. Mag. 13* (1882): 427-429.
________. "On the Dimensions of a Magnetic Pole in the Electrostatic System of Units," *Phil. Mag. 14* (1882): 225-226.
________. "On the Vibrations of a Vortex Ring, and the Action upon each other of two Vortices in a Perfect Fluid," *Phil. Trans. 173* (1882): 493-521.
________. "On the Determination of the Number of Electrostatic Units in the Electromagnetic Unit of Electricity," *Phil. Trans. 174* (1883): 707-721.
________. "On a Theory of the Electric Discharge in Gases," *Phi. Mag. 15* (1883): 427-434.
________. "On Electrical Oscillations and the Effects produced by the Motion of an Electrified Sphere," *Proceedings of the London Mathematical Society 15* (1884): 197-218.
________. "On the Chemical Combination of Gases," *Phil. Mag. 18* (1884): 233-267.
________. "On some Applications of Dynamical Principles to Physical Phenomena," *Phil. Trans. 176* (1885): 307-342.
________. "The Vortex Ring Theory of Gases. On the Law of the Distribution of Energy among the Molecules," *Proc. Roy. Soc. 39* (1885): 23-36.

________ & H. F. Newall. "On the Formation of Vortex Rings by Drops falling into Liquids, and some allied Phenomena," *Proc. Roy. Soc. 39* (1885): 417-436.

________. "Report on Electrical Theories," *B. A. Report* (1885): 97-155.

________. "Note on the Rotation of the Plane of Polarization of Light by a Moving Medium," *Proc. Camb. Phil. Soc. 5* (1885): 250-254.

________. "Some Experiments on the Electric discharge in a Uniform Electric Field, with some Theoretical Considerations about the Passage of Electricity through Gases," *Proc. Camb. Phil. Soc. 5* (1886): 391-409.

________. "Electrical Oscillations on Cylindrical Conductors," *Proceedings of the London Mathematical Society 17* (1886): 310-328.

________ & R. Threlfall. "Some Experiments on the Production of Ozone," *Proc. Roy. Soc. 40* (1886): 340-342.

________ & R. Threlfall. "On an Effect Produced by the Passage of an Electric Discharge through Pure Nitrogen," *Proc. Roy. Soc. 40* (1886): 329-340.

________. "Some Applications of Dynamical Principles to Physical Phenomena-Part II," *Phil. Trans. 178* (1887): 471-526.

________. "The Bakerian Lecture; On the Dissociation of some Gases by the Electric Discharge," *Proc. Roy. Soc. 42* (1887): 343-345.

________ & H. F. Newall. "On the Rate at which Electricity leaks through Liquids which are bad Conductors of Electricity," *Proc. Roy. Soc. 42* (1887): 410-429.

________ & H. F. Newall. "Experiments on the Magnetization of Iron Rods, especially on the Effect of Narrow Crevasses at Right Angles to their Length," *Proc. Camb. Phil. Soc. 6* (1887): 84-90.

________. *Applications of Dynamics to Physics and Chemistry* (London: MacMillan, 1888).

________. "Electrical Oscillations on Cylindrical Conductors," *Proceedings of the London Mathematical Society 19* (1888): 520-549.

________ & J. Monckman. "The Effect of Surface Tension on Chemical Action," *Proc. Camb. Phil. Soc. 6* (1888): 264-268.

________. "On the Magnetic Effects produced by Motion in the Electric Field," *Phil. Mag. 28* (1889): 1-14.

________. "The Resistance of Electrolytes to the Passage of very rapidly Alternating Currents, with some Investigations on the Times of Vibration of Electrical Systems," *Proc. Roy. Soc. 45* (1889): 269-290.

________. "Note on the Effect produced by Conductors in the Neighborhood of a Wire on the Rate of Propagation of Electrical Disturbances along it, with a Determination of this Rate," *Proc. Roy. Soc. 46* (1889): 1-13.

________. "Specific Inductive Capacity of Dielectrics when acted on by very rapidly alternating Electric Forces," *Proc. Roy. Soc. 46* (1889): 292-295.

________. "The Application of the Theory of the Transmission of Alternating Currents along a wire to the Telephone," *Proc. Camb. Phil. Soc. 6* (1889): 321-325.

________. "On the Effect of Pressure and Temperature on the Electric Strength of Gases," *Proc. Camb. Phil. Soc. 6* (1889): 325-329.

________ & G. F. C. Searle. "A Determination of 'v,' the Ratio of the Electromagnetic Unit of Electricity to the Electrostatic Unit," *Phil. Tans. 181* (1890): 583-621.

________. "On the Passage of Electricity through Hot Gases." *Phil. Mag. 29* (1890): 358-366 & 441-449.

________."Some Experiments on the Velocity of Transmission of Electric Disturbances, and their Application to the Theory of the Striated Discharge through Gases," *Phil. Mag. 30* (1890): 129-140.

________. "Note on the Conductivity of Hot Gases," *Phil. Mag. 31* (1891): 135-136 & 515.

________. "On the Illustration of the Properties of the Electric Field by Means of Tubes of Electrostatic Induction," *Phil. Mag. 31* (1891): 149-171.

________. "On the Discharge of Electricity through Exhausted Tubes without Electrodes," *Phil. Mag. 32* (1891): 321-336 & 445-464.

________. "On the Rate of Propagation of the Luminous Discharge of Electricity through a Rarefied Gas," *Proc. Roy. Soc. 49* (1891): 84-100.

________. "On the Electric Discharge through Rarefied Gases without Electrodes," *Proc. Camb. Phil. Soc., 7* (1891): 131.

________. "Some Experiments on Electric Discharge," *Proc. Camb. Phil. Soc. 7* (1889-92): 314.

________. "On the Pressure at which the Electric Strength of a Gas is Minimum," *Proc. Camb. Phil. Soc. 7* (1892): 330.

________. *Notes on Recent Researches in Electricity and Magnetism* (Oxford: Clarendon Press, 1893).

________. "On the Effect of Electrification and Chemical Action on a Steam-Jet, and of Water-Vapour on the Discharge of Electricity through Gases," *Phil. Mag. 36* (1893): 313-327.

________. "The Electrolysis of Steam," *Proc. Roy. Soc. 53* (1893): 90-110.

________. "On the Electricity of Drops," *Phil. Mag. 37* (1894): 341-358.

________. "On the Velocity of the Cathode-Rays," *Phil. Mag. 38* (1894): 358-365.

________. "Electric Discharge through Gases," *Not. Proc. Roy. Inst. 14* (1894): 239-247.

________. "The Connection between Chemical Combination and the Discharge of Electricity through Gases," *B. A. Report* (1894): 482-493.

________. *Elements of the Mathematical Theory of Electricity and Magnetism* (Cambridge: Cambridge University Press, 1895).

________ "The Relation between the Atom and the Charge of Electricity carried by it," *Phil. Mag. 40* (1895): 511-544.

________. "On the Electrolysis of Gases," *Proc. Roy. Soc. 58* (1895): 244-257.

________. "A Method of comparing the Conductivities of badly conducting Substances for rapidly Alternating Currents," *Proc. Camb. Phil. Soc. 8* (1895): 258-269.

________. "On the Discharge of Electricity produced by the Röntgen Rays, and the Effects produced by these Rays on Dielectrics through which they pass," *Proc. Roy. Soc. 59* (1896): 274-276.

________. "Presidential Address (Section A in the British Association Annual Meeting)," *B. A. Report* (1896): 699-706.

________ & E. Rutherford. "On the Passage of Electricity through Gases exposed to Röntgen Rays," *Phil. Mag. 42* (1896): 392-407.

________. "Longitudinal Electric Waves and Röntgen's X-Rays," *Proc. Camb. Phil. Soc. 9* (1896): 49-61.

________ & J. A. McClelland. "On the Leakage of Electricity through Dielectrics traversed by Röntgen Rays," *Proc. Camb. Phil. Soc. 9* (1896): 126-140.

________. "Cathode Rays," *Not. Proc. Roy. Inst. 15* (1897): 430-432; also *Phil. Mag. 44* (1897): 293-316.

________. "On the Cathode Rays," *Proc. Camb. Phil. Soc. 9* (1897): 243-244.

________ & S. Skinner. "On a Chemical Effect produced by the Impact of Kathode Rays," *Proc. Camb. Phil. Soc. 9* (1897): 371-372.

________. "On the Effect of Zinc on a Photographic Plate," *Proc. Camb. Phil. Soc. 9* (1897): 372.

________. "A Theory of the Connection between Cathode and Röntgen Rays," *Phil. Mag. 45* (1898): 172-183.

________. "On the Charge of Electricity carried by the Ions produced by Röntgen Rays," *Phil. Mag. 46* (1898): 528-545.

________. "On the Diffuse Reflection of Röntgen Rays," *Proc. Camb. Phil. Soc. 9* (1898): 393-397.

________. "On the Connection between the Chemical Composition of a Gas and the Ionisation produced in it by Röntgen Rays," *Proc. Camb. Phil. Soc. 10* (1898): 10-14.

________. *The Discharge of Electricity through Gases* (New York: Archibald Constable & Co., 1898).

________. "On the Theory of the Conduction of Electricity through Gases by Charged Ions," *Phil. Mag. 47* (1899): 253-268.

________. "On the Masses of the Ions in Gases at Low Pressures," *Phil. Mag. 48* (1899): 547-567.

________. "On the Motion of a Charged Ions in a Magnetic Field," *Proc. Camb. Phil. Soc. 10* (1899): 49-52.

________. "On the Conductivity of Gases exposed to Entladungsstrahlen," *Proc. Camb. Phil. Soc. 10* (1899): 74-77.

________. "On the Genesis of the Ions in the Discharge of Electricity through Gases," *Phil. Mag. 50* (1900): 278-283.

________."Ionisation of Gases in the Electric Field," *Proc. Camb. Phil. Soc. 10* (1900): 380.

________. "Indications relatives à la constitution de la matière," *Rapports Congrès de physique 3* (1900): 138.

________. "On a kind of easily absorbed Radiation produced by the Impact of slowly moving Cathode Rays; together with a theory of the Negative Glow, the Dark Space, and the Positive Column," *Phil. Mag. 1* (1901): 361-376.

________. "The Existence of Bodies Smaller than Atoms," *Not. Proc. Roy. Inst. 16* (1901): 574-586.

________. "On the question as to whether or not there are any Free Charged Ions produced during the Combination of Hydrogen and Chlorine; and of the effect produced on the Rate of the Combination by the presence of such Ion," *Proc. Camb. Phil. Soc. 11* (1901): 90-91.

________. "On the Theory of Electric Conduction through Thin Metallic Films," *Proc. Camb. Phil. Soc.* 11 (1901): 120-123.

________. "On the Effect of a traverse Magnetic Field on Metallic Resistance," *Phil. Mag. 3* (1902): 353-356.

________. "On some of the consequences of the Emission of Negatively Electrified Corpuscles by Hot Bodies," *Phil. Mag. 4* (1902): 253-262.

________. "Experiments on Induced-Radioactivity in Air, and on the Electrical Conductivity produced in Gases when they pass through Water," *Phil. Mag. 4* (1902): 352-367.

________. "On Induced Radio-activity," *Proc. Camb. Phil. Soc. 11* (1902): 504.

________. "On the increase in the Electrical Conductivity of Air produced by its passage through Water," *Proc. Camb. Phil. Soc. 11* (1902): 505.

________. "On the Charge of Electricity carried by a Gaseous Ion," *Phil. Mag.* 5 (1903): 346-355.

________. "Radium," *Nature 67* (1903): 602.

________. "The Magnetic Properties of Systems of Corpuscles describing Circular Orbit," *Phil. Mag. 6* (1903): 673-693.

________. "On the Dynamics of the Electric Field," *Proc. Camb. Phil. Soc. 12* (1903): 83.

________. "On the Existence of a Radio-active Gas in the Cambridge Tap-water," *Proc. Camb. Phil. Soc. 12* (1903): 172-174.

________. "Experiment to show that Negative Electricity is given off by a Metal exposed to Röntgen Rays," *Proc. Camb. Phil. Soc. 12* (1903): 312.

________. "On the Structure of the Atom: an Investigation of the Stability and Periods of Oscillation of a Number of Corpuscles arranged at equal Intervals around the Circumference of a Circle; with Application of the Results to the Theory of Atomic Structure," *Phil. Mag. 7* (1904): 237-265.

________. "On Momentum in the Electric Field," *Phil. Mag. 8* (1904): 331-356.

________. "On the presence of Radio-active Matter in Ordinary Substances," *Proc. Camb. Phil. Soc. 12* (1904): 391-397.

________. "On the Vibrations of Atoms containing 4, 5, 6, 7 and 8 Corpuscles and on the Effect of a Magnetic Field on such Vibrations," *Proc. Cam. Phil. Soc. 13* (1904): 39-48.

________. "On the Positive Electrification of α Rays, and the Emission of slowly moving Cathode Rays by Radio-active Substances," *Proc. Camb. Phil. Soc. 13* (1904): 49-54.

________. *Electricity and Matter* (New York: Charles Scribner's Sons, 1904).

________. "On the Emission of Negative Corpuscles by the Alkali Metals," *Phil. Mag. 10* (1905): 584-590.

________. "The Structure of the Atom," *Not. Proc. Roy. Inst. 18* (1905): 49-63.

________. "Do the γ Rays carry a Charge of Negative Electricity? *Proc. Camb. Phil. Soc.* 13 (1905): 121-123.

________. "On the Absence of excited Radio-activity due to temporary exposure to γ Rays," *Proc. Camb. Phil. Soc. 13* (1905): 124.

________. "The Rate of Recombination and the Size of Gaseous Ions," *Proc. Camb. Phil. Soc. 13* (1905): 170-173.

________."Some experiments on Canal-strahlen," *Proc. Camb. Phil. Soc. 13* (1905): 212-214.

________. "On the Number of Corpuscles in an Atom," *Phil. Mag. 11* (1906): 769-781

________. *Conduction of Electricity through Gases,* second edition (Cambridge: Cambridge University Press, 1906).

________. "A Theory of widening of Lines of Spectra," *Proc. Camb. Phil. Soc. 13* (1906): 318-321.

________. "On Secondary Röntgen Radiation," *Proc. Camb. Phil. Soc. 13* (1906): 322-324; *14* (1906): 109-113.

________. "Electrification produced by heating Salts," *Proc. Camb. Phil. Soc. 14*(1906): 105-108.

________. "Some Applications of the theory of Electric Discharge to Spectroscopy," *Not. Proc. Roy. Inst. 18* (1906): 193-206.

________. "On Rays of Positive Electricity," *Phil. Mag. 13* (1907): 561-575.

________. "Rays of Positive Electricity," *Phil. Mag. 14* (1907): 295-359.

________. "On the Theory of the Rotation of the Plane of Polarization by Solutions," *Proc. Camb. Phil. Soc. 14* (1907): 313-317.

________. "On the Ionisation of Gases by Ultra-Violet Light and on the evidence as to the Structure of Light afforded by its Electrical Effects," *Proc. Camb. Phil. Soc. 14* (1907): 417-424.

________. "Rays of Positive Electricity," *Not. Proc. Roy. Inst. 18* (1907): 577-592.

________. *The Corpuscular Theory of Matter* (London: Archibald Constable & Co., 1907).

________. "Positive Rays," *Phil. Mag. 16* (1908): 657-691.

________. "The Carriers of Positive Electricity," *Not. Proc. Roy. Inst. 19* (1908): 171-201.

________. "The Nature of the γ Rays," *Proc. Camb. Phil. Soc. 14* (1908): 540.

________. "On the Velocity of Secondary Cathode Rays from Gases," *Proc. Camb. Phil. Soc. 14* (1908): 541-545.

________. "On the Carriers of the Positive Charges of Electricity emitted by Hot Wires," *Proc. Camb. Phil. Soc. 15* (1908): 64.

________. "On the Electric Theory of Gravitation," *Proc. Camb. Phil. Soc. 15* (1908): 65-69.

________. "On the Distribution of Electric Force along the Striated Discharge," *Proc. Camb. Phil. Soc. 15* (1908): 70.

________. *On the Light thrown by recent Investigations on Electricity on the Relation between Matter and Ether* (The Adamson Lecture) (Manchester: University of Manchester Press, 1908).

________. "On Striations in the Electric Discharge," *Phil. Mag. 18* (1909): 441-451.

________. "Positive Electricity," *Phil. Mag. 18* (1909): 821-845.

________. "Electrical Striations," *Not. Proc. Roy. Inst. 19* (1909): 577-586.

________. "The Dynamics of a Golf Ball," *Not. Proc. Roy. Inst. 19* (1909): 795-810.

________. "On the Theory of the Motion of Charged Ions through a Gas," *Proc. Camb. Phil. Soc. 15* (1909): 375-380.

________. "On a Theory of the Structure of the Electric Field and its application to Röntgen Radiation and to Light," *Phil. Mag. 19* (1910): 301-313.

________."Rays of Positive Electricity," *Phil. Mag. 19* (1910): 424-435.

________. "On the Theory of Radiation," *Phil. Mag. 20* (1910): 238-247.

________. "Rays of Positive Electricity," *Phil. Mag. 20* (1910): 752-767.

________. "A new Method of investigating the Positive Rays," *Proc. Camb. Phil. Soc. 16* (1910) 120.

________. "Radiant Energy and Matter," *Engineering 91* (1911): 319-321, 353-354, 386-388, 421-422, 454-455.

________. "Rays of Positive Electricity," *Phil. Mag. 21* (1911): 225-249.

________. "A New Method of Chemical Analysis," *Not. Proc. Roy. Inst. 20* (1911): 140-148.

________. "Application of Positive Rays to the study of Chemical Reactions," *Proc. Camb. Phil. Soc. 16* (1911): 455.

________. "Ionisation by Moving Electrified Particles," *Phil. Mag. 23* (1912): 449-457.

________. "Further Experiments on Positive Rays," *Phil. Mag. 24* (1912): 209-253.

________. "Multiple-charged Atoms," *Phil. Mag. 24* (1912): 668-672.

________."The Unit Theory of Light," *Proc. Camb. Phil. Soc. 16* (1912): 643-652.

________. "Some Further Applications of the Method of Positive Rays," *Not. Proc. Roy. Inst. 20* (1913): 591-600.

________. "On the Structure of the Atom," *Phil. Mag. 26* (1913): 792-799 & 1044.

________. "Further Applications of Positive Rays to the study of Chemical Problems," *Proc. Camb. Phil. Soc. 17* (1913): 201.

________. *Rays of Positive Electricity and Their Application to Chemical Analyses* (London: Longmans, Green and Co., 1913).

________. "The Forces between Atoms and Chemical Affinity," *Phil. Mag. 27* (1914): 757-789.

________. "The Production of Soft Röntgen Radiation by the impact of Positive and Slow Cathode Rays," *Phil. Mag. 28* (1914): 620-625.

________. "Rays of Positive Electricity (Bakerian Lecture)," *Proc. Roy. Soc. 89* (1914): 1-20.

________. "Further Researches o Positive Rays," *Not. Pro. Roy. Inst. 21* (1914): 263-268.
________. *The Atomic Theory* (The Romanes Lecture, 1914) (Oxford: Clarendon Press, 1914).
________. "Conduction of Electricity through Metals," *Phil. Mag. 30* (1915): 192-202.
________. "The Mobility of Negative Ions at Low Pressures," *Phil. Mag. 30* (1915); 321-328.
________. "Experiments on Slow Cathode Rays," *Not. Proc. Roy. Inst. 21* (1915): 406-410.
________. "The Genesis and Absorption of X-rays," *Not. Proc. Roy. Inst. 21* (1916): 726-734.
________. "Industrial Application of Electrons," *Not. Proc. Roy. Inst. 22* (1917): 175-178.
________. "Lord Rayleigh, O. M., F. R. S," *Nature 103* (1919): 365-366.
________. *The Structure of Light: The Fison Memorial Lecture 1925* (Cambridge: Cambridge University Press, 1925).
________. "The Cavendish Laboratory," *Supplement to Nature 118* (1926): 41-44.
________ et al. *James Clerk Maxwell: A Commemoration Volume, 1831-1879* (Cambridge: Cambridge University Press, 1931).
________ & G. P. Thomson. *Conduction of Electricity through Gases, third edition* (Cambridge: Cambridge University Press, 1928, 1933).
________. *Recollections and Reflections* (New York: MacMillan & Co., 1937).
Threlfall, R. "On an Automatic Mechanical Arrangement for maintaining a Constant High Potential." *Proc. Camb. Phil. Soc. 5* (1885): 226-229.
________. "On the Theory of Explosions," *Phil. Mag. 21* (1886): 165-179.
Todd, G. W. "The Mobility of the Positive Ion in Gases at Low Pressures," *Proc. Camb. Phil. Soc. 16* (1910): 21-25.
________. "The Mobility of Positive Ions produced from heated Aluminum Phosphate in Gases at Low Pressures," *Phil. Mag. 22* (1911): 791-804.
________. "On the Mobility of the Negative Ion at Low Pressures," *Proc. Camb. Phil. Soc. 16* (1912): 653-657.
________. "Further Experiments on the Mobility of the Positive Ion at Low Pressures," *Phil. Mag. 25* (1913): 163-171.
Tolstoy, I. *James Clerk Maxwell* (Chicago: University of Chicago Press, 1981).
Tomas, D. G. *Tradition, Context of Use, Style and Function; Expansion Apparatus used at the Cavendish Laboratory during the period 1895-1912* (Master thesis, University of Montreal, 1979).
Tooley, P. "The Terling Laboratory," *Chemistry in Britain 15* (1979): 284-285.
Topper, D. R. *J.J. Thomson and Maxwell's Electromagnetic Theory*, unpublished Ph.D. dissertation (Cleveland: Case Western Reserve University, 1970).
________. "Commitment to Mechanism: J.J. Thomson, the Early Years," *Archive for History of Exact Sciences 7* (1971): 393-410.
________. "To Reason by Means of Images: J.J. Thomson and the Mechanical Picture of Nature," *Annals of Science 37* (1980): 31-57.
Townsend, J. S. "Magnetisation of Liquids," *Phil. Trans. 187* (1896): 533-549.
________. "On Electricity in Gases and the Formation of Clouds in Charged Gases," *Proc. Camb. Phil. Soc. 9* (1897): 244-258.
________. "Electrical Properties of Newly Prepared Gases," *Proc. Camb. Phil. Soc. 9* (1897): 345-371.
________. "Applications of Diffusion to Conducting Gases," *Phil. Mag. 45* (1898): 469-480.
________. "The Formation of Clouds with Ozone," *Proc. Camb. Phil. Soc. 10* (1899): 52-58.
________. "Second Röntgen Rays," *Proc. Camb. Phil. Soc. 10* (1899): 217-226.
________. "The Diffusion of Ions produced in Air by the Action of a Radio-active Substance, Ultra-Violet Light and Point Charge," *Phil. Trans. 195* (1901): 259-278.
Turner, G. L. E. (ed.) *The Patronage of Science in the Nineteenth Century* (Leiden: Noordhoff International, 1976).
[University of Cambridge]. *A Short Account of University and College Finance* (Cambridge: Cambridge University Press, 1881).
________. *The Student's Guide to the University of Cambridge* (Cambridge: Cambridge University Press, 1893).
________. *The Historical Register of the University of Cambridge: being a Supplement to the Calendar with a Record of University Offices Honours and Distinctions to the Year 1910*, edited by J.R. Tanner (Cambridge: Cambridge University Press, 1917).

________. *The Historical Register of the University of Cambridge, Supplement, 1911-20* (Cambridge: Cambridge University Press, 1922).

________. *Index to Tripos Lists 1748-1910* (Cambridge: Cambridge University Press, 1923).

________. *The Book of Matriculations and Degrees: A Catalogue of those who have been Matriculated or Admitted to any Degree in the University of Cambridge from 1851-1900* (Cambridge: Cambridge University Press, 1902).

Varley, W. M. "On the Photo-Electric Discharge from Metallic Surface in different Gases," *Phil. Trans. 202* (1903): 439-458.

________. "On the Absorption of Ultra-Violet Light in different Gases," *Proc. Camb. Phil. Soc., 12* (1904): 510-516.

Vegard, L. "Researches upon Osmosis and Osmotic Pressure," *Phil. Mag. 16* (1908): 247-271 & 396-419.

________. "On the Electric Discharge through the Gases, HCl, HBr, and HI," *Phil. Mag. 18* (1909): 465-483.

________. "On the Free Pressure in Osmosis," *Proc. Camb. Phil. Soc. 15* (1909): 13-23.

________. "On some general Properties of Mixed Solutions," *Proc. Camb. Phil. Soc. 15* (1909): 275-274.

________. "An Experiment on Ionisation with γ Rays," *Proc. Camb. Phil. Soc. 15* (1909): 78-81.

Vincent, J. H. "On the Photography of Ripples—Second Paper," *Phil. Mag. 45* (1898): 191-197.

________. "On the Photography of Ripples—Third Paper," *Phil. Mag. 46* (1898): 290-296.

________. "On the Construction of a Mechanical Model to Illustrate Helmholtz's Theory of Dispersion," *Phil. Mag. 46* (1898): 557-563.

________. "On the Use of Logarithmic Coordinates," *B. A. Report* (1898): 159-178.

________. "On the Photography of Ripples—Fourth Paper," *Phil. Mag. 47* (1899): 338-344.

________. "On Some Photographic Phenomena connected with the Colson-Russell Effect," *Chemical News 79* (1899): 302-303.

________. "The Density and Coefficient of Cubical Expansion of Ice," *Phil. Trans. 198* (1902): 463-481.

________. "On a General Numerical Connexion between the Atomic Weight," Phil. Mag. 4 (1902): 103-115.

________. "The Action of Ultra-Violet Light on Moist Air," *Proc. Camb. Phil. Soc. 12* (1903): 305-311.

________. "On the Influence of Magnetic Field on Thermal Conductivity," *Phil. Mag. 5* (1903): 529-537.

________. "The Action of Ultra-Violet Light on Moist Air," *Proc. Camb. Phil. Soc. 12* (1903): 305-311.

Wade, E. B. H. "On a new Method of Determining the Vapour Pressures of Solutions," *Proc. Roy. Soc. 61* (1897): 285-287: *Proc. Roy. Soc. 62* (1897): 376-385.

Walker, G. W. "On the Dependence of the Refractive Index of Gases on Temperature," *Phil. Trans. 201* (1903): 435-455.

________. "On the Theory of the Quadrant Electrometer," *Phil. Mag. 6* (1903): 238-250.

________. "On Unsymmetrical broadening of Spectral Lines," *Phil. Mag. 6* (1903): 536-540.

Warwick, A. "Exercising the Student Body: Mathematics and Athleticism in Victorian Cambridge," in Christopher Lawrence and Steven Shapin (eds.) *Science Incarnate: Historical Embodiments of Natural Knowledge* (Chicago: University of Chicago Press, 1998), 288-326.

________. "A Mathematical World on Paper: Written Examinations in Early 19th Century Cambridge," *Studies in History and Philosophy of Modern Physics, 29* (1998): 295-319.

________. "The World of Cambridge Physics," in R. Staley (ed.), *The Physics of Empire: Public Lectures* (Cambridge: Whipple Museum of the History of Science, 1994), 57-86.

________. "Cambridge Mathematics and Cavendish Physics: Cunningham, Campbell and Einstein's Relativity, 1905-1911. Part I: The Uses of Theory," *Studies in History and Philosophy of Science 23* (1992): 625-656.

________. "Cambridge Mathematics and Cavendish Physics: Cunningham, Campbell and Einstein's Relativity, 1905-1911. Part II: Comparing Traditions in Cambridge Physics," *Studies in History and Philosophy of Science 24* (1993): 1-25.

Watson, H. E. "On Regularities in the Spectrum of Neon," *Proc. Camb. Phil. Soc. 16* (1910): 130-135.

________. "Some Experiments on the Electrical Discharge in Helium and Neon," *Proc. Camb. Phil. Soc. 17* (1912): 90-107.

Wellisch, E. M. "The Passage of Electricity through Gaseous Mixtures," *Proc. Roy. Soc. 82* (1909): 517.

________. "An Electric Detector for Electromagnetic Waves," *Proc. Camb. Phil. Soc, 15* (1909): 337-339.
________. "The Mobility of Ions produced by Röntgen Rays in Gases and Vapours," *Phil. Trans. 209* (1909): 249-279.
________. "The Passage of Electricity through Gaseous Mixtures," *Proc. Roy. Soc. 82* (1909): 500-517.
________. "The Laws of Mobility and Diffusion of the Ions formed in Gaseous Media," *Proc. Camb. Phil. Soc. 15* (1909): 1-10.
________. "An Electric Detector for Electromagnetic Waves," *Proc. Camb. Phil. Soc. 15* (1909):337-339.
________. "The Theory of the small Ion in Gases," *Phil. Mag. 19* (1910): 201-203.
Wheaton, B. R. *The Tiger and the Shark: Empirical Roots of Wave-Particle Dualism* (Cambridge: Cambridge University Press, 1983).
Whetham, W. C. D. "On the Alleged Slipping at the Boundary of a Liquid in Motion," *Phil. Trans. 181* (1890): 559-582.
________. "Note on Kohlrausch's Theory of Ionic Velocity," *Phil. Mag. 32* (1891): 43-44.
________. "Illustrations of a Method of Measuring Ionic Velocities," *Proc. Camb. Phil. Soc. 7* (1891). 250.
________. "Ionic Velocities," *Phil. Trans. 184* (1893): 337-359.
________. "On the Velocities of the Ions and the Relative Ionisation-Power of Solvents," *Phil. Mag. 38* (1894): 392-396.
________. "On the Velocities of the Ions," *Phil. Trans. 186* (1895): 507-522.
________. "The Theory of the Migration of Ions and of Specific Ionic Velocities," *B. A. Report* (1897): 227-245.
________. "The Ionizing Power of Solvents," *Phil. Mag. 44* (1897): 1-9.
________. "The Coagulative Power of Electrolytes," *Phil. Mag. 48* (1899): 474-477.
________. "The Ionisation of Dilute Solutions at the Freezing Point," *Phil. Trans. 194* (1900): 321-360.
________. "The Theory of Electrolytic Dissociation," *Phil. Mag. 5* (1903): 279-290.
________. "The Electrical Conductivity of Solutions at the Freezing-Point of Water," *Proc. Roy. Soc. 71* (1903): 332-338.
________. "The Electrical Conductivity of Dilute Solutions of Sulphuric Acid," *Proc. Roy. Soc. 76* (1905): 577-583.
________ & H. H. Paine. "The Electrolytic Properties of Dilute Solutions of Sulphuric Acid," *Proc. Roy. Soc. 81* (1908): 58-80.
Whiddington, R. "Some Fatigue Effects of the Cathode in a Discharge Tube," *Proc. Camb. Phil. Soc. 15* (1909): 183-188.
________. "Note on the Electrical Behaviour of Fluorescing Iodine Vapour," *Proc. Camb. Phil. Soc. 15* (1909): 189.
________. "Preliminary Note on the Properties of easily absorbed Röntgen Radiation," *Proc. Camb. Phil. Soc. 15* (1910): 574-575.
________. "The Production and Properties of Soft Röntgen Radiation," *Proc. Roy. Soc. 85* (1911): 99-118.
________. "The Production of Characteristic Röntgen Radiations," *Proc. Roy. Soc. 85* (1911): 323-332.
________. "The Transmission of Cathode Rays through Matter," *Proc. Camb. Phil. Soc. 16* (1911): 150-154.
________. "The Velocities of the Cathode Particles ejected by Characteristic Röntgen Radiations," *Proc. Camb. Phil. Soc. 16* (1911): 326-330.
________. "The Transmission of Cathode Rays through Matter," *Proc. Roy. Soc. 86* (1912): 360-370.
________. "The Velocity of the Secondary Cathode Particles ejected by the Characteristic Röntgen Rays," *Proc. Roy. Soc. 86* (1912): 370-378.
________. "The Transmission of Cathode Rays through Matter," *Proc. Roy. Soc. 89* (1913): 554-560.
Wilberforce, L. R. "On Some Experiments on the Measurement of the Capacity of a Condenser," *Proc. Camb. Phil. Soc. 5* (1884): 169-175.
________. "Note on an Electrostatical Problem," *Messenger of Mathematics 13* (1884): 152-154.
________. "On a New Method of producing the Fringes of Interference," *Proc. Camb. Phil. Soc. 5* (1885): 325.

________. "On a New Method of determining Specific Inductive Capacity," *Proc. Camb. Phil. Soc. 6* (1886): 24-26.
________. "On the Calculation of the Coefficient of Viscosity of a Liquid from its Rate of Flow through a Capillary Tube," *Phil. Mag. 31* (1891): 407-414.
________. "An Experiment on the Vibrations of a Spiral Spring," *Proc. Camb. Phil. Soc. 8* (1894): 61.
________. "On the Vibrations of a Loaded Spiral Spring," *Phil. Mag. 38* (1894): 386-392.
Williamson, R. (ed.) *The Making of Physicists* (Bristol: Adam Hilger, 1987).
Willows, R. S. "On the Variation of the Resistance of certain Amalgams with Temperature," *Phil. Mag. 48* (1899): 433-456.
________. "On the Effect of a Magnetic Field on the Discharge through a Gas," *Phil. Mag. 1* (1901): 250-260.
________. "On the Absorption of Gas in a Crookes Tubes," *Phil. Mag. 1* (1901): 503-517.
Wills, R. L."Effect of Temperature on the Hysteresis Loss in Iron," *Phil. Mag. 5* (1903): 117-133.
Wilson, C. T. R. "On the Formation of Cloud in the Absence of Dust," *Proc. Camb. Phil. Soc. 8* (1895): 306.
________. "The Effect of Röntgen's Rays on Cloudy Condensation," *Proc. Roy. Soc. 59* (1896): 338-339.
________. "On the Action of Uranium Rays on the Condensation of Water Vapour," *Proc. Camb. Phil. Soc. 9* (1897): 333-338.
________. "Condensation of Water-Vapour in the Presence of Dust-Free Air and Other Gases," *Phil. Trans. 189* (1897): 265-307.
________. "On the Condensation Nuclei produced in Gases by the Action of Röntgen Rays, Uranium Rays, Ultra-Violet Light, and other Agents," *Phil. Trans. 192* (1899): 403-453.
________."On the Comparative Efficiency as Condensation Nuclei of Positively and Negatively," *Phil. Trans. 193* (1899): 289-308.
________. "On the Leakage of Electricity through Dust-Free Air," *Proc. Camb. Phil. Soc. 11* (1900): 32.
________. "On the Ionisation of Atmospheric Air," *Proc. Roy. Soc. 68* (1901): 151-161.
________. "On the Spontaneous Ionisation of Gases," *Proc. Roy. Soc. 69* (1901): 277-282.
________. "On Radio-active Rain," *Proc. Camb. Phil. Soc. 11* (1902): 428-430.
________. "Further Experiments on Radio-activity from Rain," *Proc. Camb. Phil. Soc. 12* (1902): 17.
________. "On Radio-activity from Snow," *Proc. Camb. Phil. Soc. 12* (1903): 85.
________."On a Sensitive Gold-Leaf Electrometer," *Proc. Camb. Phil. Soc. 12* (1903): 135-139.
________. "The Condensation Method of demonstrating the Ionisation of Air under Normal Conditions," *Phil. Mag. 7* (1904): 681-690.
________. "On a Portable Gold-leaf Electrometer for Low or High Potentials, and its Application to Measurements in Atmospheric Electricity," *Proc. Camb. Phil. Soc. 13* (1905): 184-189.
________. "On the Measurement of the Earth-Air Current and on the Origin of Atmospheric Electricity," *Proc. Camb. Phil. Soc. 13* (1906): 363-382.
________. "A Curvature Method for measuring Surface Tension," *Proc. Camb. Phil. Soc. 14* (1907): 206.
________. "On the Measurement of the Atmospheric Electric Potential Gradient and the Earth-Air Current," *Proc. Roy. Soc. 80* (1908): 537-547.
________. "On the Thunderstorm Electricity," *Phil. Mag. 17* (1909): 634-641.
________. "On a Method of making Visible the Paths of Ionising Particles through a Gas," *Proc. Roy. Soc., 85* (1911): 285-288.
________. "On an Expansion Apparatus for making Visible the Tracks of Ionising Particles in Gases and Some Results obtained by its Use," *Proc. Roy. Soc. 87* (1912): 277-292.
________. "On some Determinations of the Sign and Magnitude of Electric Discharge in Lightning Flashes," *Proc. Roy. Soc. 92* (1916): 555-574.
Wilson, D. *Rutherford: Simple Genius* (Cambridge, Mass.: MIT Press, 1983).
Wilson, D. B. "Experimentalists among the Mathematicians: Physics in the Cambridge Natural Sciences Tripos, 1851-1900," *HSPS 12* (1982): 325-371.
________. Kelvin and Stokes: A Comparative Study in Victorian Physics (Bristol: Adam Hilger, 1987).
Wilson, H. A. "On the Influence of Dissolved Substances and of Electrification on the Re-formation of Clouds," *Phil. Mag. 45* (1898): 454-459.
________. "On Velocities of Solidification," *Proc. Camb. Phil. Soc. 10* (1898): 25-35.

________. "On the Electrical Conductivity of Flames containing Salt Vapour," *Phil. Trans. 192* (1899): 499-528.

________. "On the Electric Conductivity of Air and Salt Vapour," *Phil. Trans. 197* (1901): 415-441.

________. "Note on the Magnetic Deflection of Cathode Rays," *Proc. Camb. Phil. Soc. 11* (1901): 179-182.

________. "On the Hall Effect in Gases at Low Pressures," *Proc. Camb. Phil. Soc. 11* (1901): 249-263; *Proc. Camb. Phil. Soc. 11* (1902): 391-395.

________. "On the Coefficient of Mutual Induction for a Circle and a Circuit with Two parallel sides of infinite Length," *Proc. Camb. Phil. Soc. 11* (1902): 398-406.

________. "The Laws of Electrolysis of Alkali Salt-Vapours," *Phil. Mag. 4* (1902): 207-214.

________. "The Current-Density at the Cathode in the Electric Discharge in Air," *Phil. Mag. 4* (1902): 608-614.

________. "A Determination of the Charge on the Ions produced in Air by Röntgen Rays," *Phil. Mag. 5* (1903): 429-441.

________. "The Electric Intensity in the Uniform Positive Column in Air," *Phil. Mag. 6* (1903): 180-188.

________. "On the Discharge of Electricity from Hot Platinum," *Phil. Trans. 202* (1903): 71-121.

________. "On the Electric Effect of rotating a Dielectric in a Magnetic Field," *Phil. Trans. 204* (1904): 121-137.

________. "On Convection of Heat," *Proc. Camb. Phil. Soc. 12* (1904): 406-423.

________. "The Electrical Conductivity of Flames," *Phil. Mag. 10* (1905): 476-485.

Wilson, W. "On the β-Particles reflected by Sheets of Matter of different Thickness," *Proc. Roy. Soc. 87* (1912): 100-108.

________. "On the Absorption and Reflection of Homogeneous β-particles," *Proc. Roy. Soc. 87* (1912): 310-325.

Winstanley, D. A. *Later Victorian Cambridge* (Cambridge: Cambridge University Press, 1947).

Wise, Norton M. (ed.), *The Values of Precision* (Princeton: Princeton University Press, 1995).

Worthington, A. M. "On the Stretching of Liquids," *B. A. Report* (1888): 583-584.

________. *A Study of Splashes* (London: Longmans, Green, 1908).

Wood, A. "Effect of Screening on Ionisation in closed Vessels," *Proc. Camb. Phil. Soc. 12* (1904): 477-480.

________."Spontaneous Ionisation of Air in closed Vessels and its Causes," *Phil. Mag. 9* (1905): 550-576.

________ & N. R. Campbell. "Diurnal Periodicity of the Spontaneous Ionisation of Air and Other Gases in Closed Vessels," *Phil. Mag. 13* (1907): 265-276.

________. *The Cavendish Laboratory* (Cambridge: Cambridge University Press, 1946).

Woodall, A. J. & A. C. Hawkins, "Laboratory Physics and its Debt to G. F. C. Searle," *Physics Today 4* (1969): 283-285.

Zeleny, J. "On the Ratio of the Velocities of the Two Ions produced in Gases by Röntgen Radiation; and on some Related Phenomena," *Phil. Mag. 46* (1898): 120-154.

________. "On Convection Currents and on the Fall of Potential at the Electrodes, in Conduction produced by Röntgen Rays," *Proc. Camb. Phil. Soc. 10* (1898): 14-25.

________. "The Velocity of the Ions produced in Gases by Röntgen Rays," *Phil. Trans. 195* (1901): 193-234.

________. "On the conditions of Instability of Electrified Drops, with applications to the Electrical Discharge from Liquid Points," *Proc. Camb. Phil. Soc. 18* (1914): 71-83.

INDEX

Archimedes

NEW STUDIES IN THE HISTORY AND PHILOSOPHY OF SCIENCE AND TECHNOLOGY

1. J.Z. Buchwald (ed.): *Scientific Credibility and Technical Standards. In 19th and Early 20th Century Germany and Britain.* 1996 ISBN 0-7923-4241-0
2. K. Gavroglu (ed.): *The Sciences in the European Periphery During the Enlightenment.* 1999 ISBN 0-7923-5548-2
3. P. Galison and A. Roland (eds.): *Atmospheric Flight in the Twentieth Century*, 2000 ISBN 0-7923-6037-0
4. J.M. Steele: *Observations and Predictions of Eclipse Times by Early Astronomers.* 2000 ISBN 0-7923-6298-5
5. D-W. Kim: *Leadership and Creativity.* A History of the Cavendish Laboratory, 1871-1919. 2002 ISBN 1-4020-0475-3

KLUWER ACADEMIC PUBLISHERS – DORDRECHT/BOSTON/LONDON